FOREST

CONSERVATION

in the

ANTHROPOCENE

FOREST

CONSERVATION

→ *in the* ←

ANTHROPOCENE

Science, Policy, and Practice

edited by

V. Alaric Sample,

R. Patrick Bixler,

and Char Miller

UNIVERSITY PRESS OF COLORADO
Boulder

© 2016 by University Press of Colorado

Published by University Press of Colorado
5589 Arapahoe Avenue, Suite 206C
Boulder, Colorado 80303

 The University Press of Colorado is a proud member of
Association of American University Presses.

The University Press of Colorado is a cooperative publishing enterprise supported, in part,
by Adams State University, Colorado State University, Fort Lewis College, Metropolitan State
University of Denver, Regis University, University of Colorado, University of Northern Colorado,
Utah State University, and Western State Colorado University.

∞ This paper meets the requirements of the ANSI/NISO Z39.48–1992 (Permanence of Paper).

ISBN: 978-1-60732-458-4 (cloth)
ISBN: 978-1-60732-521-5 (paper)
ISBN: 978-1-60732-459-1 (ebook)

Library of Congress Cataloging-in-Publication Data

Names: Sample, V. Alaric, editor. | Bixler, R. Patrick (Richard Patrick), editor. | Miller, Char,
    1951– editor.
Title: Forest conservation in the Anthropocene : science, policy, and practice / edited by V. Alaric
    Sample, R. Patrick Bixler, and Char Miller.
Description: Boulder : University Press of Colorado, 2016.
Identifiers: LCCN 2015035743 | ISBN 9781607324584 (cloth) | ISBN 9781607325215 (pbk) | ISBN
    9781607324591 (ebook)
Subjects: LCSH: Forest microclimatology. | Forest conservation. | Climatic changes.
Classification: LCC SD390.7.C55 F6635 2016 | DDC 634.9—dc23
LC record available at http://lccn.loc.gov/2015035743

Cover photograph from US Department of the Interior Bureau of Land Management (CC BY 2.0)

Contents

Acknowledgments

Many thanks to the authors whose contributions appear in this volume, some of whom are committed scientists at the forefront of research on the unprecedented array of environmental changes affecting forest ecosystems, and others of whom are pioneering resource managers in the field striving to sustain the ecological, economic, and social values of forests in the face of uncertainty. Special thanks to David Peterson, David Cleaves, Sam Foster, and Linda Joyce (all with the US Forest Service), Jessica Halofsky (University of Washington), and Joe MacCauley and Mark Shaffer (US Fish and Wildlife Service) for their support and leadership. Merci beaucoup to Jean-Luc Peyron of ECOFOR-Paris for his international leadership on forest management adaptation to climate change and his contributions to this effort, and to Marc Magaud, Pierre Michel, and their colleagues in the office of the Science and Technology Attaché at the Embassy of France for their partnership in support of international scientific and technical exchange on climate issues. We are grateful to Jessica d'Arbonne at the University Press of Colorado for her vision and encouragement in the development of this book, to Laura Furney and Kelly Lenkevich for their patience and many editorial improvements, and to the anonymous readers for the press who provided such constructive suggestions. Special thanks to Judi Lipset for her considerable talent and expertise as a technical editor. Major support for this book was provided by the Kelley Family Foundation, Doris Duke Charitable Foundation, and the US Forest Service.

V. ALARIC SAMPLE
R. PATRICK BIXLER
CHAR MILLER

FOREST

CONSERVATION

in the

ANTHROPOCENE

Introduction

Forest Conservation and Management in the Anthropocene

V. Alaric Sample

Throughout Earth's history, its climates have been changing and its biotic systems have mutated, migrated, and otherwise adapted as tectonic shifts have reconfigured the continents and as polar ice caps have ebbed and flowed across the latitudes through glacial cycles. There is substantial evidence that climatic changes that in the past have taken place over the course of millennia are now happening in a matter of decades, and that these rapid transformations are human caused. That has led researchers to coin a new term for this era, the Anthropocene—the Age of Humans. These accelerated alterations in climate challenge the ability of human civilization and the natural systems on which it depends to adapt quickly enough to keep pace. Through efforts like the Intergovernmental Panel on Climate Change and the United Nations Framework Convention on Climate Change, leading scientists from around the world have focused their energies on understanding the nature and implications of these changes, and the world's governments are striving to develop the institutions and resources to enable timely and effective actions to mitigate and adapt to changes that are anticipated or already under way.

DOI: 10.5876/9781607324591.c000

The people and organizations charged with the conservation and sustainable management of the world's forests and their associated renewable natural resources are at the forefront of efforts to understand and address these challenges. "Rapid climate change is the defining conservation issue of our generation," notes a report by a group of federal natural resource management agencies and nonprofit organizations. "Indeed, preparing for and coping with the effects of climate change—an endeavor referred to as climate change adaptation—is emerging as the overarching framework for conservation and natural resource management" (Glick et al. 2011). Traditional wildlife and biodiversity conservation strategies have relied heavily on the establishment of reserves and other protected areas to conserve habitat, but as climate changes, optimal habitat zones shift to different places on the landscape as well. So a basic question that has arisen for conservation biologists is whether protected areas that are fixed and static on the landscape can still play a useful role in protecting plant and animal species that are in the process of relocating.

As Gary Tabor and his coauthors note in chapter 13, landscape-scale habitat conservation strategies, originally developed to address the issue of habitat fragmentation, are now being pressed into service as climate adaptation strategies. Corridors and linkages that can connect habitat across several degrees of latitude are becoming critically important to facilitate the emigration of some plant and animal species and the immigration of others. However, some species within a given ecological community are more mobile than others: some are able to migrate and others are left behind, disassembling existing communities of interdependent species. At the same time, a region will experience the immigration of mobile species from elsewhere, developing species assemblages that may never have existed before. How to regard these "novel ecosystems" is a topic of considerable ongoing debate among conservation biologists. From one perspective, many of these novel ecosystems are highly biologically productive and may also exhibit a high level of species diversity, so they may represent a significant biodiversity resource in themselves. This migratory process, according to Tim Caro and his coauthors (chapter 6), will increase the importance of large protected areas with well-buffered interior regions that are more resistant to immigration by species from distant locales.

This still leaves the question of whether something can be done to minimize the emigration of species from such reserves and the dismantling of the existing ecological community. Anderson and Johnson (chapter 9) describe characteristics that can help scientists define biological and geological characteristics that allow the identification of "resilient sites" that tend to resist the influence of climate change and hold their ecological communities intact. These sites tend to have certain characteristics of geology, soils, and topography. Identifying, mapping, and then protecting a sufficient number of these resilient sites across large landscapes can be an important component in a comprehensive, portfolio approach to adapting biodiversity conservation to the effects of climate change.

There are significant additional challenges associated with actually implementing such a strategy on large landscapes predominantly characterized by private ownership and comprising many small tracts. These tracts are typically managed for objectives as diverse as the private owners themselves, who may or may not understand or share a commitment to biodiversity conservation. Once again, large landscape conservation strategies originally developed for other purposes can be repurposed to help achieve biodiversity conservation objectives in regions characterized by mixed public-private or predominantly private ownerships.

One of those key complications is the complex relationship between climate change, water, and forests, and its direct, indirect, and induced effects. Regions that experience prolonged drought and elevated temperatures will obviously face challenges resulting from lower precipitation and higher evapotranspiration, and areas that depend upon high-elevation snowpack to maintain late-season flows will more often find themselves in extreme water emergencies. This will be a major issue for aquatic habitat, especially when combined with higher water temperatures and lower dissolved oxygen levels. Cold-water species such as West Slope cutthroat trout and Dolly Varden (bull trout) may face particular environmental stress, and localized populations unable to migrate to more suitable habitat may die out (Williams et al. 2009; Wenger et al. 2011).

Intact forests can mitigate all of these influences on water supply, quality, and temperature, but as forests themselves begin to show the effects of climate change their ability to do so will be sharply reduced. Forests are remarkably efficient at absorbing precipitation, storing it, and gradually releasing it as streamflow. Forests in higher elevations can be managed for optimal snow interception by maintaining crown cover that is open enough not to intercept too much snow, where it will sublimate back into the atmosphere, but closed enough to provide shade to the snow that does penetrate to the ground, slowing spring snowmelt and helping to maintain late-season flows. Climate-induced environmental stress that results in tree mortality from drought, insects, or disease diminishes each of these functions (Allen et al. 2010).

The most extreme effects are from wildfire. Extensive crown fires in Colorado's Front Range in 1996 and 2002 caused major damage to the Strontia Springs and Cheesman Reservoirs, threatening the municipal water supply for Denver and communities along the Front Range. A decade later, local water authorities were still spending millions of dollars annually for additional water treatment and the removal of tons of sediment and debris from check dams installed upstream from these reservoirs after the fires. By leaving slopes barren of trees and other vegetation, and reducing the ability of soils to absorb water during major storms, recent wildfires in Colorado, California, and other parts of the western United States have been shown to contribute to flash flooding for many years subsequent to the fire itself (Cannon et al. 2008).

The decisive steps that Denver took to reduce the likelihood of wildfire damage to its other reservoirs provide a model that other cities and communities are taking up, especially as the changing climate is raising the stakes. Denver Water and several water authorities serving other Front Range communities sought and received permission to add a small surcharge to customers' regular water bills, creating a fund that was used to accomplish hazardous fuels treatments and forest health thinnings on forest lands upstream from municipal reservoirs. Most of these lands are national forests, and Denver Water and the US Forest Service subsequently entered a cooperative agreement in which each party would contribute more than $16 million to accelerate treatments on thousands of acres of forest (Denver Water Authority 2013).

The lessons learned in Denver were not lost on other western communities surrounded by fire-prone forests that, should a wildfire occur, would cause substantial damage to the local water supply. Laura McCarthy (chapter 11) analyzes one such proactive initiative, a study that The Nature Conservancy conducted for the city of Santa Fe, New Mexico. The study estimated the economic losses should a major fire occur in the city's primary watershed on the Santa Fe National Forest and demonstrated that the probability of such a fire could be significantly reduced through hazardous fuels treatments and forest health thinnings. The cost for these interventions would be a fraction of the projected damages. As a result, voters approved and the city enacted a modest surcharge on local water customers, and used this to create a water fund that underwrites the necessary forest management activities.

In regions of the country where the changing climate is expected to bring higher levels of precipitation and more of it in the form of extreme storm events, intact forests are becoming a high-value asset. Hurricane Irene in 2011 dumped an extraordinary volume of rain on the Mid-Atlantic states and New England in a very short period, and satellite photos from a few days after the storm showed the Susquehanna River in full flood stage, choked with sediment and debris, which was flushing into Chesapeake Bay and turning its northern portion an opaque brown. Municipal water supplies were interrupted for nearly two weeks in Harrisburg, Pennsylvania, and other communities drawing their drinking water from the Susquehanna, and power plants drawing cooling water from the river were either shut down or operating at reduced output (Water Research Foundation 2012).

In the same satellite photo, the next major watershed to the east, the Delaware River, can be seen running clear and blue, sparkling in the sunlight. One major reason for this is the fact that the headwaters of the Delaware River are roughly 80 percent forested, whereas forest cover has been reduced to less than 40 percent in the headwaters of the Susquehanna. There is a major effort now under way to restore thousands of acres of riparian forest in the upper Susquehanna watershed—a valuable initiative but one that will take years to begin having a meaningful effect. Meanwhile the upper Delaware River watershed continues to lose forest cover at an average of more than 100 acres (40 ha) a week. The Pennsylvania, New York, and New Jersey counties that come together in the upper Delaware are the fastest developing counties in their respective states. Will Price and Susan Beecher (chapter 14) explain the collaborative efforts to create the Common Waters Fund, an innovative mechanism developed to give private forestland owners a financial incentive to conserve their forest, and to manage it in ways that will enhance its watershed-protection capabilities.

But most of the communities and businesses downstream on the Delaware have yet to be convinced that it is in their interest to invest in keeping the forested headwaters intact. Water supply and water quality have been good recently, and many water users seem willing to take a chance that the continued loss of forest cover to development will not have any significant impact on them. Yet the growing prospect of more extreme storm events like Hurricanes Irene and Sandy may be changing that benefit/cost ratio. The economic impacts of a severe flood event on the Delaware would be enormous, and the forested headwaters play an important role in flood mitigation and buffering the effects of extreme storm events. Unlike the Rio Grande or the Susquehanna, whose headwaters forests are in need of costly restoration, the headwaters forests of the Delaware simply need to be maintained as they are. Currently there are more private forestland owners in the upper Delaware waiting to participate than there is money in the Common Waters Fund to enlist them—and the development pressure continues. As the effects of climate change become more pronounced, it will be clear that what the headwaters forests provide in terms of water supply, water quality, flood mitigation, and buffering extreme storm events is well worth the modest investment needed to keep them intact.

A different set of challenges confronts the wood products industry. Certain high-value hardwood species, for example, are likely to become more susceptible to exotic pests and pathogens such as the emerald ash borer (*Agrilus planipennis*), Asian long-horned beetle (*Anoplophora glabripennis*), and Sudden Oak Death (*Phytophthora ramorum*). Hopefully this will not have the impacts that the chestnut blight (*Cryphonectria parasitica*) has had on the American chestnut (*Castanea dentata*) that once dominated the eastern hardwood forests, but it is not something that even the best scientists are able to predict with confidence.

In the dry conifer forests of the Southwest and central Rockies, native forests have already been fundamentally altered by widespread mortality from the mountain pine beetle (*Dendroctonus ponderosae*) and other naturally endemic species, the result of a perfect storm in which warmer winters have fostered the survival of extraordinarily high populations of bark beetles and other agents, and drought stress has drastically reduced the ability of trees to resist and survive insect attacks. Even in fire-adapted forest types such as Ponderosa pine (*Pinus ponderosa*), the unnatural and all-consuming crown fires that often follow leave vast areas of forests with no means to regenerate and restore themselves. Many areas, especially in the American Southwest, will not return to forest within the foreseeable future and are even now in the process of converting from forest ecosystems to grassland or shrubland ecosystems. As Craig Allen (chapter 4) and Anthony L. Westerling and others (chapter 3) note, this is profoundly changing water regimes, wildlife habitat, and biodiversity across immense areas of forests, challenging local communities as well as resource managers to quickly develop new strategies for resistance, resilience, or the readjustment of their future expectations in light of climate change.

In the intensively managed forests in regions such as the US South and Pacific Northwest, there seems to be a sense that the short rotations typical of commercial plantation forests will allow forest managers to stay ahead of the accelerating pace of climate change. Research is producing more drought-resistant varieties of important commercial tree species, which presumably will replace existing forests as they are harvested. Genetic modification may offer opportunities to attune certain tree species to new and evolving climate characteristics, but the acceptance of widespread use of such techniques is far from certain. A strategy based simply on more frequent opportunities to replace existing trees may not fully account for other climate-related effects such as more intense storms like Hurricanes Katrina and Hugo that can destroy millions of acres of forest very quickly. Prolonged drought and elevated temperatures also reduce resistance to pests such as bark beetles, which can still kill large expanses of forest in a relatively short time. All of these factors contribute to increases in wildfire activity, a trend that is already being documented even in the South (Vose et al. 2012).

Getting a handle on some of these trends requires the development of vulnerability assessments that encompass terrestrial and aquatic habitat, biodiversity, vegetation management, hydrology, and forest road systems. As Jessica Halofsky and her colleagues argue (chapter 12), these tools are also essential to understanding the potential effects of climate change on forest ecosystems as a whole, and the implications for the range of environmental, economic, and social values and services that forests provide. These authors are optimistic that climate change can be mainstreamed in the policies and management of the Forest Service and other federal agencies by the end of this decade by considering climate change as one of many risks to which natural resources are subjected, and by considering adaptation as a form of risk management.

Resource managers need decision-support tools that allow them to integrate vulnerability assessments with action strategies to establish reasoned priorities and make the best-informed decisions possible (Peterson et al. 2011; Halofsky et al. 2011). Resource managers must also be able to utilize these tools to determine what they need to do differently in the future, and what existing practices will continue to be the best approach as part of an overall strategy for mitigating and adapting to climate change. The catch: budgets and human resources will never be unlimited for managers of public or private forest lands, so managers will need to be as adaptive and resilient as the forested landscapes they manage.

As early as 2020, US forests are expected to decline in their ability to serve as the nation's largest terrestrial "carbon sink." Today, these forests store enough carbon to absorb roughly 14 percent of total US greenhouse gas emissions. Before 2050, this very significant carbon offset could drop to zero, and the nation's forests would themselves become net sources of carbon emissions (USDA Forest Service 2012a). As Westerling et al. (chapter 3), Allen (chapter 4), and Stephens (chapter 5) point out, this is largely due to two factors: the increasing size, frequency, and intensity of wildfires, as most of the western United States continues on a trend of elevated temperatures and extended drought, and the continuing loss of private forests for development. It is still possible to avoid or at least mitigate this projected future, but it will require decisive actions and a substantial strengthening of current conservation and sustainable forest management efforts to change the trajectory US forests are now following. These actions include:

1. Increase afforestation and decrease deforestation:

   - Stem the conversion of forests to development and other land uses; the loss of forests and open space to development was recently estimated at approximately 6,000 acres (2,400 ha) per day, or roughly 4 acres (1.6 ha) per minute.

   - Increase the resistance and resilience of dry forests in the western United States to minimize the conversion to grassland ecosystems in the wake of major insect or disease outbreaks and wildfires.

2. Increase substitution of wood for fossil fuels in energy production, and for other building materials to maximize long-term carbon storage:

   - Increased biomass energy from the current 2 percent of US energy use to 10 percent would prevent the release of 130–190 million metric tons/

year of carbon from fossil fuels (Perlack et al. 2005, Zerbe 2006); commitment to conservation and reforestation of harvested sites is critically important to this net gain.

- Use of 1 metric ton of carbon in wood materials in construction in place of steel or concrete can result in 2 metric tons in lower carbon emissions, due to lower emissions associated with production processes (Sathre and O'Connor 2008, Schlamadinger and Marland 1996). Using wood from fast-growing forests for long-lived wood products and bioenergy can be more effective in lowering atmospheric carbon than storing carbon in the forest, where increased wood production is sustainable (Baral and Guha 2004, Marland and Marland 1992, Marland et al. 1997).

3. Manage carbon stocks in existing forests:

- Increase forest carbon stocks through longer harvest intervals and protect forests with high biomass.

- Manage forest carbon with fuel treatments: carbon emissions from wildland fires in the coterminous United States have averaged 67 million metric tons/year since 1990 (US Environmental Protection Agency 2009, 2010). Stand treatments intended to reduce fire intensity, especially crown fires that result in near-total tree mortality, have the potential to significantly reduce carbon emissions.

Integrating these domains of science and forging a new interdisciplinary science of the Anthropocene cannot be developed in isolation. This book is intended to break down some of these barriers and open up necessary discussions among scientists, land managers, policy makers, and citizens. For this new knowledge and engaged conversations to make a difference on the ground, it must be developed in the context of actual resource management planning and decision-making (USDA 2008; USDA 2010).

The objectives of this book are to (1) summarize recent advances in the scientific understanding of the projected effects of climate change on forest ecosystems and their responses to natural disturbance and human interventions, (2) describe strategies for adapting current resource management practices to sustain these evolving ecosystems and the array of social, economic, and environmental services they provide, and (3) identify opportunities to evolve the existing institutional and policy framework to support timely and effective implementation of adaptation strategies for public and private forest lands.

This book examines existing constraints to timely and effective implementation of adaptation strategies, and steps that can be taken in the near term to accelerate the evolution of policies and institutional frameworks to address these constraints. These include:

- Public awareness. There is a lack of public awareness of how climate change affects natural resources. and this information gap influences the level and nature of adaptation by public institutions.

- Resource manager awareness. The lack of experience and understanding of climate science by resource managers can lead to low confidence in taking management action in response to climate threats (GAO 2007); similar limitations through the chain of supervision and decision-making constrain appropriate efforts (GAO 2009).

- Monitoring and adaptive management. Adaptive management has been understood as a core component of ecosystem management for more than two decades, but climate change is necessitating an even more central role for real-time monitoring, reporting, and incremental adjustments in land and resource management plans and activities (Cleaves and Bixler, chapter 16; Peterson et al. 2011; Swanston and Janowiak 2012). The effectiveness of adaptive management on public lands as well as private has been limited by the weak institutional framework for monitoring, by inadequate funding, and by the lack of analyst capacity.

- Policy and planning. Public agencies and private organizations alike are constrained by hierarchies of laws, regulations, and policy direction developed before the effects of climate change were recognized or well understood; they are based on the assumption of stable and predictable climate, and thus provide limited authority for resource managers to accommodate the dynamics of climate change. Forest management organizations of all kinds confront operational challenges in working at spatial and temporal scales compatible with climate change adaptation.

- Budget and fiscal barriers. Significant additional funding will be needed for education and training; development of science-management partnerships; vulnerability assessments; and development of adaptation strategies. Collaboration across organizational as well as geographic boundaries, leveraging of institutional capacities, and other innovative solutions will be needed to address the budget challenge.

The characteristics that define the Anthropocene are about more than just the changing climate. It is about more than 7 billion people occupying virtually every biome on the planet, and human infrastructure that influences our ability to mitigate climate change and to adapt to it (Zalasiewicz et al. 2010). We know that climate change is already affecting forests around the world, and strongly influencing their ability to provide the environmental, economic, and societal values and services on which society depends. These effects are already evident today in extraordinarily destructive wildfires and floods, unprecedented epidemics of insects and pathogens, and other manifestations of forest ecosystems already under high levels of environmental stress. As Curt Stager notes (chapter 1), the combined results of numerous climate models suggest that these climate changes will strengthen and accelerate over the next several decades and perhaps centuries.

Significant progress has been made in developing the science and management approaches needed to understand, prepare for, and ultimately to adapt to these changes. There is much more we need to learn, however, *Forest Conservation in the Anthropocene* is an important first step in developing the ideas that will drive the conversation—and the resulting policy-making—forward. But we know enough now to begin taking decisive actions at the local, regional, and national level to implement adaptation strategies on public and private forest lands. As noted at the end of this book (Cleaves and Bixler, chapter 16; Shaffer, chapter 15), the bottlenecks we are now encountering are not based so much on the limitations of our science as on limitations in the policies and the existing institutional framework within which forestry is practiced.

CHANGING CLIMATIC REGIMES AND FOREST ECOSYSTEMS

I

Climate Change in the Age of Humans

CURT STAGER

Human-driven climate change is only one of many challenges that forests must face during the twenty-first century and beyond. Even without adding more heat-trapping carbon dioxide to the atmosphere than would be available should all the planet's volcanoes erupt at once (Gerlach 2011), the presence of billions of human beings on Earth represents a major source of environmental change. We have become so numerous, our technologies so powerful, and our societies so interconnected that we have become a force of nature on a geological scale.

There is no consensus yet on when the Anthropocene epoch began (Crutzen and Stoermer 2000; Stager 2011). Most definitions date it to the Industrial Revolution, but human impacts on what were previously thought to be "untouched" landscapes have long affected forests through megaherbivore extinctions, land clearance, fires, grazing, and cultivation (Willis et al. 2004; Willis and Birks 2006; Lorenzen et al. 2011). Although its authorship and timing are difficult to pin down, the Anthropocene concept nevertheless provides a useful context for ecosystem management.

DOI: 10.5876/9781607324591.c001

With approximately one-quarter of the planet's carbon dioxide reservoir now attributable to fossil fuel emissions, our behavior has become an integral part of global ecology. Our artificial nitrogen fixation now matches or exceeds natural production of available nitrogen worldwide; we change the appearances of continents through land use practices, rising sea levels, and shrinking ice masses; we disperse some species widely while driving others to extinction; and we guide evolution through changes in gene flow, selective breeding, and genetic engineering. The human presence affects the distribution, reproduction, and community structure of forests as well as their very survival, and it will make the ecological consequences of future climatic changes unique in the history of the planet.

Theoretical modeling provides possible examples of what may lie ahead in terms of climate, but proxy records of geologic history can also help to show which scenarios are most realistic and provide examples of biotic responses to climatic shifts in the past.

Today's anthropogenic climatic effects are superimposed on a background of variability that includes cyclic and irregular fluctuations on multiple spatial and temporal scales. Long, high-resolution records from ice cores, tree rings, cave formations, and aquatic sediments show that abrupt and extreme climate events are not limited to human causes, and that many of today's tree taxa have experienced such changes before.

The last 50 million years of the Cenozoic era was dominated by cooling from the high- CO_2 hothouse of the Eocene "climatic optimum" (figure 1.1). The reasons for this are still unclear, but tectonism, weathering of the continents, and sequestration of carbon in marine sediments are likely contributors to the cooling trend (Garzione 2008). Temperatures fell low enough for an Antarctic ice cap to form between 45 and 34 million years ago, and during the last 3 million years temperatures have dropped far enough to trigger several dozen ice ages.

The overall cooling pattern of the Cenozoic was also punctuated by abrupt warming events. One of the most commonly cited examples was the PETM (Paleocene-Eocene Thermal Maximum) that occurred 56 million years ago and lasted roughly 200,000 years (figure 1.1; Dickens 2011). Atmospheric CO_2 concentrations are thought to have reached or exceeded 3,000 ppm following the release of several thousand gigatons (GT; billions of metric tons) of carbon-rich gases into the atmosphere, possibly through volcanism in the Atlantic

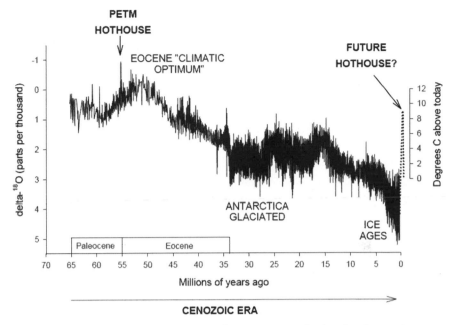

FIGURE 1.1. Deep-sea oxygen isotopes and temperatures during the Cenozoic era (after Zachos et al. 2008).

basin as well as other factors (Pearson and Palmer 2000; Dickens 2011). Global average temperatures rose 5°–10°C above their already warm states within 20,000 years or less, plant species migrated poleward, and insect herbivory on foliage increased, possibly in response to higher temperatures (Wing et al. 2005; Currano et al. 2008). Deciduous redwood forests encircled the Arctic Ocean, *Nothofagus* beech forests covered Antarctica, and ice-free, richly vegetated continents and land bridges facilitated the rapid migration of species (Bowen et al. 2002; Smith et al. 2006; Williams, Mendell, et al. 2008; Cantrill and Poole 2012).

Millennial-scale periodicities in the tilt, wobble, and orbital path of the Earth have been primary pacemakers of ice ages during the last 3 million years. Between cold glacial (longer) and stadial (shorter) periods, seasonal insolation cycles triggered warm interglacials and interstadials, as well. Sediment core evidence suggests that summers became wetter and 8°C or more warmer than today in Arctic Russia during insolation peaks between 3.5 and 2.5 million years ago that included repeated expansion of boreal forest over tundra (Brigham-Grette et al. 2013).

The last such warm period, often referred to as the Eemian Interglacial, produced regional temperatures 1°–3°C higher than today between 130,000 and 117,000 yr BP (years before present). The Arctic Ocean was partially ice free but most of the Greenland and Antarctic ice sheets remained intact despite occasional surges that lifted sea levels at least 7 m higher than today (Blanchon et al. 2009; Clark and Huybers 2009; Nørgaard-Pedersen et al. 2009). Conifers invaded Siberian tundra north of Lake Baikal, large stands of spruce, pine, and birch developed in southern Greenland, and woodlands in the Adirondack mountains of upstate New York resembled those of today's Blue Ridge, with pollen records from Eemian-age lake deposits revealing the prevalence of oak, hickory, and black gum (Muller et al. 1993; de Vernal and Hillaire-Marcel 2008; Granoszewski et al. 2004). Rainfall intensified abruptly over 200 years or less in monsoonal Asia, and greener, moister conditions in tropical Africa and the Middle East helped Stone Age peoples to migrate through what are now the Sinai and Negev Deserts (Schneider et al. 1997; Chen et al. 2003; Yuan et al. 2004; Vaks et al. 2007).

More rapid and short-lived disruptions also occurred during glacials (figure 1.2), including Dansgaard-Oeschger cycles and Heinrich events associated with ice sheet surges and extreme climate fluctuations. Around 17,000 yr BP, massive ice-rafting and cooling in the North Atlantic basin contributed to a sudden, catastrophic collapse of the Afro-Asian monsoon system that desiccated Lakes Victoria, Tana, and Van, and produced genetic bottlenecks in human populations in India ("Heinrich Stadial 1"; Stager et al. 2011). Around 13,000 yr BP, the Younger Dryas stadial represented an abrupt return to glacial-type conditions in much of the northern hemisphere that began within less than a decade in some locations and caused severe aridity in much of tropical Africa and southern Asia (Mayewski et al. 1993; Stager et al. 2002). The end of the Younger Dryas 11,700 years ago represented a rapid shift to the warmer conditions that have dominated the Holocene epoch to modern times.

During the last 11,700 years, the fluctuations preserved in ice core records were not as dramatic as they were during the preceding glacial period, leading to a common misperception that climates of the Holocene were stable before the Industrial Revolution. In fact, ecologically significant instability was still common, even at the poles (O'Brien et al. 1995; Mayewski et al. 2004). High summer insolation in the northern hemisphere during the early Holocene contributed to ice retreat on the Arctic Ocean and to

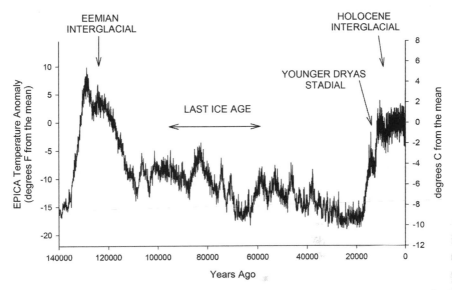

FIGURE 1.2. EPICA (European Project for Ice Coring in Antarctica) ice core record of Antarctic temperatures from the Eemian Interglacial to the present (after Augustin et al. 2004).

the expansion of lakes and forests throughout tropical Africa (DeMenocal et al. 2000; Stager et al. 2003; Kaufman et al. 2004), the effects of El Niño-Southern Oscillation (ENSO) increased notably after about 5,000 yr BP (Moy et al. 2002), and other rapid climate changes also occurred throughout the Holocene (Mayewski et al. 2004). Within the last millennium, regional warming during the Medieval Climate Anomaly (ca. 1,000–700 yr BP) brought severe drought to much of North America and East Africa, and more widespread cooling occurred during the Little Ice Age (ca. 600–200 yr BP), triggering alpine glacial advances in Europe (Mayewski et al. 2004; Maasch et al. 2005; Verschuren et al. 2009). During the last century, nonhuman sources of variability including ENSO, the North Atlantic Oscillation, shifting westerly wind tracks, and the eleven-year solar cycle have repeatedly disturbed temperature and precipitation regimes over wide areas of the planet (IPCC 2007b; Stager et al. 2007, 2012).

In sum, high-resolution paleoclimate records reveal far more natural variability than was once assumed from earlier work that failed to sample geological archives in sufficient detail. The rapidity of recent climatic changes is not, as has sometimes been suggested, by itself sufficient evidence to identify humans as the cause.

The ancestors of today's forests experienced numerous climatic shifts in the past, so change alone is not a unique threat in and of itself. However, these records also offer stern warnings about what may lie ahead as a result of human activities in the Anthropocene. The idea of an ice-free Earth, acidified oceans, and massive, rapid climatic disruptions due to greenhouse gas buildups are not merely extremist perspectives; we now know that these conditions existed before, even without major human impacts. And although a facile interpretation of geological history might lead one to conclude that climate change is not a threat because it is "natural and ongoing," a more careful reading of paleoecological records shows that extreme climatic shifts of the past would be unwelcome in today's world, which is now inhabited by more than 7 billion human beings. What does the future hold? Climates will continue to change as they always have, but the effects of human presence will redefine the baselines upon which natural variability plays out. We are essentially loading the world's weather dice through a hotter, more vigorously circulating atmosphere. The limitations of models along with uncertainties regarding human behavior and technology forbid precise portrayals of what lies ahead, but the general direction and nature of global-scale changes are clear. The more greenhouse gases that we release, the higher global mean temperatures will become and the farther inland oceans will advance. Paleoclimate records also show that, in general, large-scale warming has tended to increase the water content and extent of monsoon systems, to shift mid-latitude storm tracks poleward, and to reduce the extent of ice sheets, glaciers, and sea ice.

Surprises can also emerge from such a complex system. For example, although much of tropical Africa often became more arid during northern hemisphere cool periods and today tends to experience more intense rainfall in years just prior to solar maxima, an as-yet unexplained reversal of the cool-dry, warm-wet relationship produced dramatic lake level rises in East Africa during a prolonged solar minimum of the cool Little Ice Age, thereby weakening confidence in our understanding of how tropical climates operate (Stager et al. 2005, 2007; Verschuren et al. 2009). It has also been proposed that retreat of Arctic sea ice during recent years has contributed to rapid, erratic, and extreme swings in regional climates of the northern hemisphere (Francis and Vavrus 2012). We will not be able to model our way into complete pre-paredness for everything that the climate system may do in the future, but reasonable generalizations can nonetheless be made from long-term per-spectives on the nature and causes of climate change.

One source of insights into future carbon dioxide dynamics is the work of scientists such as David Archer, whose pioneering research at the University of Chicago has been corroborated by other investigators as well (Archer 2005; Archer and Brovkin 2008; Eby et al. 2008; Schmittner et al. 2008). The astoundingly long-term views of the future that these studies provide show that we are setting in motion a much larger and longer-lasting array of dis-ruptions than the relatively short-term global temperature rise that currently occupies our attention (Stager 2011).

At the heart of these findings is a simple question: "where does carbon dioxide go when it leaves our smokestacks and tailpipes?" Roughly three-quar-ters of it will dissolve directly into the oceans during the next several cen-turies to millennia, leaving slow weathering of carbonate and silicate min-erals to wash the airborne remainder into the sea over tens of thousands of years (figure 1.3). When fossil fuel emissions inevitably level off and decline, whether by design or by depletion, marine uptake will cause atmospheric CO_2 concentrations to level off and then to drop nearly as steeply as they rose until the oceans can absorb no more and mineral reactions more slowly consume the leftovers. During the relatively brief turnaround phase of "cli-mate whiplash," many of the selection pressures that operated in the con-text of rising temperatures may swing into reverse during the cooling that follows (Stager 2011).

The form and timing of the peak, whiplash, and the long tail of the cooling-recovery curve will largely depend upon how much carbon dioxide we release during the next century or so. In a relatively moderate emissions scenario such as B1 (IPCC 2007b) in which non-fossil energy sources quickly replace coal, oil, and gas, approximately1,000 GT of carbon will have been emitted since the Industrial Revolution. If instead we burn all remaining fos-sil fuel reserves in a scenario more like A2 (IPCC 2007b), then a total dis-charge of closer to 5,000 GT is more likely. This would lead to a higher, later, and more protracted peak and a much longer recovery (figure 1.3).

In one moderate scenario in which emissions decline after AD 2050, atmo-spheric concentrations of CO_2 reach 550–600 ppm by AD 2200 (figure 1.3;

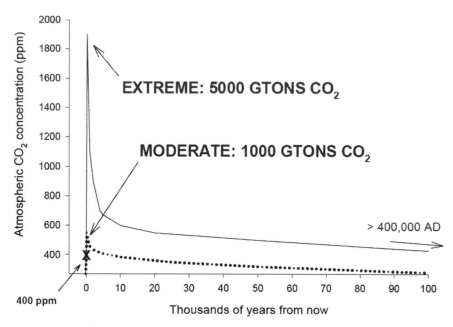

FIGURE 1.3. Carbon dioxide concentrations in the atmosphere under two emissions scenarios (after Archer 2005).

Stager 2011). At thermal maximum around AD 2200–2300, global average temperatures are 2°–4°C higher than today. After a whiplash stage lasting several centuries, CO_2 concentrations decrease steeply for several millennia due to marine uptake, and then fall within the range of preindustrial conditions after tens of millennia, possibly as long as 100,000 years. Even in this relatively mild case, the thermal effects of the excess carbon dioxide in the atmosphere are likely to prevent the next ice age, which orbital cycles could otherwise trigger around AD 50,000 (Berger and Loutre 2002; Archer and Ganopolski 2005).

In a more extreme scenario, CO_2 concentrations peak close to 2,000 ppm around AD 2300 and take at least 400,000 years to recover (figures 1.3, 1.4; Stager 2011). The whiplash stage lasts for several thousand years, producing a seemingly stable plateau of PETM-style warmth 5°–9°C warmer than today that could persist long enough for ecosystems and cultures to coevolve with before the long recovery period destabilizes them again. In both scenarios, the staggered responses of temperature and sea level to changing CO_2

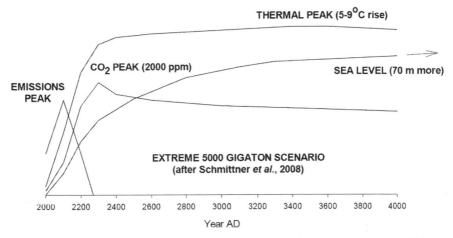

FIGURE 1.4. Sequential environmental changes expected in an extreme 5,000 GT carbon emission scenario (after Schmittner et al. 2008).

concentrations further complicate environmental settings for future forests as well as for human beings. In figure 1.4, for example, global mean temperature continue to climb for several centuries after the CO_2 peak, and sea levels continue to rise for thousands of years after the thermal peak because the temperatures are still high enough to melt continental ice masses.

What does the geological record reveal about possible consequences of such scenarios for future forests? The more moderate case has much in common with the Eemian interglacial. Although it was not caused by greenhouse gas buildups, it produced conditions warmer than today in many locations, particularly during summers in the northern hemisphere. Even though it lasted for about 13,000 years, it failed to de-ice the planet entirely, and polar bears and other arctic biota survived it. Extensive poleward migrations of forests and animals resulted, and rapid changes in sea level followed sporadic destabilizations of ice sheets, eventually submerging much of Florida.

The more extreme case has much in common with the PETM, which resulted from a greenhouse gas release comparable to our own. Fossil carbon is depleted in the stable isotope, ^{13}C, and a dramatic global decline in delta–1[3]C in PETM sediments due to enormous geological inputs of fossil carbon into the air and oceans is a diagnostic marker for that event. A similar global dilution of ^{13}C content of the world and its inhabitants is currently under way as a result of our own fossil carbon emissions, and its isotopic signal is being preserved in the geologic record as an anthropogenic sequel to the PETM. Warming during the PETM increased the intensity of rainfall, weathering, and runoff over most of the planet, and it left no refuge for cold habitats. Conditions similar to those of the PETM are likely to develop again in a "burn-it-all" emissions scenario, but several factors will differ in an Anthropocene reprise of former hothouse states. A modern return to the CO_2 concentrations and temperatures of 56 million years ago would involve increasing those parameters from a much lower thermal baseline that currently allows extensive cold-dependent ecosystems to exist at high latitudes and altitudes. The overall pace of the changes associated with such a return to an ice-free world, were they to occur over a span of several centuries, would outstrip those of the PETM and similar greenhouse warming of the earlier Cenozoic, which occurred when the atmosphere and oceans were already warmer than now (figure 1.1).

The responses of forests so far back in time may not be directly comparable to those of the Anthropocene, but more recent sediment records suggest that many of today's plant species may be quite resilient to climatic fluctuations if they are free to migrate in response. Pollen and other data from Arctic Russia 3.5–2.5 million years ago show that when summer temperatures there were 8°C or more higher than today, forests in the region changed their geographical distributions and abundances but still consisted of larch, pine, birch, alder, spruce, and other taxa that have also persisted through multiple glacial and interglacial periods to the present day (Brigham-Grette et al. 2013). Human presence, however, could seriously restrict such adaptive movements during the Anthropocene.

We cannot know exactly what new technologies will eventually arise or how future societies will respond to the climatic settings that we bequeath to them. Perhaps an ice-free Arctic will come to seem natural and preferable to the frozen state that we now consider to be normal, and what we would call "recovery" might be experienced as a global cooling disaster thousands of years from now. Even so, potentially important insights arise from such long-term perspectives:

  1. Human influences on the planet have become more powerful than many of us yet realize.

2. Rapidity of climatic change can be more ecologically stressful than the magnitude or direction of change. The last century's warming (ca. 0.7°C) proceeded at least twice as quickly as the onset of the Eemian interglacial, and in an extreme emissions scenario global mean temperature could rise by 2°C–5°C per century between now and AD 2300 (Figure 1.4), significantly faster than the onset of the PETM.

3. Although extreme climatic instability has occurred before, the restriction of free migration and other human impacts now make such instability more challenging for species and ecosystems of the Anthropocene.

Although the upward direction of global average temperature change is easy to anticipate, variable responses within different components of the climate system will make accurate prediction of regional- and local-scale conditions difficult. Long-term, global-scale climates are easier to simulate and predict than the more here-and-now, down-to-earth scales of change that many forest managers and urban planners deal with. The inherent limitations of global climate models can be magnified when they are downscaled to focus on relatively short time scales and specific regions, and demand for detailed projections on the regional and local scales sometimes leads people to ask more of climate models than they can reliably produce (Hulme et al. 2009; Heffernan 2010; Schiermeier 2010; Trenberth 2010). The limitations of global climate models are often magnified when they are downscaled to focus on relatively short time scales and specific regions, and linking these in turn to models of hydrology or biological processes can amplify errors further (Schiermeier 2010; Trenberth 2010; Beier et al. 2012).

A recent comparison of sixteen commonly cited models that were downscaled to the Lake Champlain watershed of Vermont and New York illustrates some of the problems that may be faced in such studies (Stager and Thill 2010). All of the models anticipated significant warming by AD 2100, but they disagreed on the magnitudes and seasonality of the changes. The question of seasonality has serious implications for forest ecology because the distribution of temperatures through the year affects ecologically important factors such as snowpack, spring runoff, and net water balance in summer. Perhaps the most reliable projection regarding future temperature in this region may simply be that it will increase as greenhouse gas concentrations rise, which one could conclude even without the aid of models.

Precipitation patterns are also important to forests, but they are more difficult to simulate and to predict (Schiermeier 2010). Precipitation can vary tremendously over small geographical areas, making it difficult to obtain accurate observational records of regional precipitation alone, much less to develop accurate predictive models. Topography, humidity, wind direction and speed, albedo, and other factors further complicate regional-scale modeling of future precipitation. ENSO also strongly influences precipitation patterns around the world, but there is as yet no consensus regarding its likely behavior in the future, and similar uncertainty obscures the future of the North Atlantic Oscillation, Pacific Decadal Oscillation, and other sources of variability as well (IPCC 2007b). In the case of the Lake Champlain study, the sixteen regional precipitation scenarios varied in magnitude, seasonality, and even in the direction of trends. Although it is common in such cases to note that the ensemble average of the models states "thus and so," it is difficult to know in advance which models are the most accurate, and sticking with the majority may mean rejecting a more reliable minority.

Despite these limitations, observational data and paleoclimate records can document regional and local patterns that have accompanied global-scale changes of the last century, and which can help to inform speculations about future changes. Such data indicate that poleward retreat of the austral westerly wind belt in a warming future could reduce winter rainfall over southwestern Australia and the fynbos region of South Africa (Biastoch et al. 2009; Stager et al. 2012). A warmer atmosphere is likely to be more energetic and turbulent and to carry more water vapor from the oceans and vegetation (IPCC 2007b), and an associated widening of the tropical rain belt has already been observed (Seidel and Randel 2007). The effects of such processes are also apparent in historical records of decreasing ice cover on Lake Champlain and a recent rise in the intensity of extreme precipitation events, which supports model projections of similar trends in a warmer future (figure 1.5; Stager and Thill 2010).

Even if climate models cannot tell us exactly what will happen in the future with complete certainty, they still provide valuable examples of what realistically could happen. High-quality, multiparameter simulations can reveal unexpected consequences from perturbations in complex systems that might otherwise be overlooked, and they offer ballpark ranges of variability that can sometimes be refined further through reference to historical and

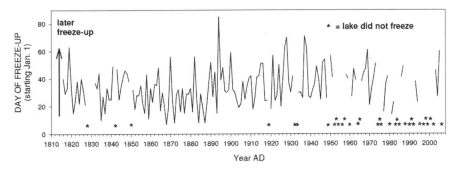

FIGURE 1.5. Freeze-up dates for the main body of Lake Champlain. Asterisks indicate winters in which the lake did not completely freeze (after Stager and Thill 2010).

geological records. The spectrum of precipitation variability suggested by diverse arrays of models may also provide helpful estimates of the possible magnitudes of such changes.

Forests have long experienced climatic instability in the past, but the most extreme, global-scale disruptions were uncommon in relation to the lifetimes of individual organisms. In addition, most of those events were not accompanied by very large increases in atmospheric carbon dioxide concentrations: Ecmian CO_2 levels, for example, remained well below the ca. 400 ppm concentrations that prevailed in 2014. Perhaps most important, none of the extreme climatic shifts of the past occurred while such a large array of additional anthropogenic factors were in operation. We are sailing into an uncertain future with no exact historical analogs to inform us and little more than educated guesswork to guide us, but it is clear that the rapid pace and global extent of Anthropocene warming will make it increasingly difficult to preserve biotic communities in their present form.

The ecological effects of long-term climatic change may be particularly abrupt on regional and local scales when physiological or physical thresholds are crossed. When winters are no longer cold enough to exclude an herbivorous insect or pathogen from a habitat, sudden changes in forest composition may result (Dale et al. 2001), and the melting point of snow and ice is a fixed boundary across which today's seasonally cold ecosystems will be pulled as the world warms. Old-growth forests that were established in western North America under cool, moist conditions of the Little Ice Age may become increasingly vulnerable to sudden replacement by other taxa through future disturbance events (Willis and Birks 2006). In addition, latitudinal shifts and meanders in major wind belts can bring rapid, extreme climatic changes to sites that lie adjacent to or beneath them.

Paleoecological records show that vegetational communities did not always move as coherent units during climatic shifts of the past, and differential migration rates and climatic tolerances of different species will likely produce new combinations in a warming future (Pitelka et al. 1997; Williams and Jackson 2007). As paleoecologist Tom Webb (1988) wrote: "plant assemblages are to the biosphere what clouds, fronts, and storms are to the atmosphere . . . they are features that come and go." Current distributions of tree species do not necessarily reflect their true potential bioclimatic niches, which may further limit our ability to predict their responses to future changes, and the effects of climatic shifts on animal communities can indirectly influence forests as well.

As warming pushes isotherms poleward in coming centuries, they may force taxa to enter regions in which their preferred soils, precipitation patterns, and/or light regimes are not available, and no-analog communities and novel environmental settings are likely to emerge in the future as they have in the geological past (Williams and Jackson 2007). If rising temperatures open the high Arctic to colonization by forests, for instance, plants that can tolerate the long darkness of circumpolar winters will be most likely to replace what is now tundra, potentially creating vegetational assemblages that have not existed since the early to mid-Cenozoic or, possibly, at any previous time in history (Sturm et al. 2005).

Although human intervention may produce new or "unnatural" ecosystems in the future, paleoecological records show that novelty is not unusual in Earth history. Humans have been affecting forest composition for tens of thousands of years, and differential rates and modes of dispersal in the face of environmental instability have often led to new combinations of species. Some of the vegetational communities that we are familiar with today are surprisingly young: oak savannas of northeastern Iowa and grassy montane parklands of northwestern Wyoming originated roughly 3,000 years ago, ponderosa pine forests of the Bighorn Basin arose 2,000 years ago, and mixed northern hardwood/conifer forests of northeastern Michigan may be only a thousand years old (Jackson 2006, 2012). In light of the ephemeral nature of many plant communities and the huge extent of the modern human footprint, a long-term Anthropocene perspective suggests that managing ecosystems in ways that include difficult tradeoffs, triage, or new combinations of species may not necessarily make them too unusual or unnatural to be desirable (Kareiva et al. 2007).

The unsteady nature of future climate will make the long-term stability of many forest communities an impossible goal, particularly for those that are confined within static and / or shrinking borders. Flexibility and resilience in the face of environmental and cultural change will become ever more important to the sound management of ecosystems in this new and unprecedented Age of Humans.

2

Invasive Plants, Insects, and Diseases in the Forests of the Anthropocene

ALEXANDER M. EVANS

The unprecedented mixing of species from across continents and ecosystems is one of the profound changes of the Anthropocene. Species introduced into completely different ecosystems are freed from the constraints that limited their growth and expansion in their home systems (Phillips et al. 2010). For example, plants can escape the herbivores adapted to feed on them, insects can escape the pathogens that limited their population growth, and newly introduced species can find new opportunities such as hosts with little resistance to their attack (Liebhold et al. 1995). The combination of fewer constraints and new opportunities allow some introduced species to flourish in their new environments to the detriment of native species; in short, to become invasive species (Torchin et al. 2003). Executive Order 13112 (1999) defines invasive species as alien species whose introduction causes economic or environmental harm or harm to human health. In many cases, the introduction of species into new ecosystems is an unintended consequence of human movement and trade (Bradley et al. 2011). Some invasives were introduced intentionally to bring useful plants and animals to new places for the benefit of humans (Reichard

DOI: 10.5876/9781607324591.c002

and White 2001). However, once introduced into a new ecosystem, invasive species are able to expand in that ecosystem without human assistance (e.g., Gibbs and Wainhouse 1986). As invasive species expand their range, they can create novel ecosystem interactions and unforeseen outcomes (Hobbs et al. 2006; Mascaro et al. 2012).

In addition to their ecological costs, exotic forest invaders have a large economic impact on forest products and ecosystems services (Pimentel et al. 2005; Holmes et al. 2009). For instance, a mere three invasive insects cause approximately $1.7 billion in damages in the United States annually (Aukema et al. 2011). By one estimate, the United States spends about $1.3 million a year on surveillance to keep just one pest, the Asian gypsy moth (*Lymantria dispar*), from invading (Work et al. 2005).

The negative impact of invasive species is likely to expand during the Anthropocene, exacerbated by the warming climate (Bradley et al. 2010), more frequent extreme climatic events (Diez et al. 2012), large and severe fires (Ziska et al. 2005), and forest fragmentation (Dewhirst and Lutscher 2009). As global trade continues to move vast cargoes across the world, the introduction of new species is almost inevitable. Work and colleagues (2005) estimate that about seven species are introduced to the United States each year via refrigerated maritime cargo alone. However, it is not just the invasive species that will thrive as the climate changes; even native insects, plants, and diseases may act more like invasive species in the Anthropocene under new climate conditions (Weed et al. 2013).

Invasive species will help define the forests of the Anthropocene, hence it is vital to understand the types of invaders we face, their impacts, and how they interact in natural ecosystems. While all ecosystems have been altered by invasive species, this chapter is limited to plants, insects, and diseases affecting forested ecosystems. Though animals such as the brown tree snake (*Boiga irregularis*) or feral pigs (*Sus scrofa*) have detrimental impacts on forested ecosystems, they are excluded from this chapter in an effort to limit an already expansive topic. For the same reason, this chapter also excludes invasion of wetland and coastal communities. While all the examples and most of the research cited here are drawn from the United States, the issue of invasives in the Anthropocene is international (e.g., Yan et al. 2001).

Humans are enthusiastic about importing new species of plants for economic benefit or aesthetic appeal, but these introductions frequently go wrong and result in exotic plants invading native forests (e.g., Forseth and Innis 2004). By one estimate, the horticultural trade is responsible for over 80 percent of invasive plants in the United States (Reichard and Hamilton 1997). Other common pathways include accidental introduction with crop seeds and purposeful introductions for soil erosion control (Reichard and White 2001). Many of the invasive plants in the United States are agricultural weeds—in other words, plants that interfere with crop production or graz-ing—but these are generally outside of the scope of this chapter.

Mapping from programs such as the Early Detection and Distribution Mapping System (www.eddmaps.org/distribution/) shows that invasive plants cover the entire United States. Although not every forested acre has been invaded by nonnative plants, they are ubiquitous at the county scale in the coterminous United States. For example, a study of twenty-four north-eastern and midwestern states found 66 percent of all plots had at least one invasive plant (Schulz and Gray 2013). Disturbed areas, particularly roadsides, accumulate invasive plants because many invasives are adept at colonizing open growing space (Aikio et al. 2012).

Invasive plants disturb ecosystems in a number of ways. Out of the 1,055 threatened plant species in the United States, about 57 percent are affected by invasive plants (though often in combination with other stressors) (Gurevitch and Padilla 2004). Invasive species outcompete and overwhelm native plant species. For example, kudzu covers some 7.4 million acres in the United States, where it shades out and crushes other plants (Forseth and Innis 2004). Similarly, stiltgrass outcompetes native plants, reduces herbaceous diversity, impedes native woody species regeneration, and creates extensive stiltgrass monocultures (Oswalt et al. 2007; Adams and Engelhardt 2009). Invasive plants can disrupt plant reproductive mutualism such as pollination or seed dispersal, causing population reductions (Traveset and Richardson 2006). An example of a less visible influence of the presence of invasive plants is the alle-lopathic effect of Tree of Heaven, which has a detrimental impact on red oak regeneration (*Quercus rubra*), an important tree economically and ecologically in the eastern US (Gómez-Aparicio and Canham 2008). Another example is melaleuca (*Melaleuca quinquenervia*), which has converted wetlands to uplands through increased litter inputs over many years (Strayer et al. 2006).

Invasive plants often negatively impact water quantity because they tend to grow fast and use more water than native species (Brauman et al. 2007).

Invasive plants alter, usually negatively, habitat for wildlife. Some reduction in habitat quality is to be expected where animals have adapted to a plant community that invasives subsequently disrupt. Birds that nest in honeysuckle (*Lonicera periclymenum*) and buckthorn (*Rhamnus cathartica*), for instance, experience higher predation rates than those nesting in native plants (Schmidt and Whelan 1999). Even when invasive species like buckthorn provide fruits for animals (birds in this case), these fruits are often less nutritious than those provided by the native species the invaders have displaced (Smith et al. 2013). About 28 percent of birds listed as threatened are negatively affected by invasive plants (Gurevitch and Padilla 2004).

There are some 455 invasive insects in US forests, though only about 62 cause significant ecosystem damage (Aukema et al. 2011). Of those insects that have a significant impact on forested ecosystems and feed on trees, about a third feed on sap, a quarter are wood borers, and the remainder feed on foliage (Aukema et al. 2010). Over the last century, an average of about 2.5 nonnative insects have been newly detected in the Umited States per year (Aukema et al. 2010) and Koch and colleagues (2011) predict new alien forest insect species establishments every five to fifteen years in select urban areas. Not every foreign insect that establishes in the United States becomes a destructive invasive, but many have. Some of these insects, such as the gypsy moth, have been in this country for over a century, and many have spread through the entire range of their new hosts. Mapping tools such as the Alien Forest Pest Explorer illustrate that at least one or more invasive forest insects infest every forested region in the United States.

Many invasive insects are specialists that feed on, or live in, one particular tree or shrub species or genus. For instance, hemlock woolly adelgid (*Adelges tsugae*) feeds only on species of hemlock. Others, such as the gypsy moth, attack a broad range of tree species. The Northeast and Appalachian forests have a particularly high number of destructive insects in part because of their proximity to busy eastern ports and in part because of the large number of tree species that can support a large number of species-specific invaders (Licbhold et al. 2013). In contrast, western interior forests have fewer different species of invasive insects (Liebhold et al. 2013).

Insect populations often expand and collapse in response to environmental conditions. For native insects, populations can be very low and individuals difficult to find until conditions are right for an outbreak. The population then eventually crashes due to declines in the host, lack of available food, climate shifts, predator response, or pathogens that spread easily at high population densities. Invasive species can likewise build large, outbreak-type populations as they invade new areas because of the lack of constraints in the new environment. Because these are novel outbreaks, native trees are ill-equipped to resist or recover from them. For example, populations of hemlock woolly adelgid can be very abundant once they have established in a new area, but even though adelgid populations decline as the health of hemlock trees decline, the outbreaks result in significant hemlock mortality (McClure 1991).

Polyphagous insects can cause a reduction in tree growth through massive defoliation, but species- or genus-specific invaders can also have disastrous impacts on forested ecosystems. By 2006, some 15 million ash trees had been killed by the emerald ash borer (*Agrilus planipennis*) (Poland and McCullough 2006). This widespread mortality has cascading effects through the ecosystems with ash trees, including the loss of native insects (Gandhi and Herms 2010b). The death of hemlocks from hemlock woolly adelgid affects herbaceous plants (Eschtruth et al. 2006), nutrient cycling (Cobb et al. 2006), stream temperatures, fish communities (Ross et al. 2003), bird diversity (Tingley et al. 2002), and habitat for deer and other mammals (DeGraaf et al. 1992). More generally, by removing important trees from US forests, invasive insects have the potential to affect fundamental forest composition, structure, and function (Ellison et al. 2005, Gandhi and Herms 2010a). The complexity of interdependencies within ecosystems makes it difficult to trace the full impact of invasive forest insects (Kenis et al. 2009).

There are likely many more nonnative disease-causing organisms in the United States than have been identified because they are often difficult to detect. As with nonnative insects, those we are most aware of are those that cause serious damage. For example, an early introduction, chestnut blight (*Cryphonectria parasitica*), functionally removed American chestnut (*Castanea dentata*) from its ecological role as a dominant tree in eastern forests by the 1950s (Tindall et al. 2004). Although the list of significant invasive forest diseases is shorter than that of insects, significant invasive forest diseases cover most forested regions of the United States (Aukema et al. 2010). Chestnut blight, Dutch elm disease (*Ophiostoma* spp.), and butternut canker (*Sirococcus clavigignenti-juglandacearum*) cover the entire range of their host trees (Evans and Finkral 2010). Beech bark disease (*Ophiostoma* spp.) has spread through forests where beech trees (*Fagus americana*) are most dense (Morin et al. 2007). Based on past spread rates, it is likely that other significant diseases including sudden oak death (*Phytophthora ramorum*), dogwood anthracnose (*Discula destructiva*), laurel wilt (*Raffaelea lauricola*), and phytophthora root rot (*Phytophthora cinnamomi*) will likewise expand to fill their ecological niche in the United States (Evans and Finkral 2010).

A lack of coevolution between host and pathogen can result in limited resistance in the host tree and excessive aggressiveness (i.e., greater host mortality) in the pathogen, which in turn causes disease outbreaks (Brasier 2001); there is, for example, very limited genetic resistance of tanoaks (*Notholithocarpus densiflorus*) to sudden oak death (Hayden et al. 2011). Diseases introduced to forests have removed dominant tree species, reduced diversity, altered disturbance regimes, and affected ecosystem function (Liebhold et al. 1995, Mack et al. 2000). The cascading effects of the removal of important trees species are similar to the effects of invasive insects and influence forest structure as well as the animals and plants connected to the diseased trees.

Humans are also a key factor in the spread of invasives. The transportation of firewood has been identified as an important vector for invasive insects, particularly long-distance dispersal (Bigsby et al. 2011; Koch et al. 2012). Human development and infrastructure also help invasive species flourish. Many invasive plants such as Asian bittersweet and multiflora rose (*Rosa multiflora*) thrive in disturbed areas and the open-edge habitat that human development creates (Yates et al. 2004; Kelly et al. 2009). The trees of these disturbed, edge habitats may also be more stressed and hence more susceptible to insects and diseases. In one Ohio study, 84 percent of new emerald ash borer infestations were within 0.6 miles (1 km) of major highways (Prasad et al. 2010)

Even low-density residential areas are associated with a greater density of invasive plants (Gavier-Pizarro et al. 2010). The effect of human land use on invasives lasts a long time, as demonstrated by a study that links invasive plants in North Carolina with historical land use and reforestation (Kuhman et al. 2010).

About one-third of the coterminous United States was human dominated by 2001, and an additional 35,600 square miles (92,200 km², or roughly the size of Indiana) are likely to be converted from natural cover to development by 2030 (Theobald 2010). About 15 percent of the current acreage of southern forests could be converted to housing and other uses by 2040 (Hanson et al. 2010).

Although the long-term trend in the Northeast during the twentieth century was one of increasing forest cover, this trend has recently reversed, and the total number of forested acres has started to decline again (Drummond and Loveland 2010). As much as 909,000 acres (368,000 ha), or about 2 percent of forest land, could convert from forest to other land uses in Maine, New Hampshire, Vermont, and New York by 2050 (Sendak et al. 2003). This growing human presence and increased fragmentation is a significant driver in the spread and domination of invasive species in US forests (Lundgren et al. 2004; Gavier-Pizarro et al. 2010; Schulz and Gray 2013). An indirect effect of fragmentation and suburbanization is the population growth of animals that thrive in human environments. For instance, white-tailed deer (*Odocoileus virginianus*) populations have grown significantly in many suburban/forest interface zones. The high deer populations help spread invasives and, at the same time, hamper the regeneration of native species (Evans 2008; Williams, Ward, and Ramakrishnan 2008).

Human-driven changes to the climate also benefit invasives. A warming climate opens new ecosystems to invaders previously limited by cold. Warming will facilitate the spread of invasive plants such as kudzu and privet (*Ligustrum sinense*) as far north as New England by 2100 (Jarnevich and Stohlgren 2009; Bradley et al. 2010).

In general, invasive plants have been far better able to respond to recent climate change in New England than native species (Willis et al. 2010). Warming will also facilitate the spread of invasive insects such as hemlock woolly adelgid (Evans and Gregoire 2007). Two or three times more forest in Canada will be at risk from gypsy moth by 2060 because of a changing climate (Régnière et al. 2009). Similarly, climate changes will modify forest pathogen dynamics and may exacerbate some disease problems (Sturrock et al. 2011). For instance, sudden oak death has the potential to expand its range under a warming climate (Venette and Cohen 2006). Increasing summer temperatures appear to exacerbate outbreaks of cytospora canker (*Valsa melanodiscus*) and mortality of alders (*Alnus incana*) in the southern Rocky Mountains (Worrall et al. 2010).

A changing climate means more than just warming temperatures. Other climate changes such as increased CO_2 concentrations and more frequent and more powerful storms will benefit invasives. Rising CO_2 concentrations commonly give invaders an extra edge in competition with native species (Manea and Leishman 2011): cheatgrass, for one, is able to take advantage of increased CO_2 concentrations by increasing productivity (Smith et al. 2000). Higher CO_2 levels help kudzu and honeysuckle tolerate cold temperatures and hence expand these species' capacity for invading new forests (Sasek and Strain 1990). Moreover, hurricanes, ice storms, wind storms, droughts, and fire can all create forest disturbances that invasive species can capitalize on (Diez et al. 2012). A study in Florida found that nearly 30 percent of the species regenerating after Hurricane Andrew were invasive and that invasive vines negatively affect the regeneration of native plants (Horvitz et al. 1998). Similarly, tufted knotweed (*Polygonum caespitosum*) and mile-a-minute weed (*Persicaria perfoliata*) were able to expand after Hurricane Isabel hit Maryland (though garlic mustard decreased because of the increased light) (Snitzer et al. 2005).

The warming and, in many regions, drying predicted for the United States will increase the area burned in the United States over the next century (Moritz et al. 2012). Some invasive species contribute to the increase in fire activity. Cheatgrass provides surface fuel that spreads fire more frequently than before the grass's invasion (Ziska et al. 2005). Sudden oak death also encourages fire by killing trees and creating more heavy fuel (Valachovic et al. 2011). This synergy between sudden oak death and fire has caused a fourfold increase in the mortality risk for redwood trees (*Sequoia sempervirens*) (Metz et al. 2013). While many native species are adapted to fire, altered fire regimes (more frequent or more severe fires) can benefit invasives. Uncharacteristically severe fire kills dominant vegetation that would have survived more natural fire and thus can create growing space for invasives.

Is there any hope for native forest ecosystems in the Anthropocene? For conservationists, ecologists, foresters, wildlife biologists, and all those who work in the woods, the answer must be yes. The first key element in any response to invasive species should be concerted effort to limit new introductions (Hayes and Ragenovich 2001; Lodge et al. 2006). Increased surveillance at ports and other introduction pathways can limit the growth of the invasive problem. Improved early detection strategies directed at a quarter of US agricultural and forest land would likely be able to detect 70 percent of invaded counties (Colunga-Garcia et al. 2010). If an invasive species avoids detection, a rapid response can help limit establishment (Anderson 2005). Similarly, policy or management actions that limit fragmentation and carbon emissions will rein in the negative interactions between invasives and these other forest stressors. There are steps that forest land owners and managers can take to increase ecosystem resistance to the effects of climate change and resilience to negative impacts of invasive pests and plants (Waring and O'Hara 2005). Eradication is impossible for many invasives and management should focus on those invasives that cause the most damage or those that can be effectively removed (Ellum 2009). A cornerstone of forest management in the face of the uncertainties of the Anthropocene is maintaining species diversity (Linder 2000). Maintaining or restoring species diversity on a site can increase the likelihood that some native species will flourish in this new epoch. Intact, diverse forest ecosystems may be more resistant to invasion (Jactel et al. 2005; Huebner and Tobin 2006; Mandryk and Wein 2006). For example, the impact of sirex wood wasp has been less dramatic in the diverse forests of the United States than in the single-species plantations in the southern hemisphere (Dodds et al. 2010).

Even in the Anthropocene, invasives are not invincible. Much of their competitive advantage comes from escaping the predators, pests, and pathogens of their region of origin. When those predators, pests, and pathogens catch up with an invader in a new region, the invader is less able to cause unusual damage or disrupt ecosystems. For example, *Entomophaga maimaiga*, a fungus that attacks gypsy moth, appears to have begun to limit the extent and impact of outbreaks in the areas longest infested by gypsy moth (Andreadis and Weseloh 1990). Similarly, a leaf blight has been discovered on stiltgrass that can cause reduced seed production, wilting, and, in some cases, death of stiltgrass plants (Kleczewski and Flory 2010). In a third example, an insect pest that can significantly retard the growth of kudzu has recently been found in Georgia (Zhang et al. 2012). As with biological control of invasive plants and insects, human intervention may also be able to change the dynamics of some invasive pathogens. New transgenic techniques hold promise for engineering resistance into tree such as elm and chestnut to battle exotic diseases (Merkle et al. 2007). Because genetic resistance to invasive diseases may vary in a native tree population, identifying and protecting potential resistant individuals is an important management response (Schwandt et al. 2010). For example, selection and breeding present a possible route to increasing resistance to beech bark disease in American beech populations (Koch et al. 2010).

As climate change alters ecosystems, there is the possibility that new restoration opportunities may emerge; canopy openings created by hurricanes and other storm events, for instance, could provide ideal planting sites for the restoration of American chestnut (Rhoades et al. 2009). In addition, climate change may render some areas unfavorable to invasives that previously seemed entrenched. Models suggest that cheatgrass will no longer be viable in some areas of the western United States as the climate warms (Bradley and Wilcove 2009). In these locations, cheatgrass could be replaced with native species. Managers should be ready to seize these novel restoration opportunities if and when they emerge during the Anthropocene.

Though it can be considered heresy, invasive species may not be all bad. Some can provide ecosystem services, while others might fill novel ecological niches created by climate change and inaccessible to native species. Invasive tamarisk provides habitat for the endangered willow fly catcher (*Empidonax traillii*) (Shafroth et al. 2005). With the recent introduction of the tamarisk leaf beetle (*Diorhabda carinulata*), which reduces the competitive advantages of tamarisk (Pattison et al. 2011), it is worth reconsidering tamarisk's potential positive role in riparian ecosystems. A study in Hawaii demonstrates that though invasives caused the decline of native tree species, the new species were able to maintain some ecosystem functions (Mascaro et al. 2012). While protecting against new invasives and fighting the spread of existing invasives are important, it may be time to accept some nonnative species.

Protecting refuges, such as parks and preserves, where threatened native species face fewer stressors may help those native species survive through the Anthropocene. Outside of parks and preserves, management that fosters diversity at the stand and landscape scales can help minimize the threat of invasives. Managers must be ready to embrace any opportunities for proactive restoration that may emerge because of a warming climate, species shifts, or disturbances. For entrenched invasives, conservationists may have to move from denial to acceptance and adapt forest management to a new mix of species. Though invasives are a significant threat to forests in the Anthropocene, all is not lost.

3

Climate and Wildfire in Western US Forests

Anthony L. Westerling, Timothy J. Brown,
Tania Schoennagel, Thomas W. Swetnam,
Monica G. Turner, and Thomas T. Veblen

Climate change is generating higher temperatures and more frequent and intense drought (Cayan et al. 2010; Peterson et al. 2013). Globally, the 1980s, 1990s, and 2000s have each in turn been the warmest decade in history (Arndt et al. 2010). In the United States, 2014 was the warmest year on record (Blunden and Arndt 2013), and drought has become more widespread across the western United States since the 1970s (Peterson et al. 2013). Climate projections suggest increased likelihood of heat waves in the western United States and droughts in the Southwest (Wuebbles et al. 2013), and the fire season and area burned are expected to increase substantially by midcentury across the western United States due to expected climate change (Yue et al. 2013a).

Climate—primarily temperature and precipitation—influences the occurrence of large wildfires through its effects on the availability and flammability of fuels. Climatic averages and variability over long (seasonal to decadal) time scales influence the type, amount, and structure of the live and dead vegetation that comprises the fuel available to burn in a given location (Stephenson

1998). Climatic averages and variability over short (seasonal to interannual) time scales determine the flammability of these fuels (Westerling et al. 2003).

The relative importance of climatic influences on fuel availability versus flammability can vary greatly by ecosystem and wildfire regime type (Westerling et al. 2003; Littell et al. 2009; Krawchuk and Moritz 2011). Fuel availability effects are most important in arid, sparsely vegetated ecosystems, while flammability effects are most important in moist, densely vegetated ecosystems. Climate scenarios' changes in precipitation can have very different implications than changes in temperature in terms of the characteristics and spatial location of wildfire regime responses (viz., changes in fire frequency, average area burned, and fire severity). While climate change models generally agree that temperatures will increase over time, changes in precipitation tend to be more uncertain, especially in arid mid-latitude regions (Dai 2011; Moritz et al. 2012; Gershunov et al. 2013). In ecosystems where wildfire risks have been strongly affected by variations in precipitation, there is less certainty about how these wildfire regimes may change. Yet in ecosystems where wildfire risks have been sensitive to observed changes in temperature, climate change is likely to lead to substantial increases in wildfires. As climate change alters the potential spatial distribution of vegetation types, ecosystems and their associated wildfire regimes will be transformed synergistically.

The type of vegetation (i.e., fuels) that can grow in a given place is governed by moisture availability, which is a function of precipitation (via its effect on the supply of water) and temperature (via its effect on evaporative demand for water) (Stephenson 1998). As a result, the spatial distribution of vegetation types and their associated fire regimes is strongly correlated with long-term average precipitation and temperature (e.g., Westerling 2009). Climatic controls (temperature and precipitation) on vegetation type along with successional stage largely determine the biomass loading in a given location, as well as the sensitivity of vegetation in that location to interannual variability in the available moisture. These factors in turn shape the response of the wildfire regime in each location to interannual variability in the moisture available for the growth and wetting of fuels. Cooler, wetter areas (forests, woodlands) have greater biomass, and wildfires there tend to occur in dry years. Warmer, drier areas (grasslands, shrublands, pine savannas) tend to have less biomass and wildfires there tend to occur after one or more wet seasons or years (Swetnam and Betancourt 1998; Westerling et al. 2003; Crimmins and Comrie 2004).

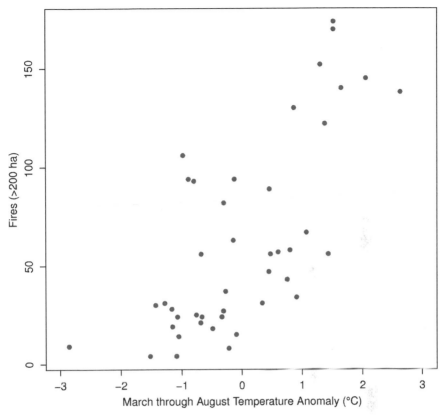

FIGURE 3.1. Scatter plot of annual number of large (> 200 ha) forest wildfires versus average spring and summer temperature for the western United States. Forest Service, Park Service, and Bureau of Indian Affairs management units reporting 1972–2004. Fires reported as igniting in forested areas only. (Source: Westerling 2009.)

Some locations are much more sensitive to variability in temperature than others. In the western United States, cool, wet, forested locations tend to be at higher elevations and latitudes where snow can play an important role in determining summer moisture availability (Sheffield et al. 2004). Above-average spring and summer temperatures in these forests can have a dramatic impact on wildfire, with a highly nonlinear increase in the number of large wildfires above a certain temperature threshold (figure 3.1). Westerling et al. (2006) concluded that this increase is due to earlier spring snowmelt and a longer summer dry season in warm years. They found that years with early arrival of spring account for most of the forest wildfires in the western United States (56 percent of forest wildfires and 72 percent of area burned, as opposed to 11 percent of wildfires and 4 percent of area burned occurring in years with a late spring).

Fire severity tends to be highest, with large infrequent stand-replacing fires that burn in the forest canopy, in cooler, more moist forests at generally higher elevations and/or latitudes, such as the lodgepole pine forests in the northern and central Rocky Mountains (Baker 2009). Prior to the era of extensive/intensive livestock grazing (post-1850s) and active fire suppression by government agencies (post-1900s), warmer, drier forests tended to have mixed or low severity; more frequent fire, with more of the fire concentrated in surface fuels (grass, shrub, forest litter); and less tree mortality (Allen et al. 2002). Increased fuel loads due to historic fire suppression and ongoing land use changes, combined with more extreme climatic conditions, have resulted in high-severity fire in some forests where it was rare prior to the twentieth century (Miller et al. 2009). The frequency of large (> 1,000 acre) forest fires and the area burned in those fires has continued to increase steadily over the last three decades as temperatures have risen throughout the region.

Forests of the northern and central Rocky Mountains where fire typically burned with high severity but was infrequent have been the most sensitive to changes in temperature, accounting for the largest share of the increase in burnt forest area (Westerling et al. 2006). As discussed below, projections of additional increases in future temperatures imply further increases in fire activity. However, warming and fire frequency may increase past critical thresholds, with some forests no longer able to sustain large high-severity fires. That is, fuel availability may become a limiting condition on fire in areas where climatic controls on fuel flammability were recently the dominant constraint on fire.

Climate in the Rockies is generally semiarid with summer-dry conditions in the northwest of the range, and summer-wet in the southeast (Bailey 1996), and generally moister and cooler conditions relative to regions at lower latitudes. Elevation ranges from 3,000 to 7,000 feet in the southern and central portions, and 3,000 to over 9,000 feet in the northern portion. Mixed evergreen-deciduous forests dominate montane and subalpine elevations in the north, with strong topographic controls on moisture fostering diverse forest vegetation zones to the south (Bailey 1996; Cleveland 2012). Forests with characteristically infrequent high-severity, stand-replacing fires account for the largest area (mixed spruce-fir, lodgepole pine), with significant forest area character-ized by mixed- (e.g., Douglas fir) and low- (e.g., ponderosa pine) severity fire regimes prior to the historical fire suppression era (Schoennagel et al. 2004).

Some northern Rockies ponderosa pine forests, usually associated with low-severity surface fire regimes, may have experienced occasional high-severity, stand-replacing fires during extended droughts of past millennia, as inferred from sedimentary charcoal studies (Pierce et al. 2004). However, the patch sizes of these ancient high-severity fires within ponderosa pine–domi-nant or mixed forests are unknown for almost all forests of these types, and it is possible that current large, high-severity patch sizes and subsequent geo-morphic responses may be unique over the late Holocene, as similar sedimen-tary charcoal studies in Colorado pine and mixed conifer forested watersheds suggest (Bigio et al. 2010). In the only detailed, highly systematic study of tree age structures and fire scar evidence at landscape scales in northern stands of ponderosa pine (i.e., in the Black Hills of South Dakota), Brown et al. (2008) found that about 3 percent of the landscape experienced high-severity fires during the three and one-half centuries prior to 1893, and overall, frequent, low-severity surface regimes appear to have dominated those landscapes.

In the Rockies' northern forests where infrequent, large high-severity fires occurred, these events likely were driven by extended drought associated with high-pressure atmospheric blocking patterns (Romme and Despain 1989; Renkin and Despain 1992; Bessie and Johnson 1995; Nash and Johnson 1996; Baker 2009). Paleofire studies support a strong influence of climate on fire-return interval (e.g., Whitlock et al. 2003, 2008; Millspaugh, Whitlock, and Bartlein 2004), with fuel controls playing a much lesser role (Higuera et al. 2010).

Burned area is historically concentrated in a relatively small number of very large fire events (Balling et al. 1992; Schoennagel et al. 2004; Baker 2009). From 1972 to 1999, 66 percent of burned area in the northern Rockies occurred in only two years (1988 and 1994), and 96 percent of burned area in the Greater Yellowstone area occurred in one fire year (1988) (Westerling et al. 2011b). This pattern is consistent with climatic controls on the flammability of plentiful fuels being the dominant constraint on the occurrence and spread of large wildfires (Littell et al. 2009): namely, large areas burn in rare dry years.

The effect of changes in the timing of spring on wildfire has been particu-larly pronounced in the higher-latitude (> 42° north), mid-elevation (1,680–2,590 m) forests of the Rocky Mountains, which account for 60 percent of the increase in forest wildfires in the western United States (Westerling et al. 2006). Higher-elevation forests in the same region had been buffered against these effects by available moisture, while lower elevations have a longer summer dry season on average and were consequently less sensitive to changes in the timing of spring.

The frequency and extent of wildfire is projected to continue to increase in coming decades until fuel availability and continuity become limited and supplant climatic controls on flammability as the dominant constraint on the spread of large wildfires by midcentury in the Greater Yellowstone region (Westerling et al. 2011b) and in the Rockies more generally. Increased burned area of similar magnitude has been projected by the National Research Council (2011), applying models from Littell et al. (2009) (see also Climate Central 2012).

Colorado and Utah also experience high geographic and interannual variability in temperature and precipitation due to elevation, topography, and latitude. A number of low-elevation forests (e.g., below 2,100 m in the central Colorado Front Range; Sherriff and Veblen 2008) with grass or other fine-fuels in the understory record regional fires during dry summers when preceded by increased spring-summer moisture availability up to four years prior, which enhances fine-fuel accumulation and contributes to fire spread when subsequently cured (Donnegan et al. 2001; Grissino-Mayer et al. 2004; Brown et al. 2008; Sherriff and Veblen 2008; Gartner et al. 2012). Moister, higher-elevation forests lacking grass understories do not record this wet-dry signature in the fire record (Sibold and Veblen 2006; Brown et al. 2008; Schoennagel et al. 2011). Documentary records of area burned in ecoregions encompassing Colorado and Utah showed that moist antecedent conditions are associated with greater area burned (and were more important than warmer temperatures or drought conditions in the year of fire) in grasslands, shrublands, and arid low-elevation woodlands with grass or shrub understories, but only fire-year conditions were significant in moister high-elevation and/or west-slope forests (Knapp 1995; Westerling et al. 2003; Collins et al. 2006; Littell et al. 2009).

Littell et al. (2009) found that area burned in the southern Rockies (1977–2003) was positively related to winter temperature and negatively related to spring temperature, along with spring and summer precipitation and lagged drought (r^2 = 0.77; Littell et al. 2009). Predictions for Utah-Nevada mountains were linked to lagged spring temperature but were much less robust

(r^2 = 0.33). The southern Rockies only accounted for less than 1 percent of recent increase in wildfire activity since 1985, in contrast to the northern Rockies, which accounted for 60 percent, primarily related to longer fire seasons and snowpack reduction (Westerling et al. 2006).

Average annual summer and winter temperatures are expected to increase dramatically in Colorado and Utah by 2050, yet models show low agreement for precipitation (figure 5.1 in Ray et al. 2010). However, Seager et al. (2007) predict that the Southwest (125°W–95°W, 25°N–40°N, which includes most of Colorado and Utah) will become more arid during the next century as annual mean precipitation minus evaporation becomes more negative. Similarly, Gutzler and Robbins (2011) predict that higher evaporation rates due to positive temperature trends will exacerbate the severity and extent of drought in the semiarid West.

Brown et al. (2004) predict that reduced relative humidity will increase the number of days of high fire danger at least through the year 2089, compared to the base period, however, the Colorado Rockies and Front Range showed no change in predicted fire risk thresholds, suggesting little change in wildfire activity. This contrasts with a study (Spracklen et al. 2009) that predicts higher temperature will increase annual mean area burned by 54 percent by the 2050s, relative to the 1980–2004 period, with the entire Rocky Mountains showing large increases (78 percent) and high interannual variability.

The National Research Council (2011) predicts that burn area in parts of western North America may increase by 200 to 400 percent for each degree (°C) of global warming relative to 1950–2003, adapting methods developed by Littell et al. (2009) to use temperature and precipitation as the predictor variables. Across Colorado and Utah, the southern Rocky Mountain Steppe Forest is predicted to experience the greatest increase in mean annual area burned (> 600%), with the least in the Nevada-Utah Mountains (only 73%).

The Southwestern United States (Arizona and New Mexico) is generally a semiarid region. Considerable topographic relief, however, results in a very diverse biotic landscape and consequent differences in vegetation and wildfire. These differences are often expressed along relatively short distances (tens of kilometers) and elevational gradients from desert basins to forested mountains. Natural fire regimes along these gradients vary from essentially no spreading wildfires in the pre-twenty-first-century historical record (e.g., lower Sonoran Desert), to frequent, low-severity surface fires (e.g., mid-elevation ponderosa pine forests, with intervals between widespread fires ranging from two to twenty years), to low-frequency, high-severity, stand-replacing fires (high-elevation spruce-fir forests, with intervals between large crown fires ranging from 150 to more than 300 years) (Swetnam and Baisan 1996, 2003; Margolis et al. 2007; Margolis and Balmat 2009).

Seasonal climate of the Southwest is characterized by bimodal precipitation, with winter-cool season and summer-warm season maxima, with a pronounced dry season during most years in late spring to early summer. The peak of fire activity tends to occur in this warm/dry season with a maximum area burned in the driest weeks of June, and the maximum number of fire ignitions in July when monsoonal moisture and convective activity generates large numbers of lightning strikes (Crimmins 2006; Keeley et al. 2009). Human-set fires are also important in Southwestern landscapes, in the distant past (i.e., by Native Americans), and in the modern era. During some seasons and years human-set fires exceed areas burned by lightning-set fires, especially during some recent years when extraordinarily large fires were set accidentally or purposely during spring-summer droughts. Paleo- and modern records of fire and climate show the strong importance of prior cool-season and current spring-through-summer moisture indices to fire activity in this region (especially regionally synchronized fire events in the paleorecord and total area burned/fire season/year in the modern record (Swetnam and Betancourt 1998; Westerling et al. 2002; McKenzie et al. 2004; Crimmins and Comrie 2004; Crimmins 2006; Holden et al. 2007; Littell et al. 2009; Williams et al. 2013).

Because comprehensive documentary evidence of wildfire goes back only a few decades, proxy paleorecords of past fire and climate activity have been developed to provide annual- to millennial-scale perspectives on fire, vegetation, and inferred climate variability (Swetnam and Baisan 1996; Swetnam and Brown 2010; Falk et al. 2011; International Multiproxy Paleofire Database (http://www.ncdc.noaa.gov/paleo/impd/impd_data_intro.html); Anderson et al. 2008; Frechette and Meyer 2009; Bigio et al. 2010). These paleorecords demonstrate the following specific findings:

1. Widespread surface fires were ubiquitous in ponderosa pine forests and mixed-conifer forests across the Southwest region before the advent of extensive livestock grazing in the late nineteenth century and active fire suppression by government agencies beginning about 1910. High-severity,

stand-replacing crown fires occurred in some dense pinyon-juniper wood-lands (Romme et al. 2009), shrublands, and higher-elevation spruce-fir for-ests (Margolis et al. 2007; Margolis and Balmat 2009) in the pre-1900 period, but large, high-severity fires were rare in ponderosa pine forests. Although some evidence of high-severity fires in ponderosa pine and mixed-conifer forests has been found in charcoal sediments (e.g., Frechette and Meyer 2009; Bigio et al. 2010), and small patch size (< 200 ha) high-severity fires have been reconstructed in a few tree-ring studies (Swetnam et al. 2001; Iniguez, Swetnam, and Baisan 2009), we lack any clear evidence at this time that large patch size (> 200 ha) high-severity fires occurring in ponderosa pine-dominant forests in the past were as extensive as those occurring today (Cooper 1960; Allen et al. 2002).

2. Extreme droughts and regional fire activity in the Southwest are highly correlated over the past four centuries in the available tree-ring record. Lagging patterns are evident in lower-elevation forests and woodlands, with wet conditions in the prior one to three years, coupled with dry conditions during the current year, often leading to extensive regional fire years in the past (Swetnam and Betancourt 1998).

3. Decadal-scale variation in past fire activity is evident in parts of the Southwest, with occasional periods of one to two decades of either decreased or increased local to regional fire activity (Swetnam and Betancourt 1998; Grissino-Mayer and Swetnam 2000; Brown and Wu 2005; Margolis and Balmat 2009; Roos and Swetnam 2011). Many studies have shown some association between these annual-to-decadal-scale patterns and climatic variations (e.g., Swetnam and Betancourt 1990, 1998; Kitzberger et al. 2007; Brown and Wu 2005).

4. There are relatively few long-term, sedimentary charcoal-based records of fire activity in the Southwest compared to other more mesic regions with more lakes and bogs. The available records do show, however, decadal-to-centennial-scale variations in fire and vegetation that are likely associated with climatic variations on those time scales (e.g., Anderson et al. 2008). One striking finding in a comparison of tree-ring and charcoal-based fire histories is the unprecedented lack of fire in the most recent century (due to livestock grazing and fire suppression) in a record of more than 7,000 years (Allen et al. 2008).

The longest modern records for the Southwest show a similar pattern to that observed in some other forests across the western United States during the twentieth and twenty-first centuries, namely, some large fires occurred during early decades of the twentieth century, there were lower levels of fire activity during the mid-twentieth century (but with several large events > 5,000 ha during the 1950s drought), and after the late 1970s a rather sharp rise in numbers of large fires and area burned occurred (e.g., Rollins et al. 2001; Holden et al. 2007).

The post-2000 period includes several fires in forested landscapes that exceed in area any other wildfire in this two-state region over at least the past 100 years (e.g., most notably, the 189,651 ha [468,640 acre] Rodeo-Chediski Fire in central Arizona in 2002, the 217,741 ha [538,049 acre] Wallow Fire in east-central Arizona and west-central New Mexico in 2011, and the 63,000 ha [156,593 acre] Las Conchas Fire in New Mexico in 2011). Since the late 1990s, large areas of forest and woodland have experienced extensive tree mortality due to a combination of direct drought-induced physiological stress and mortality, and attacks by phloem-feeding bark beetles (Allen and Breshears 1998; Breshears et al. 2005). Williams et al. (2010) summarize the mortality extent across the Southwest by these agents (drought, fire, bark beetles) and they estimate that nearly 20 percent of forested areas experienced high levels of tree mortality between 1984 and 2010.

In California, about 13 percent of the forest area is composed of forest types with naturally high-severity (30%–80% crown-burned) fire regimes with mean fire-return intervals (MFRI) of 15 to 100 years (predominately cedar/hemlock/Douglas-fir, red fir), while nearly 70 percent is composed of forest types that experienced frequent, low-severity prehistoric fire regimes (MFRI ≤ 10 yr, crown burned ≤ 5%; predominately mixed conifer, mixed California evergreen, redwood, and ponderosa pine) (Stephens et al. 2007). A policy of fire suppression and land use changes reduced the annual burned area in California forests from presettlement levels by more than 90 percent in the twentieth century (Stephens et al. 2007). Miller et al. (2009) document trends toward increasing fire severity in the Sierra Nevada and hypothesize that fire suppression and increased precipitation over the twentieth century increased fuel densities, contributing to increased fire severity. The frequency of large fires, total area burned, mean fire size, and fire severity have all increased in northern California forests since the mid-1980s (Westerling et al.

2006; Miller et al. 2009) (figure 3.1). Because a large portion of the interannual variability in northern California forest wildfire burned area is due to variability in ignitions from clustered lightning strikes, only a modest fraction of observed interannual variability in burned area can be explained by climate alone (Preisler et al. 2011; Westerling et al. 2011a).

Wildfire is predicted to increase substantially in northern California forests in the Sierra Nevada, southern Cascades, and Coast Ranges under some climate change scenarios. Westerling and Bryant (2008) project 100 percent to 400 percent increases in the probability of large fire occurrence over much of the Sierra Nevada, southern Cascades, and Coast Ranges under a relatively warm, dry climate scenario (GFDL SRES A2). A study by the National Research Council (2011), applying regression methods from Littell et al. (2009) for fire aggregated by ecosystem provinces similarly found increases exceeding 300 percent for a 1°C temperature increase. Westerling et al. (2011a) find increases in burned area ranging from 100 percent to over 300 percent for much of northern California's forests across a range of climate and growth and development scenarios using three climate models (NCAR PCM1, CNRM CM3, GFDL CM 3.1) for the SRES A2 emissions scenario. Spracklen et al. (2009) find increases in burned area on the order of 78 percent by midcentury for the GISS GCM under the SRES A1b emissions scenario, which is similar in magnitude to Westerling et al. (2011a) for midcentury northern California forests under GFDL SRES A2 scenarios. Conversely, increases in California forest wildfire frequency and burned area are more modest under a lower (SRES B1) emissions scenario, with end of century burned area roughly the same as midcentury (Westerling and Bryant 2008; Westerling et al. 2011a; Yue et al. 2013b).

The direct effects of anthropogenic climate change on wildfire are likely to vary considerably according to current vegetation types and whether fire activity is currently more limited by fuel availability or flammability. In the long run, climate change is likely to lead to changes in the spatial distribution of vegetation types, implying that transitions to different fire regimes will occur in locations with substantial changes in vegetation. Most long-term projections of changing wildfire activity have not successfully incorporated dynamic changes in vegetation types and fuels characteristics in response to climate and disturbance. Using existing fire-climate-vegetation interactions we can understand the likely direction and magnitude of climate-driven changes in fire activity over the next few decades. Beyond that, we may be able to use these models and our understanding of current ecosystems to assess when changes in climate and disturbance regimes will begin to lead to qualitative changes in ecosystems. Given the lack of analogues to projected climate changes—especially the substantial changes in the latter half of the twenty-first century that are projected to result from continued high emissions of greenhouse gases—precise modeling of future changes in vegetation and disturbances like wildfire becomes significantly more challenging for later in this century and beyond.

The overall direction and spatial pattern of changes in precipitation under diverse climate change scenarios varies considerably across future greenhouse gas emissions scenarios and global climate models (Dettinger 2006). In ecosystems where climatic influences on fire risks are dominated by precipitation effects, this implies greater uncertainty about climate change impacts on wildfire in those locations (Westerling and Bryant 2008). Overall, greater warming will lead to more evaporation of moisture from soils and from the live and dead vegetation that fuels forest wildfires. Given the substantial interannual variability in precipitation characteristic of western US climate, it is likely that fire activity will at least increase in drought years in coming decades, across a broad range of future climate scenarios.

Climate scenarios (even those with rapid reductions in global greenhouse gas emissions) project increases in temperature substantially greater than those observed in recent decades (IPCC 2001, 2007a, 2007b, 2013a, 2013b), which have been associated with substantial increases in wildfire activity in western US forests (Gillett et al. 2004; Westerling et al. 2006; Soja et al. 2007; Williams et al. 2013; figures 1 and 2). Strategies for adapting to a warmer world will therefore need to consider the impacts of climate change on wildfire.

The effects of climatic change on wildfire will depend on how past and present climates have combined with human actions to shape extant ecosystems. Climate controls the spatial distribution of vegetation, and the interaction of that vegetation and climate variability largely determines the availability and flammability of the live and dead vegetation that fuels wildfires. In moist forest ecosystems where snow plays an important role in the hydrologic cycle and fuel flammability is the limiting factor in determining fire risks, anthropogenic increases in temperature may lead to substantial increases in fire activity.

In dry ecosystems where fire risks are limited by fuel availability, warmer temperatures may not increase fire activity significantly. Warmer temperatures and greater evaporation in some places could actually reduce fire risks over time if the result is reduced growth of grasses and other surface vegetation that provide the continuous fuel cover necessary for large fires to spread. The effect of climate change on precipitation is also a major source of uncertainty for fuel-limited wildfire regimes. However, in some places these are the same ecosystems where fire suppression and land uses that reduce fire activity in the short run have led to increased fuel loads today as formerly open woodlands have become dense forests. This increases the risk of large, difficult-to-control fires with ecologically severe impacts.

4

Forest Ecosystem Reorganization Underway in the Southwestern United States

A Preview of Widespread Forest Changes in the Anthropocene?

CRAIG D. ALLEN

Extensive high-severity wildfires and drought-induced tree mortality have intensified over the last two decades in southwestern US forests and woodlands, on a scale unseen regionally since 1900. Abundant and diverse paleoecological and historical sources indicate substantial variability in southwest fire regimes and forest vegetation patterns over the past 10,000 years, providing context for recent fire and vegetation trends. In particular, over the past 150 years regional forest landscapes and fire regimes have responded sensitively, strongly, and in understandable ways to changes in human land management, as well as to interactions with climate variability and trends. Widespread, high-frequency surface fire activity ceased on most landscapes of the Southwest in the late 1800s due to changed land use patterns, grading into increasingly vigorous active fire suppression after 1910. Fire suppression allowed woody plant establishment to explode during several wet climate windows favorable for tree regeneration and growth, particularly ca. 1905–1922 and 1978–1995. By the early 1990s many forests of the Southwest likely had reached locally maximum potential levels of tree density, leaf area,

DOI: 10.5876/9781607324591.c004

biomass and carbon storage, and surface and ladder fire-fuel loads—unsustainable levels upon the inevitable recurrence of episodic drought. Decadal-scale drought returned to the region in the late 1990s, along with historically unprecedented warmth. This warm, global-change-type drought has affected the Southwest almost continuously since 2000 through the present (February 2015). The uniquely recent combination of anomalously overgrown forests and extreme global-change-type drought has fostered more extensive and severe forest disturbance processes, driving ongoing reorganization of southwest forests into new ecosystem patterns. This chapter describes drivers and some cascading ecological effects of these interactive landscape changes, along with adaptation strategies to enhance forest ecosystem resilience in the context of ongoing and projected climate trends.

The American Southwest recently has been subject to large increases in severe wildfire activity and overall tree mortality in response to the combination of protracted drought and early twenty-first-century warmth. Research on physiological responses of diverse tree species to climate variables is providing important insights into the linked roles of drought and heat stress in driving southwest forest productivity and health, physiological thresholds of tree mortality, and forest disturbance processes (Adams et al. 2009; McDowell et al. 2011). Williams et al. (2012) recently derived a forest drought-stress index (FDSI) for the Southwest using a comprehensive tree-ring growth data set representing AD 1000–2007, driven by both warm-season temperature and cold-season precipitation (figure 4.1). Substantial warming over the past twenty-five years is significantly amplifying regional forest drought stress, likely by increasing atmospheric vapor pressure deficits during the growing season months. Strong correspondence exists between FDSI and forest productivity, tree mortality, bark-beetle outbreaks, and wildfire in the Southwest, illustrating the powerful interactions among climate, land use history, and disturbance processes in this region. If regional temperatures increase as projected by climate models, the mean FDSI by the 2050s will exceed that of the most severe droughts in the past 1,000 years.

Multiple lines of evidence now indicate ongoing changes in forest structures and compositions in the Southwest, including documented changes in the elevational distributions and dominance of many plant species, pointing toward novel patterns emerging over the course of the twenty-first century. With the onset of global-change-type drought (Breshears et al. 2005) since the

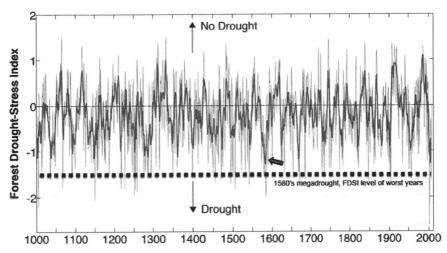

FIGURE 4.1. Reconstructed Forest Drought-Stress Index (FDSI) for the American Southwest, AD 1000–2007 (updated from Williams et al. 2012). Annual values in light gray, with a 10-year smooth in darker gray. The 1580s "megadrought" is marked by an arrow, and the dashed line indicates the upper bound of the driest 50% of years in this drought, representing tree-killing levels of drought stress. Note that the warm drought in 2002 (circled year) caused the worst year for regional forest growth in the entire tree-ring record since at least AD 1000.

late 1990s, overgrown forests in the Southwest have been subject to wildfires and tree mortality episodes of historically unprecedented extent and severity (figures 4.2–4.4), along with emergent shifts in vegetation patterns. The observed convergence of climate, human land use patterns and histories, and disturbance trends in the southwestern United States may presage widespread forest ecosystem changes more broadly in North America, and globally.

The southwest United States has an abundance of diverse paleoecological records that make this one of the best places in the world to determine past patterns of climate, vegetation, and fire, providing context to evaluate recent trends in forest and landscape change. For example, in this region scientists have used information locked in the tree-rings of ancient wood to precisely reconstruct past patterns of precipitation, temperature, stream flow, drought stress, and tree growth and death going back as much as 2,000 years (Swetnam and Betancourt 1998; Salzer and Kipfmueller 2005; Swetnam et al. 1999, 2011; Allen et al. 2008; Brown and Wu 2005; Woodhouse et al. 2010; Falk

FIGURE 4.2. View from the heart of a large (> 10,000 ha) patch with nearly no live conifer trees remaining after the initial 14-hour high-severity run of the June 26, 2011, Las Conchas Fire (Jemez Mountains, New Mexico). The area in this photo was primarily dense ponderosa pine forest prior to the 1996 Dome Fire (though snags in foreground are mostly Douglas fir on a north slope); the 2011 Las Conchas Fire severely reburned through this area, which in 2011 had much shrub cover of *Quercus* and *Robinia*. (Photo by C. D. Allen, August 2011.)

et al. 2011; Margolis et al. 2011; O'Connor et al. 2014). Dendroclimatological data from the Southwest illustrate fluctuations in precipitation and associated forest drought stress at multiple time scales (figure 4.1) that apparently are driven by atmospheric teleconnections with oscillations in ocean temperature patterns, particularly including the multiyear El Niño-Southern Oscillation (ENSO, Swetnam and Betancourt 1998), the multidecadal Pacific Decadal Oscillation (PDO), and the Atlantic Multidecadal Oscillation (AMO; McCabe et al. 2008; Pederson et al. 2013). Compared to other regions in the United States, the Southwest is characterized by relatively arid conditions and high levels of variability in precipitation at annual, decadal, multidecadal, and centennial time scales (Swetnam and Betancourt 1998; Woodhouse et al.

FIGURE 4.3. Map of high- and moderate-severity (tree-killing) fire patches in the Jemez Mountains (New Mexico), only fires with severity map data from 1977 to 2013 (fires without data on severity were excluded)—all but one fire occurred since 1996. The size of individual stand-replacing fire patches from recent fires now ranges up to > 10,000 ha in this landscape. Map data primarily from various fire-specific Burned Area Emergency Rehabilitation (BAER) reports, on file at USGS Jemez Mountains Field Station.

2010). Such climate variability drives associated large changes in southwestern forest growth patterns. This is exemplified by the recent development of a regional FDSI extending back over 1,000 years (figure 4.1), which strongly links warm growing-season temperatures to reduced growth of southwest conifers (Williams et al. 2012).

Other paleoenvironmental evidence in the Southwest extends back tens (or even hundreds) of thousands of years in the form of plant pollen, other plant remains, and charcoal deposited in layers of sediment at the bottoms of lakes and bogs (e.g., Weng and Jackson 1999; Anderson et al. 2008). These sediment records document how today's high-mountain tree species like

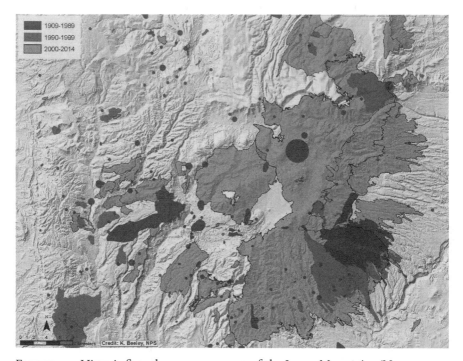

FIGURE 4.4. Historic fire atlas summary map of the Jemez Mountains (New Mexico), showing perimeters of all recorded fires larger than 0.1 acres for the period 1909–2013, color coded into three time periods of occurrence. The source of pre-1960 fires are original hand-drawn fire atlas maps (with associated original annual fire suppression records in tabular form), curated by the Santa Fe National Forest; these fires were redrawn on modern base maps and then digitized into a geographic information system (Snyderman and Allen 1997). Almost all post-1969 fires were mapped from various digital sources. Fires mapped as perfect circles represent occurrences with perimeter data lacking, but where a point location and a fire size class were available. Note large extent of fires since 2000.

spruce and fir were growing at much lower elevations during the colder climate of the last ice age, before moving upslope as the world's climate moved into the current warmer interglacial period about 11,000 years ago. Similarly, plant macrofossils preserved in the middens of ancient packrat nests directly show how much and how fast the ranges of plant species have expanded and contracted geographically, moving north and south, and locally upslope and downslope, in response to climate variations (Betancourt 1990). These pollen and macrofossil records also show that southwestern vegetation communities in the past often consisted of combinations of plant species unknown today (Betancourt 1990; Weng and Jackson 1999; Anderson et al. 2008).

Linked changes in climate, vegetation, and fire activity are evident in paleoecological records from this region. For example, documented midden evidence of ponderosa pine (*Pinus ponderosa*) is almost nonexistent in the Southwest during the last ice age, but with the early postglacial warming and the associated development of our summer monsoon climate after about 10,000 years ago this pine expanded across the region to eventually become a widespread forest species (Betancourt 1990; Weng and Jackson 1999). During this same time period, the abundance of charcoal deposited in lakes and bogs increased markedly across the region (Anderson et al. 2008; Allen et al. 2008), reflecting increased frequency and extent of fire activity on southwestern landscapes, which likely also favored the expansion of fire-adapted and fire-fostering species, like ponderosa pine (Weng and Jackson 1999). Numerous charcoal records over the past 1,000 years in the West and Southwest generally show the modulating effects of climate on fire activity, with modest increases in charcoal concentrations during the Medieval Warm Period, and also some significant decline during the Little Ice Age (Marlon et al. 2012). The world's greatest regional concentration of tree-ring studies is from the Southwest, including tens of thousands of precisely dated fire scars from hundreds of forest sites across the region—these reconstruct fine-resolution spatial and temporal patterns of fire extending back more than 400 years, documenting high levels of frequent and widespread fire activity that were closely tied to climate patterns until ca. 1900 (Swetnam et al.1999, 2011; Falk et al. 2011).

These pre-1900 fire-climate relationships are consistent with those that we see today (Swetnam and Betancourt 1998; Swetnam et al. 1999), with much higher levels of fire activity in warm dry years. For about two-thirds of the fire scars we can even date the season that the fire scar formed, recording that most pre-1900 fire activity occurred in the dry spring and early summer period, just as today, before the July onset of summer rains. Tree-ring reconstructions document that frequent, low-severity surface fires characterized the pre-1900 fire activity in the widespread ponderosa pine and dry mixed-conifer forests that predominate in much of the Southwest (Swetnam and Baisan 2003). Still, note that there is great diversity of forests and associated fire patterns across the substantial elevational and regional landscape gradients present in the Southwest (Swetnam et al. 2011; Vankat 2013). For example, mixed-severity and high-severity stand-replacing fires naturally occurred in cooler and wetter mixed-conifer and spruce-fir forests, which occupy relatively limited high-elevation portions of this region (e.g., Fulé et al. 2003; Margolis et al. 2011; Margolis and Balmat 2009; O'Connor et al. 2014).

Over the past 150 years regional forest landscapes and disturbance regimes (fire, drought stress, insect outbreaks) have responded to changes in human land use and land management in concert with patterns of climate variability (figure 4.5). The prehistoric pattern of widespread, high-frequency surface fire regimes across the Southwest initially collapsed in the late 1800s, because with the entry of railroads to this region there was a buildup of domestic livestock herds that interrupted the former continuity of the grassy surface fuels by widespread overgrazing, trampling, and trailing (Swetnam et al. 1999; Allen 2007). This mostly inadvertent suppression of surface fires by overgrazing then morphed into active fire suppression and exclusion efforts by land management agencies in the early 1900s, which have continued with ever-increasing effort and expenditure to the present (Stephens et al. 2012a). Since forest types historically characterized by high-frequency surface fire regimes (ponderosa pine and dry mixed conifer) are a substantial majority of southwest forests (about 70%, based upon vegetation area estimates from Vankat 2013), over a century of fire suppression greatly affected most forests in the Southwest.

After the late 1800s collapse of surface fire regimes in most southwestern forests, the multitude of young trees that periodically established no longer were thinned out by naturally frequent surface fires that previously had favored relatively open, grassy forest conditions. As a result, woody plant establishment and forest densification exploded during the twentieth century, particularly fostered by two favorable wet climate windows for tree regeneration (Savage et al. 1996; Brown and Wu 2005) and growth in the early and late 1900s (figure 4.5). Increasingly intensive fire suppression efforts by land managers during the twentieth century also were necessary to enable the general pattern of regional "woodification," with widespread expansion of trees into regional grasslands and meadows, along with substantial increases in the densities of most (although not all) southwestern forests and woodlands. For example, in some of the most common forest types, like various ponderosa pine and dry mixed-conifer forests, tree densities commonly increased tenfold or more, often from less than 100 to over 1,000 trees per acre (Covington and Moore 1994; Allen et al. 2002).

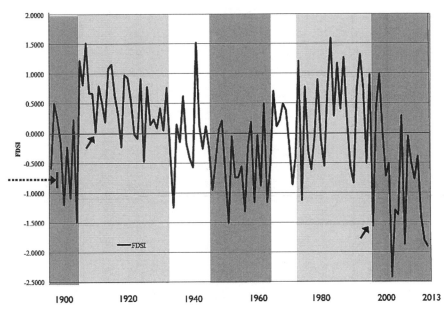

FIGURE 4.5. Historic sequence of interactions among climate, land-use, disturbance regimes, and forest change in the Southwest, showing graph of regional Forest Drought-Stress Index (FDSI) for 1896–2013 (updated from Williams et al. 2012); wet periods (light shading) and dry periods (darker shading); former period of widespread high-frequency surface fires (dashed arrow) that ends in late 1890s (black bar); onset of fully institutionalized fire suppression ca. 1910 (arrow) early in a wet period that supported abundant tree regeneration; late 1900s wet period with maximal development of dense high-biomass forests with widespread ladder fuels; and the onset of severe and persistent forest drought stress since 1996 (arrow), which has driven regionally extensive wildfire and tree mortality.

Such increases in forest density also were accompanied by huge increases in surface fuel loads and the widespread development of understory thickets of small, suppressed trees and with live crowns near the ground surface. In combination these "ladder fuels" allow surface fires to easily spread upward into tree canopies, where the high-energies being liberated through combustion can generate strong convection that drives positive feedback toward more intense fire activity (Allen 2007). Severe regional drought in the 1950s started to expose the potential for larger stand-replacing fires as more susceptible fuel structures began to emerge in ponderosa pine forests, but concurrent fire suppression advances generally kept a lid on extreme fire activity until drought stress moderated again. Generally wet conditions in the Southwest from the late 1970s through 1995 drove rapid tree growth and further buildup of forest biomass—and importantly, the wet conditions in this period also helped firefighters keep wildfires in check despite the hazardous fuel conditions that prevailed by this time (figure 4.5). Thus by the mid- 1990s many southwestern forests likely were near their maximum possible densities and levels of biomass accumulation and leaf area at landscape and stand scales; the former fire-maintained mosaic of mostly low-density forests (with interspersed patches of thicker forest and open meadows) across diverse southwest landscapes had morphed into a relatively homogenous blanket of dense forests with vertical and horizontal fuel structures that could support the initiation and extensive spread of explosive high-severity canopy fires. During this wet period, forest growth was strong and forest disturbances (e.g., fire and bark beetle mortality) were limited—southwestern forests seemed to be resilient and secure.

Drier winter conditions abruptly returned to the Southwest in 1996, with near-continuous and ongoing drought since 2000, along with historically novel warmer temperatures. As a result, over the last seventeen years southwestern forests and woodlands have been subject to much more extensive and severe fire activity (e.g., figures 4.2–4.4) and drought-induced tree mortality (including associated bark beetle outbreaks), with about 20 percent of regional forests affected by significant tree mortality from combinations of drought stress, bark beetles, and high-severity wildfire between 1984 and 2012 (updated from Williams et al. 2010). The scale of these recent tree-killing forest disturbances is unprecedented in the Southwest since historic record-keeping began around 1900, and almost certainly is unprecedented since the megadrought of late 1500s (Swetnam and Betancourt 1998); the size of recent high-severity fire patches in southwestern ponderosa pine forests (e.g., Figures 4.2, 4.3) quite possibly is also unprecedented (Stephens et al. 2013) since before modern patterns of climate, vegetation, and fire regimes established by ca. 9,000 to 6,000 years ago (Anderson et al. 2008).

Given that substantially warmer temperatures and greater drought stress are projected for the Southwest in coming years (Seager and Vecchi 2010; Williams et al. 2010, 2012), we should expect even greater increases in mortality of drought-stressed trees, high-severity fire, and ultimate conversion of current forests into different ecosystems, ranging from grasslands and shrublands to new forests dominated by different tree species (Williams and Jackson 2007; Jackson et al. 2009). Increasingly frequent and severe droughts and fires favor plant life-forms that can survive aboveground stem dieback and fire damage by resprouting from belowground tissues—many grass and shrub species can do this. After high-severity fires, successful regeneration of the main conifer tree species in the Southwest primarily depends upon the local survival of enough mother trees to serve as seed sources. Recent observations and studies document the risks of postfire vegetation type conversions from forest to nonforest ecosystems in the Southwest (Barton 2002; Savage and Mast 2005; Goforth and Minnich 2008; Savage et al. 2013). These conversions can be caused by the ever larger, high-severity fire patches where essentially all tree seed sources are killed across tens of thousands of acres, as observed in some recent fires (figures 4.2, 4.3). Such large stand-replacing fire patches greatly limit recolonization rates by some of the most common tree species such as piñon pine, ponderosa pine, and Douglas fir, allowing dense grasslands or shrublands of resprouting species to achieve dominance before conifer trees can reestablish. It is also beginning to be observed that once large areas of resprouting shrubs, like Gambel oak, become heavily mixed in and around surviving postfire conifer tree populations, a subsequent hot reburn through the shrubs can then kill nearly all of those adult tree survivors and associated young regeneration. In this way a sequence of hot burns can eliminate local tree seed sources over very large areas (figures 4.2, 4.3). In addition, millions of hectares of forest and woodland in the Southwest have been affected by high levels of tree mortality since 2000 from combinations of drought and heat stress, amplified by biotic agents, particularly various bark beetle species (Breshears et al. 2005; Raffa et al. 2008; Williams et al. 2010, 2012). The growing extent and severity of recent forest disturbances in this region, and the lack of tree regeneration on some extensive sites after severe fires, are evidence that we already may be reaching tipping points of regional forest ecosystem change, changes that are new in the historical era.

The observed recent ramp-up in the extent and severity of climate-related forest disturbances across the Southwest (Williams et al. 2010, 2012) may represent the beginning of substantial reorganization of ecosystem patterns and processes into new configurations (Barton 2002; Goforth and Minnich 2008; Jackson et al. 2009; Brusca et al. 2013; Worrall et al. 2013), as southwestern forest landscapes transition toward more open and drought-resistant

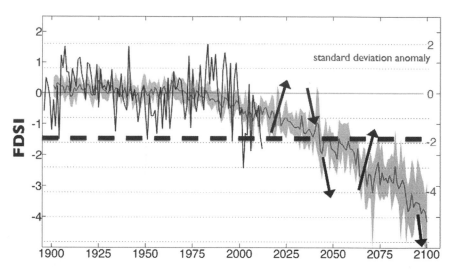

FIGURE 4.6. Climate model projections of the Forest Drought-Stress Index (FDSI) for the American Southwest, 1900–2100 (updated from Williams et al. 2012). Observed FDSI 1896–2013 (black line), mean of ensemble model projections (dark gray line), and inner quartile of ensemble projections (shaded band). Hypothetical deviations from this projection due to decadal-scale climatic oscillations (PDO, AMO) are shown by arrows. The dashed line indicates the upper bound of the driest 50% of years of the 1580s megadrought, representing tree-killing levels of drought stress.

ecosystems in response to recent climate forcing. If the climate projections of further rapid warming and drought for the Southwest are correct (e.g., Seager and Vecchi 2010), then in coming decades southwestern forests, as we know them today, will experience ever-growing levels of vegetation mortality, driving the emergence of transformed ecosystems with new dominant species (Williams and Jackson 2007).

Still, forest management practices have potential to improve forest resistance and resilience to climate stressors and associated disturbances. For example, combinations of mechanical tree harvesting, ground mulching, and managed fire treatments can reduce forest densities and hazardous fuel loadings, decreasing between-tree competition for water (Grant et al. 2013), thereby reducing overall forest drought stress and risk of high-severity fires (Finney et al. 2005; Ager et al. 2010) and providing protection to mountain watersheds (TNC 2014). Such treatments also can restore historical forest ecological conditions that were sustainable for at least many centuries prior to 1900 in many southwest forest types (Swetnam et al. 1999; Allen et al. 2002; Sisk et al. 2005; Fulé 2008; Stephens et al. 2012a; Stephens et al. 2013).

While a unique combination of geography, climate, land use, and disturbance histories have driven the recent period of high-magnitude forest stress and disturbance in the Southwest, similar patterns of forest change could emerge more broadly as projected climate changes progress at continental and global scales. Similar interactions among drought, heat, and land use are widely observed to be drivers of major fire and forest die-off episodes more broadly in western North America (Westerling et al. 2006; Raffa et al. 2008; Littell et al. 2009; Bentz et al. 2010; Allen et al. 2010; Williams et al. 2012; Hicke et al. 2013) and globally (Bowman et al. 2009; Allen et al. 2010; Pechony and Schindell 2010; McDowell et al. 2011; Matusick et al. 2013; Worrall et al. 2013). Given projections of substantial further global warming (IPCC 2013a) and increased drought stress in coming decades for much of western North America (National Climate Assessment 2013) and many areas globally (IPCC 2013a), the recent emergence of high levels of forest drought stress and disturbance (fire, die-off) in the Southwest may foreshadow future forest trends globally in the Anthropocene.

Assessing Vulnerability and Threats to Current Management Regimes

5

Increasing Resiliency in Frequent Fire Forests

Lessons from the Sierra Nevada and Western Australia

SCOTT L. STEPHENS

In forests in the western United States, fire hazard reduction treatments have become a priority as the size and severity of wildfires have been increasing in some forest types (Miller et al. 2009; Westerling et al. 2006; Miller and Safford 2012). However, the scale and implementation rate for fuel treatment projects is well behind what is necessary to make a meaningful difference across landscapes (North et al. 2012). This issue is particularly relevant as wildfire size and intensity are projected to increase in many parts of the western United States based on climate-fire modeling (Lenihan et al. 2008; Westerling et al. 2011a; Yue et al. 2013a).

In contrast to crown fire adapted ecosystems, ecosystems that once experienced frequent, low-to-moderate-intensity fires can be managed to reduce their susceptibility to high-severity wildfires (Fulé et al. 2012) and to increase their ecosystem resiliency (Stephens et al. 2012a, 2013). Research has determined that there are few unintended consequences of forest fuel reduction treatments across forests in the United States, because most ecosystem components (vegetation, soils, small mammals and song birds, bark beetles, carbon

DOI: 10.5876/9781607324591.c005

sequestration) exhibit very subtle or no measurable effects at all (Stephens et al. 2012a). Similar results were found in Western Australia forests and shrublands that were repeatedly prescribed burned over thirty years (Wittkuhn et al. 2011). In surface fire adapted ecosystems, management actions including fuel treatments and managed wildfire (lightning ignitions allowed to burn for resource benefit) can be taken today to reduce the negative consequences of subsequent wildfires (such as large, high-severity patches; Collins and Roller 2013) that also meet restoration objectives (North et al. 2009).

Of the three principal means of fuels reduction—mechanical treatment, prescribed burning, and wildfire—the latter is often the most expensive (North et al. 2012). US Forest Service (USFS) mechanical treatments (thinning, mastication) vary widely (Hartsough et al. 2008) but costs on average were 3.5 times higher than prescribed fire in large part due to expensive service contracts for removal of small, noncommercial biomass. Wildfire costs were highest but vary tremendously between burns. In general, costs per acre increased as access became more difficult but decreased with fire size (North et al. 2012). Managing fire for multiple objectives instead of a narrow focus on fire suppression is producing some positive outcomes, such as when fire exhibits self-limiting characteristics (reduced area and severity) in some ecosystems (figure 5.1). Recurring fires consume fuels over time and can ultimately constrain the spatial extent and lessen the fire-induced effects of subsequent fires. In montane forests in Yosemite National Park, California, when the amount of time between successive adjacent fires is under nine years, the probability of the latter fire burning into the previous fire area is low (Collins et al. 2009).

Our analysis of fire-severity data by ten-year periods revealed stability in the proportion of area burned over the last three decades among fire-severity classes (unchanged, low, moderate, high). This contrasts with increasing high-severity burning in many USFS Sierra Nevada forests from 1984 to 2010 (Miller and Safford 2012), which suggests that freely burning fires over time in some forests can regulate fire-induced effects across the landscape (Stephens et al. 2008; Miller et al. 2012). It should be noted that the USFS has not used prescribed fire or managed wildfire to the degree that the US National Park Service has, which has influenced current burning patterns.

Current wildfires are burning large areas, but there is some evidence that intact fire regimes (those minimally affected by fire exclusion for several

FIGURE 5.1. Jeffrey pine forests have been repeatedly burned by managed wildfire in the Illilouette Creek basin of Yosemite National Park. Reintroduction of a functioning fire regime in 1974 has produced a forest that is resilient to fire and climate change. (Photo by Scott Stephens.)

decades) can constrain fire size (Stephens et al. 2014). For example, in montane forests of Yosemite National Park, where lightning fires have been allowed to burn under prescribed conditions for forty years, a pattern of intersecting fires emerged that limited the extent of subsequent fires to less than 4,000 ha (9,884 acres) (van Wagtendonk et al. 2012). However, wildfires have grown to over 40,000 ha (98,842 acres) on areas in or adjacent to the park where fires have been routinely suppressed and the resulting burn severity patterns (especially large patch sizes) are not within desired ranges to conserve ecosystem resiliency (Miller et al. 2012).

A recent analysis determined that fuels reduction was occurring on USFS lands in the Sierra Nevada at very low rates (North et al. 2012). With less than 20 percent of USFS lands in the Sierra Nevada receiving needed fuels treatments, and the need to re-treat many areas, the current pattern and scale of fuels reduction is unlikely to ever significantly advance restoration efforts. One means of changing current practices is to concentrate large-scale strategic (Ager, Vaillant, and Finney 2010) fuels reduction efforts and then move treated areas out of fire suppression into fire maintenance. A fundamental change in the scale of fuels treatments is needed to emphasize treating entire firesheds and restoring ecosystem processes (North et al. 2009, 2012). Without proactively addressing this situation, the status quo will relegate many ecologically important areas (including sensitive species habitat) to continued degradation from either no fire or wildfire burning at high severity (North et al. 2012).

Ironically, current USFS practices intended to protect resources identified as having high ecological value often put them at a greater risk of large, high-severity fire (Collins et al. 2010; North et al. 2012). A policy focused on suppression, which ultimately results in greater wildfire intensity, means that fuels reduction becomes the principal method of locally affecting fire behavior and reducing severity (Collins et al. 2010). Forest areas identified as having high conservation value, such as riparian conservation areas (van de Water and North 2010, 2011) and protected activity centers (PAC) for threatened and sensitive wildlife, often have management restrictions and higher litigation potential, resulting in minimal or no fuels reduction treatment (North et al. 2012). Stand conditions in these protected areas often consist of multilayered canopies with large amounts of surface fuel, resulting in increased crown fire potential (Spies et al. 2006; Collins et al. 2010). Following the particularly high-intensity Moonlight wildfire in 2007 in the Sierra Nevada, riparian and PAC areas had some of the greatest percentage of high-severity effects of any area within the fire perimeters (Safford et al. 2009). In contrast, low- and moderate-severity wildfire and prescribed burning in Yosemite National Park maintained habitat characteristics and density of California spotted owls (*Strix occidentalis*) in late successional montane forest (Roberts et al. 2011).

Recent estimates determined that at current treatment rates, the deficit of forestland "in need" of treatment would be approximately 1.2 million ha (approximately 30 million acres) in the Sierra Nevada (approximately 60%

of USFS lands in the Sierra Nevada), of which 670,000 ha (16.5 million acres) are forests dominated by ponderosa (*Pinus ponderosa*) and Jeffrey pine (*Pinus jeffreyi*) (North et al. 2012). This is a very conservative estimate of the deficit because it assumes that mechanical, prescribed fire, and wildfire areas never overlap and that all wildfires are restorative in their ecological effects, which is not the case. Although current policy recognizes the importance and need for managed wildfire (National Interagency Fire Center 2001; USDA/USDI 2005; National Interagency Fire Center 2009), studies have found very low rates of implementation. In 2004, land management agencies only let 2.7 percent of all lightning ignitions burn (NIFC 2006), consistent with a recent analysis in the Sierra Nevada that less than 2 percent of USFS lands were burned under managed wildfire between 2001 and 2008 (Silvas-Bellanca 2011). The most significant factor associated with USFS use of managed wildfire was the personal commitment of district rangers, while the main disincentives were negative public perception, lack of available resources, and perceived lack of agency support (Williamson 2007).

Given that less than 20 percent of the landscape that needs fuels treatments actually receives them, and that the need to re-treat many areas every fifteen to thirty years, depending on forest type (Stephens et al. 2012b), the current pattern and scale of fuels reduction is unlikely to significantly advance restoration efforts, particularly if agency budgets continue to decline. Treating and then moving areas out of fire suppression into fire maintenance is one means of changing current patterns (North et al. 2012).

As fuel loads increase, rural home construction expands, and agency budgets decline, delays in fuel treatment implementation will only make it more difficult to expand the use of managed fire after initial treatments. Increases in the number of managed wildfires may be criticized given current constraints but at least this strategy could stimulate discussions between stakeholders, air-quality regulators, and forest managers about current and future management options (North et al. 2012). Without proactively addressing some of these conditions, the status quo will relegate many ecologically important areas to continued degradation from fire exclusion and high-severity wildfires. In some forests, revenue generated in the initial entry (Hartsough et al. 2008) may be the best opportunity to increase the scale and shift the focus of current fuels reduction toward favoring long-term fire restoration (North et al. 2012).

FIGURE 5.2. High-severity patch in upper-elevation mixed-conifer forest in the Illilouette Creek basin in Yosemite National Park. Median high-severity patch size in this forest is < 4 ha (Collins and Stephens 2010) and the amount of high-severity fire has not changed since the early 1970s (Collins et al. 2009). (Photo by Scott Stephens.)

One of the world's best examples of a fire-management program designed to reduce wildfire impacts can be found in Australia (Boer et al. 2009). In the fire-prone forests and shrublands of southwest Western Australia, prescribed burning of native vegetation is an important management strategy for achieving conservation and land management objectives (Wittkuhn et al. 2011). Prescribed burning done at the appropriate spatial and temporal scales reduces the overall flammability and quantity of fuels in the landscape, thereby reducing the intensity and spread rate of wildfires (Stephens et al. 2014).

Broad area fuel reduction burning as a key asset protection strategy has been implemented in southwest Western Australia since the mid-1950s (figure 5.3). Approximately 8,500 prescribed burns have been conducted, burning a total area of 15 million ha (37 million acres) (Stephens et al. 2014). Over this time, an inverse relationship between the area burned by prescribed fire and wildfire has been established (Boer et al. 2009); that is, prescribed burning

FIGURE 5.3. Prescribed fire in mixed eucalyptus forests near Busselton, Western Australia. Note the variety in fire severity with some areas unburned and others burned at high intensity, which is the common goal in this area. Regrowth is already occurring only two months after the prescribed fire. (Photo by Scott Stephens.)

has reduced the impact of wildfires by reducing their size and intensity. This Australian example could be of interest for regions that continue to focus solely on fire suppression.

However, the annual area of prescribed burning in the southwest Western Australian region is trending downwards since the 1980s (mainly because of the reduced area of prescribed fire), while the annual area burned by wildfires is trending upwards (Stephens et al. 2014). In recent years there has been a spate of wildfires that have not been experienced in the region since the 1960s. Key drivers of these trends are (Stephens et al. 2014):

- Climate change. Since the 1970s, the climate has become warmer and drier (Bates et al. 2008) reducing the window of opportunity for safely carrying out prescribed burning. Longer periods of hotter, drier weather result in longer periods of high fire risk.

- Population growth in the wildland-urban interface. More people are living in fire-prone settings. In many instances, local bylaws and land use planning policies do not adequately consider the risk of wildfires, or are not adequately enforced. People are building and living in dangerous locations and often are not taking adequate fire-protection measures.

- Fire-management capacity. Resources and personnel for fire management have not kept pace with the increasing demands and complexity of managing fire. Additional staff is needed and training programs are necessary to allow this new group to become familiar with prescribed fire planning and operations.

- Smoke management. Managing air quality and the impacts of smoke on adjacent land users or homeowners further narrow burning windows and reduce the size and number of prescribed burns that can be conducted.

Effective management of wildfire risk will require the incorporation of larger-scale management processes across landscapes at the scale of 10,000 to 30,000 ha (25,000–75,000 acres) (North et al. 2012, Stephens et al. 2013). This can be done with large-scale prescribed burning programs, mechanical fuel treatments, combinations of mechanical and fire treatments, or allowing wildfires to burn under desired conditions. Managed wildfire probably has the greatest ability to meet restoration and fuel management goals in the western United States because it can be implemented at moderate-to-large spatial scales with the lowest cost (North et al. 2012), whereas in Australia, the Mediterranean Basin, the US Great Plains, and the southern United States, prescribed burning is preferable (Stephens et al. 2013). Regardless of how a fire is ignited, smoke will likely be a large concern, especially its impact on human health. However, it is important to contrast the human health effects of smoke from prescribed fires and managed wildfires with those of large, severe wildfires, which can affect large regions for weeks or months.

Federal forest managers have a great challenge in promoting forests that are resilient to changing climates. Forests that once experienced frequent, low-to-moderate-intensity fire regimes can be managed today to reduce the negative impacts of subsequent wildfire. Increased use of fire and fire surrogates treatments (McIver et al. 2009; Schwilk et al. 2009; Stephens et al. 2012a) and increased use of management wildfire for resource objectives are the only possibilities for managers to achieve desired conditions. Current rates of treatments on federal lands in the western United States are inadequate to conserve forests into the future (North et al. 2012). This treatment deficit has the potential to adversely impact critical ecosystem services that are derived from US forests.

One of the largest challenges faced by US federal land managers is the continued reduction in funding for fuels and restoration programs, which consequently emphasizes managed wildfires as their primary management option. Funding for prescribed fire, thinning, and mechanical fuels treatments has been reduced in the last two years and the 2013 US federal budget reduced it by over 85 percent in comparison to resources allocated in this area in the early 2000s. Managed wildfire is appropriate in wilderness, roadless areas, and other remote areas but is not applicable to large areas of federal land because of human infrastructure. Managing wildfires for weeks or months can produce positive ecological outputs but is much riskier than performing fuels treatments over relatively short time intervals. Managers need access to all forest management options to increase resiliency in the forest areas of the United States; removing options will further increase the backlog of areas in need of restoration.

California can learn from a successful prescribed fire program in southwestern Western Australia (Sneeuwjagt et al. 2013). Although the ecosystems in these two areas are different they both have evolved with frequent fire in Mediterranean climates. Southwestern Western Australia has successfully implement a prescribed fire program that has reduced the incidence of wildfires while conserving ecosystems (Boer et al. 2009; Wittkuhn et al. 2011). The challenge is for California managers to produce a similar outcome for frequent fire ecosystems in the Sierra Nevada.

6

Protected Areas under Threat

Tim Caro, Grace K. Charles, Dena J. Clink, Jason Riggio,
Alexandra M. Weill, and Carolyn Whitesell

A protected area (PA) is defined as "an area of land and/or sea especially dedicated to protection and maintenance of biological diversity, and of natural and associated cultural resources, and managed through legal or other effective means." Protected areas have been established to achieve several goals including the preservation of species, conservation of genetic diversity, preservation of wilderness and natural and cultural features, providing for tourism and recreation, supporting the sustainable use of resources, promoting education, and maintaining environmental services and cultural attributes. According to the World Database on Protected Areas, the number of terrestrial PAs is at least 125,000, covering about 17 million km² and more than 12.7 percent of global terrestrial area, although many of these PAs are multiple-use areas (table 6.1). From a biological standpoint, the effectiveness of PAs as a conservation tool depends on its ability to incorporate biodiversity and to buffer plant and animal populations against anthropogenic forces (e.g., Bruner et al. 2001).

Plant and animal populations inside PAs are not immune to these forces and are under threat everywhere. For example, mammal populations in

DOI: 10.5876/9781607324591.c006

TABLE 6.1. IUCN (International Union for Conservation of Nature) Protected Area categories
(adapted from Chape et al. 2005)

Category	Description	Number*	Area[†]
Ia	Strict nature reserve: PA managed mainly for science	5.2	4.6
Ib	Wilderness area: PA managed mainly for wilderness protection	1.3	3.0
II	National park: PA managed mainly for ecosystem protection and recreation	3.7	20.5
III	Natural monument or feature: PA managed mainly for conservation of specific natural features	18.8	1.4
IV	Habitat/species management area: PA managed mainly for conservation through management intervention	25.2	15.1
V	Protected landscape/seascape: PA managed mainly for landscape/seascape conservation and recreation	8.2	11.8
VI	Protected areas with sustainable use of natural resource: managed mainly for the sustainable use of natural ecosystems	8.2	11.8
Other		33.8	21.8

*Percentage out of total PAs, including noncategorized PAs.
[†]In millions of km².

African PAs show substantial declines (Craigie et al. 2010), although population declines are greater in adjacent unprotected areas (Geldmann et al. 2013). Here we review the principal contemporary threats to terrestrial PAs and demonstrate how they may be changing over time based on a crude measure of numbers of published studies. We focus on national and regional public PAs, although much can be applied to private reserves as well. Our purpose is not to provide an exhaustive list of threats to PAs, but instead to offer a temporal and spatial overview.

By the end of the twenty-first century, average global temperatures are expected to increase by 1.1°C to 6.4°C, with the largest temperature rises at high northern latitudes. Many species have already exhibited range shifts in response to climate change. Meta-analyses have uncovered latitudinal shifts of between 6.1 km and 16.9 km per decade but rates of range shifts are highly variable between and within species. Movement of species out of PAs, as well as changes in species interactions within PAs, will become important issues in the future.

Since PAs principally conserve biodiversity within their boundaries, shifts in species distributions may limit their effectiveness. Many climate change projections predict that species distributions will move out of PAs, decreasing biodiversity within PAs and leaving many species with inadequate protection (Hole et al. 2009). Even if species do shift their current ranges out of PAs, existing PAs may provide suitable colonization habitat for such species. For example, a study on 139 Tanzanian savanna bird species found that local colonization events were explained by changes in climate suitability, and that colonization preferentially occurred in PAs (Beale et al. 2013). In the United Kingdom, five bird and two butterfly species exhibiting northward range expansions preferentially colonized PAs; these results are further supported by surveys of 256 invertebrate species wherein 98 percent of those surveyed were found to disproportionately colonize PAs (Thomas et al. 2012).

Altitudinal shifts have also been documented in diverse taxonomic groups within PAs, with species showing average range shifts from 6.1 m to 11 m up altitudinal gradients per decade. In Yosemite National Park (NP), United States, small mammal distributions have changed over the past century, with over half showing an upward shift in their upper-elevation limits (Moritz et al. 2008). Tingley and colleagues (2012) found that increased temperature caused upward shifts in birds in the Sierra Nevada whereas increased precipitation led to downward shifts, resulting in a heterogeneous mix of avian range shifts within this region. Fortunately PAs are often situated on mountainsides that will benefit species whose ranges contract to higher latitudes. Increasing temperatures may affect species interactions through changes in phenology or temporal mismatches. Changes in phenology have been documented in central and southern African PAs where plant seasonality has increased in response to changes in rainfall (Pettorelli et al. 2012). In West Greenland, increased caribou (*Rangifer tarandus*) calf mortality and reduced calf production were attributed to a trophic mismatch between the timing of caribou reproduction and plant growth (Post et al. 2008). For certain species, an increase in temperature will be lethal, while for others it will increase growth. In Peñalara Natural Park, central Spain, increases in temperature have driven the growth of the chytridiomycete fungus, a disease that has killed numerous amphibian species (Bosch et al. 2007).

Due to differences in response rates to climate change, species in PAs may lose or gain prey, predators, pollinators, or competitors, leading to changes in interspecific interactions and formation of novel (nonanalog) communities (see figure 6.1). Nonetheless, our understanding remains largely theoretical as there is a large degree of uncertainty regarding ecosystem and

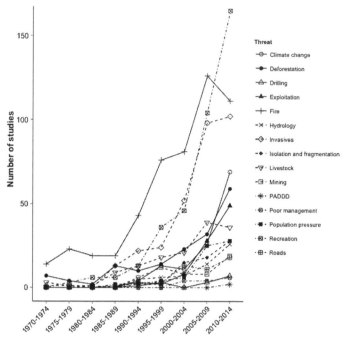

FIGURE 6.1. Number of references from Web of Knowledge™ discussing different sorts of threats in relation to PAs or NPs plotted against time in 5-year intervals. To gather a representative sample of all papers published on PAs we used the search string ("national park*" OR "protected area*") in the 'Title' field code. We excluded marine PAs by searching NOT "marine" in the 'Topic' field code. For each individual threat, we retained the previous queries and added the following search strings in a second 'Topic' field code: climate change ("climate change" OR "global warming"); fragmentation/isolation (fragment* OR isolat*); human population pressure ("human population*" OR "population growth" OR "population pressure"); PADDD (degazette* OR downgrad* OR downsiz* OR PADDD); deforestation (clear-cut* OR deforest* OR forestry); wildlife exploitation – ("poach* OR bushmeat OR medicin* OR "trophy hunt*"); invasives (alien* OR invasive* OR exotic* OR non-native*); livestock (cattle OR livestock OR livestock- wildlife OR pastorali*); fire (burn* OR campfire* OR fire* OR wildfire*); hydrology (dam OR hydrolog* OR irrigat* OR "snow pack" OR "water flow"); mining (extraction OR mine OR "mineral extraction" OR mining OR tailing*); drilling (drill* OR extraction OR "natural gas" OR oil); roads (road* OR highway*); recreation (recreation* OR trails OR tourism); and poor management ("paper park*" OR "poor management" OR underfunded OR "under staffed" OR corruption OR governance). An example query would be: ("national park*" OR "protected area*") in 'Title', NOT "marine" in 'Topic', AND ("climate change" OR "global warming") in 'Topic.'

biotic responses to climate change, posing great challenges for the management of future PAs.

Their management will be further complicated during the coming century due to a series of intensifying pressures. For example, degradation of habitat between PAs and fragmentation of PAs results in loss of connectivity and reduction in their effective size. Many parks and reserves were originally carved out of much larger wilderness areas, and as a result the available habitat for animals and plants found within these areas once extended well beyond their borders. However, PAs have become increasingly isolated due to degradation of surrounding habitat, particularly so perhaps in North America, Asia, and Africa (see table 6.2). For example, nearly 70 percent of the lands surrounding PAs in tropical forests experienced habitat loss or degradation in the past twenty years (DeFries et al. 2005). As effective reserve size decreases, there is less space available to animals and plants and their population sizes decline, resulting in increased susceptibility to extirpation. In Canada, construction of highways has led to the range contraction, habitat fragmentation, and subsequent reduction of grizzly bear (*Ursus arctos*) populations (Proctor et al. 2005). In addition, species that formally ranged outside reserve boundaries are affected by land use changes outside PAs. Woodroffe and Ginsberg (1998) showed that human-induced mortality outside reserve borders played a key role in the extirpation of large carnivores within PAs globally.

As distance between these island-like PAs increases, immigration rates decrease, leading to lower biodiversity levels. Barriers to dispersal between PAs can have the same result. Newmark (2008) found that increasing isolation of PAs in African savanna regions led to the disruption of migration and dispersal patterns, making recolonization of PAs following local extinction more difficult. Fragmentation of habitat can also lower genetic diversity of constituent populations, slow population growth rates, reduce trophic chain length of communities living in PAs, alter species interactions, and ultimately decrease biodiversity. More than 70 percent of PAs are less than 10 km² in area, which suggests that many of these reserves support small, isolated populations. For example, jaguars (*Panthera onca*) that have been restricted to fragmented PAs in South America's Atlantic Forest show significant loss of genetic diversity in individual populations (Haag et al. 2010). Furthermore, recolonization of patches by Holarctic breeding birds

TABLE 6.2. Citations addressing threats to protected areas, by region. High numbers of citations may reflect a growing problem in a particular region or may indicate that a threat had not yet been well-studied.

	CC	IF	PP	PAD	DF	EXP	INV	LS	F	H	M	D	RD	R	PM	TOTAL
North America	60	16	9	0	21	1	115	20	223	28	10	2	25	105	1	636
Central America	0	5	3	0	10	2	2	5	7	1	0	0	0	9	0	44
South America	8	5	10	0	17	7	28	16	28	1	2	2	2	10	1	137
Europe	25	7	10	0	24	9	28	17	35	12	26	1	7	117	2	320
Asia	4	14	15	0	24	25	21	35	22	5	5	0	9	53	5	237
Australia	6	0	0	0	0	0	38	5	80	1	3	2	1	32	0	168
Africa	13	12	47	0	37	44	72	45	118	9	7	5	4	47	0	460
Oceania	0	0	0	0	20	0	16	0	3	1	1	1	0	1	1	44
Caribbean	0	3	0	0	2	0	1	0	0	0	0	0	0	2	0	8
Unspecified	2	0	0	0	0	2	0	0	0	0	0	0	0	20	0	24
Global	6	3	15	4	13	1	0	1	2	0	0	1	0	2	4	46
TOTAL	118	65	109	4	168	91	321	144	518	58	54	14	48	399	14	2,125

Key: CC = climate change, IF = isolation and fragmentation, PP = population pressure, PAD = PADDD, DF = deforestation, EXP = exploitation, INV = invasives, LS = livestock-wildlife conflict, F = fire, H = hydrology, M = mining, D = drilling, R = recreation, RD = roads, PM = poor management.

is directly related to the patches' degree of isolation (Opdam 1991). In addition, extirpation of the Iberian lynx (*Lynx pardinus*) over the last thirty-five years has been linked to reduced immigration due to habitat fragmentation (Rodríguez and Delibes 2003).

Specific effects of isolation and fragmentation on species are idiosyncratic and difficult to predict. Even within the carnivore guild, isolation can have dissimilar effects on different species: isolation of PAs in the northern Rocky Mountain region of the United States has had a greater impact on grizzly bears than on wolves (*Canis lupus*), which have larger home ranges and superior dispersal abilities compared to bears (Carroll et al. 2004). Time since isolation also plays an important role in determining biodiversity remaining in PAs, as populations requiring larger areas will disappear only slowly.

Human populations are growing quickly near borders of many reserves (Zommers and MacDonald 2012), though whether this growth is significantly greater than in other rural areas is disputed (Joppa and Pfaff 2010). Population growth near PA borders may be a result of migrants being "pushed" into areas near reserves due to lack of resources elsewhere, especially arable land ("frontier engulfment"). Alternatively, people may move to these areas because they are attracted to features of the PA, such as job opportunities in ecotourism, clean water, or the very resources that are being protected, although there is less evidence for this. Whatever the cause of population growth, increasing population pressure at reserve borders, when coupled with climate disruption, may exacerbate PA isolation and other threats to these areas.

Protected area downgrading, downsizing, and degazettement (PADDD) is a developing threat to PAs (Mascia and Pailler 2011). PADDD usually occurs for the extraction of resources for human needs, such as recent efforts to allow oil drilling in the Arctic National Wildlife Refuge. In Brazil, an increase in PADDD events is attributed to increased demand for electricity throughout Amazonia. Mascia and colleagues (2014) identified 543 instances of PADDD across 57 countries in the last century. Some conservationists see PADDD as a positive conservation strategy because funds can be reallocated from poorly performing Pas, but PADDD doesn't necessarily occur in such reserves. If conservation embraces PADDD, it may be easier for PAs to be downgraded or degazetted for resource extraction without any corresponding conservation benefit.

Deforestation is an ever-present issue (figure 6.1) and poses an increasingly large threat to PAs throughout all regions of the world except for Caribbean Pas, which are greatly deforested already (table 6.2). While a meta-analysis of forty-nine locations from twenty-two countries showed that the majority of PAs had significantly lower levels of deforestation than non-PAs, PAs are not always effective at decreasing deforestation and extraction activities. In Brazil's 1.8 million km² Amazonian PA network, only 1.4 percent of natural forest had been lost between 2000 and 2006 yet the performance of individual PAs was highly variable (Barber et al. 2012). In the Wolong Nature Reserve for giant pandas (*Ailuropoda melanoleuca*), levels of forest degradation due to woodfuel extraction, timber harvesting, and other activities increased after the area was designated as a reserve (Liu et al. 2001). Despite an increase in PAs throughout Romania, old-growth forest cover continues to decline, with the greatest losses occurring within NPs (Knorn et al. 2013). In a PA complex spanning the China-North Korea border that contains important habitat for the Siberian tiger (*Panthera tigris altaica*), over half of the primary forest has been degraded by exploitative activities (Tang et al. 2010).

Deforestation in PAs occurs through clear cutting forests for activities such as oil palm cultivation, ranching, logging, fuelwood collection, and charcoal production, and PAs in Asia have the highest rates of forest loss. Secondary and regenerating forests that have undergone extraction activities have consistently lower levels of biodiversity than primary forests. Selective logging, the removal of particular species from a forest stand, can lead to forest degradation in and around PAs and can have differential impacts on organisms that occur within the PAs. For example, in the (now) strictly protected Danum Valley Conservation Area in Northern Borneo, selectively logged forests have higher floral species richness, but lower faunal species richness than primary forest (Berry et al. 2010).

Legal and illegal exploitation of wildlife occurs outside and inside PAs on every continent (table 6.2) and is driven by demand for medicine, luxury items (e.g., pets and fashion), trophy hunting, and food. Increased global wealth has driven an upsurge in wildlife exploitation for medicine and luxury items. An estimated 80 percent of the world's people depend on traditional medicine, most of which comes from plants. The $1 billion global trade of between 35,000 and 70,000 species of medical plants represents the largest single human use of biodiversity.

The lucrative trade in wildlife for luxury markets involves large volumes of animals. Between 2000 and 2005, more than 6.7 million live birds, 7.9 million live reptiles, and 30 million reptile skins were traded globally. Because it is often impossible to determine if an animal is wild-caught or captive bred, species listed by the IUCN (International Union for Conservation of Nature) as Critically Endangered or on CITES (Convention on International Trade in Endangered Species of Wild Fauna and Flora) Appendix I can still be easily bought and sold in certain pet markets. From 2002 to 2006 in Italy, officials confiscated the equivalent of 13 percent of the world's remaining wild Egyptian tortoise (*Testudo kleinmanni*) population from pet traders. Fashionable, and often illegal, wildlife items such as ivory can result in the dramatic decline or loss of species from PAs. For example, over the past 40 years elephant (*Loxodonta africana*) populations in West and Central African PAs have declined by 33 and 76 percent respectively, largely due to the demand for ivory (Bouché et al. 2011).

Trophy hunting is a valuable industry for many countries. In sub-Saharan Africa, trophy hunting annually brings in over $200 million in gross revenue from at least 18,500 hunters, while 45,000 to 60,000 foreign hunters visit Eurasian countries, contributing $40 million to $60 million each year. However, trophy hunting can have negative effects on wildlife populations in PAs when quotas are set unsustainably high. For example, in East Africa overharvesting of leopards (*Panthera pardus*) and lions (*Panthera leo*) has led to their decline in PAs (Packer et al. 2011), and sport hunting of bighorn sheep (*Ovis canadensis*) in North America has caused a significant decline in ram body weight and horn size (Coltman et al. 2003). Bushmeat consumption occurs on a vast scale. For example, the extraction of mammal bushmeat from the Congo basin is an estimated 4.9 billion kg/year, while 150 million kg are extracted from the Amazon annually (Fa et al. 2002). Bushmeat hunting is a severe threat to savanna ecosystems, too (e.g., 78,000–110,000 wildebeest are poached in Serengeti NP annually).

All types of extraction can lead to negative consequences for species and ecosystems within PAs, particularly during the Anthropocene. Large-bodied animal species are particularly vulnerable because they have wide-ranging behavior, a low rate of reproduction, and are specifically targeted by hunters. Altered population structures can lead to reduced population growth rates and can result when wildlife exploitation disproportionately targets individuals based on size, age, or sex. All forms of exploitation can reduce genetic variation. Nonnative plants, animals, and pathogens have been introduced, intentionally and unintentionally, to new habitats worldwide. Invasive species are a major cause of modern global extinctions of species: 50 percent of bird extinctions, 48 percent of fish extinctions, and 48 percent of mammal extinctions. Nonnative species are often difficult and costly to eradicate, particularly in the case of plants. Most PAs have at least one documented invasive (90% of PAs surveyed); for instance, sixteen invasive mammalian species are established in more than three-quarters of Argentina's NPs (Merino et al. 2009).

Increased predation is a common result of introduced animals in PAs. For example, the Burmese python (*Python bivittatus*) has led to a dramatic decline in frequency of observations of raccoons (*Procyon lotor*; 99.3% decrease) and opossums (*Didelphis virginiana*; 98.9% decrease), and a complete disappearance of once-common rabbits (*Sylvilagus* spp.) in the Florida Everglades NP (Dorcas et al. 2012). The effect of predation is often more pronounced on islands, where many species have few natural predators. On Gough Island, a protected reserve in the South Atlantic Ocean, introduced house mice (*Mus musculus*) predate juvenile Tristan albatrosses (*Diomedea dabbenena*) and have nearly driven them to extinction (Wanless et al. 2009). Conversely, introduced prey can have negative impacts on their predators. In Kakadu NP in northern Australia, the poisonous invasive cane toad (*Bufo marinus*) colonized the entire reserve within two years, leading to the rapid decline of the quoll (*Dasyurus hallucatus*), a carnivorous marsupial native to Australia (Woinarski et al. 2010). Introduced parasites and disease can also have detrimental effects on native populations inside and outside PAs. Avian malaria and avian poxvirus were introduced to Hawaii in 1826, likely leading to the extinction of at least thirteen bird species. Avian malaria continues to drive population declines in Hawaiian PAs as illustrated by its increased prevalence in Hawaiian honeycreepers (Aves: Drepanidinae) in Hakalau Forest National Wildlife Refuge (Freed et al. 2005). Domesticated animals can be vectors for the spread of invasive diseases in PAs. For example, anthrax (*Bacillus anthracis*) brought in by domestic animals was implicated in a large number of chimpanzee (*Pan troglodytes*) deaths in Tai NP, Ivory Coast (Leendertz et al. 2004). Invasive species also lead to declines of native species in PAs through competition. In Yellowstone NP and surrounding areas, the invasive plant *Linaria vulgaris* has dramatically reduced the cover of native plants (Pauchard et al. 2003). In the Ol Pejeta Conservancy in Kenya, the fast establishing, unpalatable, and fire-resistant invasive bush *Euclea divinorum* has outcompeted native plants to cover 27 percent of the reserve's area and is encroaching into the woody habitat favored by black rhinoceros (*Diceros bicornis*) (Wahungu et al. 2012).

Protected areas can limit the spread of nonnative plants, as in Kruger NP, where the number of nonnative plants inside the park drops off dramatically 1500 m from the border, due to reduced disturbance within the park (Foxcroft et al. 2011). The ability of PAs to buffer against invasive species is limited, however, because waterways, roads, and in-park disturbances allow invasive plants to spread more easily. Recreation can also contribute to the spread of invasive plants within PAs, as can the expansion of invaders' potential ranges due to climate change. In general, human density around PAs is also a major driver of invasions of plant and animals species. As global travel and human population increase during the rest of the twenty-first century, invasive species will become an increasingly large problem.

While pastoralists and their livestock have coexisted with wildlife for thousands of years in some areas, recent changes in land use and resource availability have brought these groups into conflict. Livestock grazing has been implicated in environmental degradation, water shortages, and forage scarcity in and around PAs, and incursions of livestock into reserves are common perhaps especially in Africa and Asia (table 6.2). Bruner and colleagues (2001) found that over 40 percent of 93 parks were ineffective at mitigating the impacts of grazing. Livestock grazing can increase competition with wildlife for food, decrease nutrient availability in soil and water, and cause environmental degradation. Numerous studies have documented the negative relationship between livestock density and wildlife density, including in Kilombero Game Controlled Area in Tanzania (Bonnington et al. 2007) and in and around Kgalagadi Transfrontier Park, Botswana (Wallgren et al. 2009).

Competition between wild herbivores and livestock is context dependent, however. For example, wild ungulates and cattle compete for food during the dry season when resources are scarce but may enhance each other's diet quality during the wet season when resources are abundant. Spatial partitioning of wildlife and livestock may lessen the effects of competition, although this partitioning may not be possible in dry years. The impacts of grazing may vary by taxa too. A study of wildlife densities inside Kenya's Masai Mara National Reserve and adjoining pastoral areas found that small-sized herbivores were more abundant in pastoral areas whereas large-bodied herbivores were more abundant in the park (Bhola et al. 2012). Piana and Marsden (2014) showed that cattle grazing in a Peruvian PA increased the diversity and density of certain raptor species. Livestock may even provide unexpected benefits to PAs by promoting seed dispersal and increasing plant diversity. Thus, the effects of livestock in PAs need not always be negative but in general they have adverse consequences for native fauna and flora. Fire is a powerful ecological disturbance that shapes ecosystem structure and can maintain biodiversity. Fire activity can also dramatically alter habitat structure and affect nutrient and particle content of soil, water, and air, and species around the world have evolved in response to particular fire regimes. The threat of fire in PAs as a result of human-imposed deviations from natural fire regimes has been well recognized for twenty-five years (figure 6.1) and can be divided into locations where it is uncommon in nature, or where fire has been excluded from regions where it was once a frequent occurrence.

The first case is often observed in tropical forests. The natural fire-return interval for most tropical forests is quite long but human influence has altered this through clearing land for agriculture. Fires set on nearby land can spread into PAs or can result from unattended campfires. Isolation and fragmentation can cause forest edges to dry out, increasing fire susceptibility, as can warming climates and invasive grasses. In the Brazilian Amazon, fires occurred in at least 20 percent of reserves in most years, with more fires in dry years, near roads, and in forests with high levels of human impact (Adeney, Christensen, and Pimm 2009). West African PAs saw fire density increase more than 22 percent from 2004 to 2009 (Grégoire and Simonetti 2010). However, a recent study of PAs in the Colombian Andes found that although the majority of PAs experienced at least one fire in the past twenty years, PAs had significantly lower rates of fire than their respective buffer zones, suggesting that PA borders can act as effective barriers (Rodríguez et al. 2013).

For subtropical and temperate forests, grasslands, and shrublands that are fire adapted, the question of fire in PAs is more an issue of maintaining regular fire in that ecosystem. Fire suppression and livestock grazing have reduced fire intervals in fire-adapted ecosystems, leading to fuel buildup and woody species recruitment. Climate change and invasive species have increased susceptibility to fire, promoting severe fires in regions where fuel load has built up due to suppression policies. Management plans that maintain healthy fire regimes are now established in many PAs, yet implementing fire-friendly policies in PAs can be difficult when people live near PA boundaries and can be at odds with species-focused approaches such as the US Endangered Species Act.

Decreasing water availability will have large impacts on PAs this century as worldwide water demand grows. Already, excessive water use means that many major rivers run dry during all or parts of their dry seasons. These include the Ganges in South Asia, the Nile in Africa, and the Colorado River in the United States. Many PAs act as dry-season water sources for wildlife and massive die-offs due to water shortages have been documented for a wide range of taxonomic groups in PAs, including migrating birds in Klamath National Wildlife Reuge (NWR) ("Water Shortage Blamed for Massive Bird Die-Off At NorCal Wildlife Refuge" 2012), piñon pines (*Pinus edulis*) in New Mexico (Breshears et al. 2008), and mammals in the South African Kalahari Gemsbok NP (Knight 1995) some of which are exacerbated by PA isolation. Sometimes artificial waterholes can mitigate these problems to a limited extent. Certain PAs have already been strongly impacted by anthropogenic water use. For example, overdrawing river water for irrigation and cattle watering outside Ruaha NP, Tanzania, has caused the Great Ruaha River to become seasonally dry, shifting the ecosystem from a traditionally wet-tropical to a dry-tropical zone (Mtahiko et al. 2006). In Serengeti NP, the Mara River, a vital source of water to this ecosystem, has decreased its flow by 68 percent in the last forty years as a result of deforestation of the Mau Forest in Kenya as well as increased extraction of water upstream (Gereta et al. 2009). Damming also alters water flow inside PAs and can destroy aquatic habitats.

Analysis of a century of hydrologic records from thirty-one North American rivers revealed flow declines for 67 percent of those measured. These rivers provide water for a large number of North American PAs as confirmed by a raft of studies in this region (table 6.2). Decrease in flow results from a combination of urbanization, irrigation, damming, and reduced snow pack due to climate change. Already, reductions in snow pack as a result of warming have led to decreases in seasonal water availability for PAs throughout the western United States. Runoff from agriculture and mining activities pollutes watersheds and threatens the survival of wildlife within PAs. Widespread use of fertilizers, herbicides, and pesticides in agriculture has negatively impacted PAs such as Kruger, Keoladeo (in India), and Everglades NPs. Pollution from agricultural systems can accelerate eutrophication in freshwater systems and has been linked to declines in numerous freshwater species.

Legal and illegal mining around PAs and accidental mining spills pollute water sources, destroy habitat, and threaten biodiversity, with much attention centered on Europe (table 6.2). Certainly, small-scale mining (mineral extraction characterized by low levels of mechanization and high labor intensity) occurs in or around many PAs. Drainage and tailings from mining activities can contaminate watersheds with lethal levels of chemicals such as arsenic, mercury, and lead. In the Coto Doñana, a protected estuarine marsh ecosystem in Spain, the accidental upstream release of 5 million m³ of acid waste from the processing of pyrite ore led to severe declines in fish, invertebrate, and bird species (Pain et al. 1998). In addition, degazettement and downsizing of PAs for mining threatens ecosystems. For example, a license for uranium mining in Selous Game Reserve, Tanzania, reduced the reserve size by 200 km² (Wippel 2012). As global demand for certain metals increases, so too will mining activity. High prices of coal have driven illegal mining in the protected Bukit Saharto Forest, Indonesia (Mattangkilang 2013), and global gold prices have driven increased mining activity and subsequent deforestation in the Peruvian Amazon (Swenson et al. 2011).

Reliance on fossil fuels has sparked unprecedented levels of oil and gas exploration and extraction and will increase in coming decades in and around PAs. In the federally owned section of the Arctic NWR in Alaska, there are approximately 7.69 billion barrels of recoverable oil, roughly equal to US oil consumption for 2007, and the possibility of opening this region for oil exploration has been intensely debated (Snyder 2008). In the western Amazon, there are approximately 180 oil and gas blocks (geographic areas zoned for hydrocarbon extraction activities) covering some 688,000 km² of forest; these blocks overlap some of the most species-rich areas of the globe and many are located within PA boundaries in Ecuador, Bolivia, and Peru. Oil and gas extraction negatively impacts wildlife, as seen in North America where fossil fuel extraction activities disrupt migration patterns of caribou (*Rangifer tarandus*) and mule deer (*Odocoileus hemionus*) (Hebblewhite 2011) and have led to significant population declines of the greater sage grouse (*Centrocercus urophasianus*) (Naugle et al. 2011). More than a quarter of the 1007 UNESCO World Heritage sites worldwide are under threat from oil and gas extraction, with the Arabian Oryx Sanctuary in Oman being the first site in history to be delisted from the World Heritage list due to a significant reduction in size for oil and gas extraction (Osti et al. 2011).

Road construction and upgrading through PAs result in habitat destruction and pollution while increased vehicle speeds result in greater roadkill counts as well as noise, litter, and chemical pollution. As demand for mineral resources increases and metropolitan centers are connected, PAs are being bisected by roads. The problem is increasingly recognized (Laurance, Goosem, and Laurance 2009) and compromise solutions to the problem of roads are being developed (Caro et al. 2014). Roads inside and outside PAs increase human traffic, exacerbating threats from resource exploitation, immigration, and recreation, reducing effectiveness of PAs and adjacent buffer zones. Worldwide, PAs have seen a substantial rise in nonconsumptive wildlife recreation and nature-based tourism (figure 6.1) with positive trends in total number of visitors in Africa, Asia, Latin America, and Europe but stable trends in Australasia and North America. Creation of roads, trails, and facilities leads to direct habitat destruction and to altered hydrologic processes, increased erosion, and damage to tree roots. Trampling by hikers, bicycles, cross-country skiers, off-road vehicles, and horses causes soil compaction and can result in decreased plant diversity. Recreational activities can also negatively affect animals in PAs by causing direct mortality, altering animal behavior, or introducing diseases (e.g., human-primate disease transmission). Species can be differentially sensitive to disturbance: wolves (*Canis lupus*) in Canadian Rockies NPs avoid areas with the highest levels of human disturbance (< 50 m from trails), but elk (*Cervus canadensis*) preferentially use areas of relatively high human impact (50–400 m from trails) as predation refuges (Rogala et al. 2011). Even quiet, nonconsumptive recreational activities that seem to have low impact can have detrimental effects on wildlife populations. For example, wildlife viewing reduces foraging efficiency in birds and causes higher nest predation or abandonment of young (Boyle and Samson 1985).

Yet nature tourism is responsible for $28.8 billion per year in spending, greatly exceeding annual conservation spending by aid agencies and contributing to government conservation budgets. Nature-based tourism can also be vital for establishing and managing PAs globally. Revenue generated by tourism in PAs contributes to the conservation of wildlife, although there is a risk that the reliance of species conservation on funds from tourism and recreation management might divert resources from other management activities. Poor management can be considered an additional threat to conservation success in the Anthropocene. Jim and Xu (2003) identified six major problem areas in protected area management: site selection, funding, people-park conflict, paper parks, cooperation of management agents, and lack of international experience among managers. While their study is limited to Chinese PAs, examples of poor management from across the globe fit one or more of these categories. For example, the locations of many PAs around the world were not selected with maximum conservation value in mind. Fuller and coworkers (2010) have argued that funds used for the least-cost-effective Australian parks could be better used in other locations. Management of people-park conflict has focused on community-based conservation programs, but many such programs have failed to produce joint solutions to biodiversity decline and poverty. Unfortunately, the paper park syndrome (parks in name only) occurs in Africa, the Americas, Russia, and China.

While there is abundant evidence that PAs are effective at conserving species and landscapes globally, reserves are still vulnerable to many human activities (figure 6.1), a vulnerability that may accelerate during an era of changing climates. Here we have examined key threats to PAs although there are many we have omitted, such as effects of civil war, as for example in the Democratic Republic of Congo or Rwanda, or clandestine airstrips built by narcotraffickers in Guatemala, so our analysis is broad but incomplete. Nonetheless, recognition of the key challenges to PAs surely allows us to begin to address them effectively and consider their role in a suite of climate adaptation strategies.

7

Mitigating Anthropocene Influences in Forests in the United States

CHADWICK DEARING OLIVER

Based on the evidence from climate temperature cycles that have occurred since the Pleistocene, North American plant and animal species can be expected to move to cooler climates to the north and higher elevations as the climate warms. During these previous periods, the species' contiguous ranges were able to move northward at a rate of about twenty-five miles per century (Davis 1981) or to higher elevations at the same latitude at about 130 feet per century (Hopkins 1920). However, plant species do not migrate as communities. That is, species found together as a community at one latitude now may not have occurred together at a lower latitude when the climate was cool and will probably not remain together at a higher latitude in the future (Davis 1981). Plant species migrate by individual seeds becoming established and outcompeting other species at a given time and place. Plant movement occurs primarily after a disturbance or at range edges, when growing space is available for the plant to become established in the new location. Plant survival during germination and initial growth seems especially sensitive to climate, but long-lived trees that begin growing under one climate can survive in the same place even after the climate has changed. Many tree species may actually be living in places where they can no longer become established (Brubaker 1986; figure 7.1).

Species establishment and migration can be strongly influenced by topography—sunny south slopes in the cool northern end of the range and shady north slopes in the southern end. Some species may be able to become established at slightly higher and lower elevations on these slopes in response to climate changes. The opportunistic nature of plant establishment and migration means a species' range may fragment if some individuals move to higher elevations, others move northward, and some remain as isolated "refugia" on favorable microsites. These fragmented habitats can lead to the isolated groups evolving along such different trajectories that they become different species—the "species pump" concept (Huston 1994).

Northward movement of a species' home range is facilitated by gentle terrain and by mountain ranges that lie in a north-south orientation (figure 7.2), as in much of North America (Flannery 2001). Other factors tend to keep species positions more stationary (figure 7.2). A mountain range can keep a species nearly stationary in terms of latitude as it finds opportunity to reestablish in cooler climates at higher elevations (a change of 400 feet in elevation can have the same effect as moving seventy miles of latitude; Hopkins 1920). Some species that move up mountains to cooler climates may eventually become isolated in "sky islands" similar to those in the southwestern United States (Warshall 1995) and possibly in the Great Smoky Mountains in the southeastern United States. Species may eventually be eliminated if the mountain is not tall enough to provide suitably cool climates.

The generally north-south mountain orientation in the United States allows species to move latitudinally quite readily with climate change; however, steep east-west valleys along the Pacific Coast have prevented—and may continue to prevent—species migrations northward (A, figure 7.2). Mountain ranges also affect rainfall patterns, cold periods, and ice storms. The orographic precipitation belt at the southeastern base of the Great Smoky Mountains (B, figure 7.2) will probably not move with climate change and freezing rainstorms typical on the Cumberland Plateau will probably continue (C, figure 7.2). The Mississippi River and its floodplains will continue to be a barrier to the east-west migration of species (D, figure 7.2). And, the crest of the Cascade and Sierra Ranges (E, figure 7.2) will continue to create a division between humid ecosystems to the west and arid ecosystems to the east.

FIGURE 7.1. The unusual mixture of Douglas firs (*Pseudotsuga menziesii* [Mirb.]
Franco), mountain hemlocks (*Tsuga mertensiana* [Bong.] Carriere), and Pacific silver
firs (*Abies amabalis* [Douglas ex Loudon] Douglas ex Forbes) growing together at
3,000 feet of elevation in the North Cascades of Washington is a result of the different climates when each species became established (from Oliver et al. 1985). The
small tree near lower right corner is a Pacific silver fir (5 feet tall, 110 years old) that
became established at a slightly cooler climate than present. Right of center: much
taller mountain hemlock, about 200 years old (established during the Little Ice Age),
growing about 1,000 feet lower than where it usually germinates at present. Center
distance: the large Douglas fir that the person is leaning on typically becomes established about 1,000 feet lower at present; it is about 800 years old and was established
during the Medieval Warm Period.

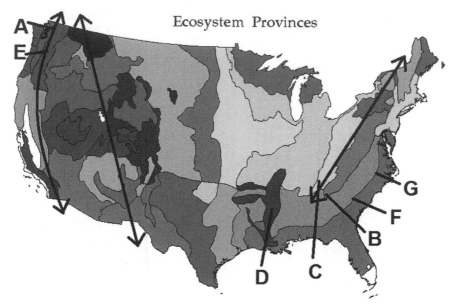

Ecosystem Provinces

FIGURE 7.2. Features affecting species movement and stability in the contiguous United States, overlain on Bailey's Ecosystem Provinces (Bailey 1995). (Letters refer to discussions in text. Arrows show north-south orientations of mountains.)

On the other hand, the freezing rain belt that defined the northern limit of slash pine's (*Pinus elliottii* Engelm) northern range in South Carolina (F, figure 7.2) may expand its range northward (Baldwin 1973). The sandy soils (Sand Hills) of the upper coastal plain (G, figure 7.2) in the southeastern United States will not move, and so longleaf pines (*Pinus palustris* Mill.), scrub oaks, and other species well adapted to those soils will probably not be able to move northwestward into the adjacent, clay soils.

Some species changes can be anticipated; however, it is impossible to predict all changes—especially behaviors of insect and disease pests such as the pine butterfly and bark beetle outbreaks in the Pacific Northwest (Scott 2012) or the native bark beetle outbreaks with warming climate in central British Columbia (Astrup et al. 2008).

Ideally, we would monitor the forests and facilitate the migration of each species to more suitable areas; however, the overwhelming numbers of species would make such an endeavor impossible in practice. Strategies can focus on the viability of all ecosystems—the "coarse filter" approach to biodiversity conservation (The Nature Conservancy 1982)—recognizing that they will reorganize with different component species in the future. As they reorganize we can reclassify the ecosystems and concentrate on maintaining their viability. If current ecosystems remain viable, it is likely that most existing species will survive, although perhaps differently arranged. We can also monitor a subset of key species of trees, herbaceous plants, and animals to ensure they are thriving—the "fine filter" approach to biodiversity conservation (The Nature Conservancy 1982).

When a unique ecosystem becomes extremely small, its component species may become exterminated by chance local events. On a coarse scale, this is not a significant concern in the eastern United States. Most ecological provinces in the eastern United States are composed of more than 50 percent forest by area (Bailey 1995; figure 7.3 and table 7.1); consequently, much of the issue will be ensuring the forests are in appropriate condition, rather than establishing or expanding forests. Exceptions are the Eastern Broadleaf Continental Forest and the Lower Mississippi Riverine Forest, which have been extensively cleared for agriculture, and the Everglades, which may always have had lesser amounts of forests. The Lower Mississippi Riverine Forest province is of special concern because it is a small area to begin with; the Ozark Broadleaf Forest-Meadow and Ouachita Mixed Forest-Meadow provinces are also small and, although currently forested, will need attention to avoid extirpation of their species. The Everglades is already receiving attention, and the Eastern Broadleaf Continental Forest and the Lower Mississippi Riverine Forest provinces may be difficult to address because increasing the forests here will shift agriculture to larger areas of less productive soils. The eastern forests are already fragmented with highways, cities, and farmlands that make migration difficult for plant and animal species. Such issues as connectivity to allow migration with climate change will need to be addressed (Redondo-Brenes 2007).

Some introduced species become benign parts of the ecosystem, while others are more harmful because they eliminate some native charismatic tree species and restrict the ranges of others. These harmful introduced species increasingly create problems in the United States and abroad as increased global trade is moving more predators and pathogens to locations where the hosts are not resistant. As the number of tree species in an area declines, the food diversity for animals also declines. Native species may also exhibit different behaviors with a changed climate (Warren and Bradford

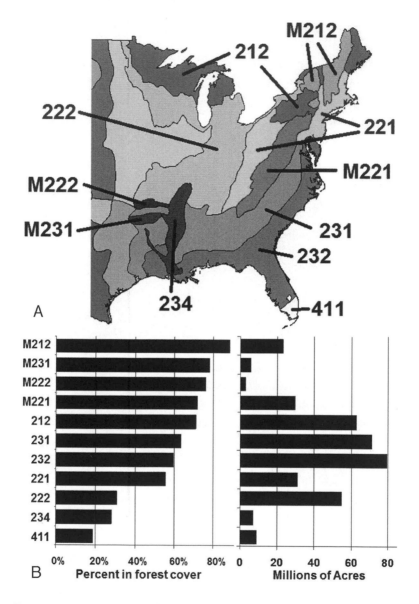

FIGURE 7.3. (A) Locations of ecosystem provinces and (B) forest proportions and areas within each province in the eastern United States (Bailey 1983, 1995). Only a few of these ecosystems have small proportions or areas of forests. (Unlike eastern forests, western forests are not continuous but are mixed with natural shrublands, grasslands, deserts, and alpine areas.) (See table 7.1 for the key to letters/numbers in Figure 7.3A.)

TABLE 7.1. Ecosystem province names and abbreviations used in Figure 7.3A (Bailey 1983).

Ecosystem province name	Abbreviation
Laurentian Mixed Forest Province	212
Adirondack-New England Mixed Forest—Coniferous Forest—Alpine Meadow Province	M212
Eastern Broadleaf Forest (Oceanic) Province	221
Central Appalachian Broadleaf Forest—Coniferous Forest—Meadow Province	M221
Eastern Broadleaf Forest (Continental) Province	222
Ozark Broadleaf Forest—Meadow Province	M222
Southeastern Mixed Forest Province	231
Ouachita Mixed Forest—Meadow Province	M231
Outer Coastal Plain Mixed Forest Province	232
Lower Mississippi Riverine Forest Province	234
Everglades Province	411

2014; Urban et al. 2012), and currently benign species may aggressively displace their long-time neighbors.

Two features of the forest are highly important to sustaining and possibly facilitating migration of species: the current and future condition of the forest and our ability to manipulate the forest. Figure 7.4 shows the recent age class and diameter distributions of the United States forests in different regions (Smith et al. 2009), the mean global temperature when each age class became established (Soon 2005), and which proportions of each region's forest are living in compatible and incompatible climates. The forty-to-eighty-year age classes became established during the warm period in the first half of the twentieth century when the climate was somewhat similar to present; consequently, plants in these forests may survive well in the warming climate. Forests less than forty years old and over eighty years may be more vulnerable to the warming climate because these forests became established when the climate was cooler. Consequently, many western forests are growing in climates much warmer than they began in, and so may not be physiologically well adapted. Older forests in the north also may be out of synchrony with their climate, but nearly all forests in the South, being young, began in warm climates similar to the present and the anticipated future.

The forests in each region will probably be most resilient to loss of functioning with climate change if they are maintained in a diversity of stand

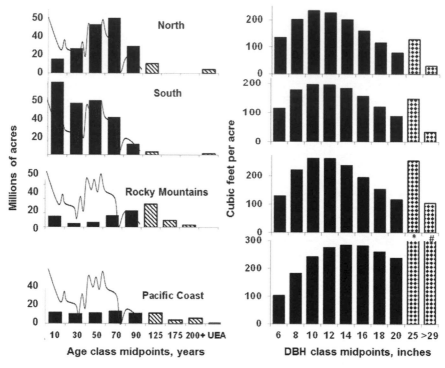

FIGURE 7.4. Distributions of forest age and diameter classes in the four major forest regions of the contiguous United States (Smith et al. 2009). The line graphs (Soon 2005) superimposed on age classes (left column) show relative temperature compared to present when each age class initiated. Hachured (left) and stippled (right) bars indicate change of horizontal (age class) scales. (UEA = uneven-age; * and # indicate even higher volumes than shown.)

structures (Oliver and Larson 1996; figure 7.5). Since each structure supports many different species (Oliver and O'Hara 2004), the diversity means that large, uniform populations of native hosts or pests are less likely to develop. The variety of structures also makes the forests less susceptible to catastrophic fires and more suitable for water infiltration. Open and savanna structures evapotranspire less water than the other structures that have more leaf area; consequently, more water flows through the soil to aquifers or streams. And, the constant need for open and savanna structures gives opportunities for available growing space where species can become established and thus migrate to more suitable climates.

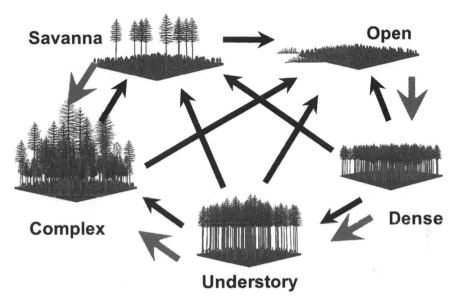

FIGURE 7.5. Forests naturally contain stands in a variety of structures that change with growth and disturbances. Each structure provides habitat for some species, so a forest maintains the most diversity—as well as being resilient to catastrophic disturbances and climate change—if all structures are maintained. Gray arrows represent growth; black arrows represent disturbances (Oliver and Larson 1996).

Each structure also will have issues that will need silvicultural attention with climate change. The older, understory and complex forests may contain trees that are physiologically weakened by the warmer climate and/or susceptible to associated storm or other weather patterns. Yet these forests are needed by some animals and plants, structures that take a long time to replace. It may be appropriate to focus especially on protecting those old forests in cool and moist topographic positions and to remove old trees where their demise may provide a human hazard, such as alongside roads where wind and ice storms can destroy power lines and block roads. It may be appropriate to begin planning areas of future old forests within the cooler parts of the current tree ranges, to serve as permanent "reserves" while the younger forests are managed to respond to climate change (Seymour and Hunter 1999). This planning may give an opportunity to promote understory and complex forests in the Southeast, where relatively few of them currently exist.

The silvicultural knowledge and techniques to manage the forests in ways that maintain their many values may change during the Anthropocene. Silviculture has changed from identifying a single "system" for managing each species and community to considering that each stand can develop naturally through any of a large number of possible "pathways" that are the result of growth and disturbances (Botkin 1991, Oliver and O'Hara 2004). A stand's pathway can be directed by using silvicultural operations that avoid and/or mimic disturbances and regeneration at key times to provide targeted structures and values at later key times. Silviculturalists can mix the many pathways of different stands within a landscape to ensure a diversity of structures and values are continuously provided. Individual structures can be placed and moved on a landscape so they are less susceptible to unwanted disturbances or pests. In the process of implementing the diversity of pathways, various species and species combinations can be regenerated in favorable habitats.

Even the best silvicultural knowledge and technology to avoid catastrophic forest problems associated with climate change are insufficient without an infrastructure of appropriate labor and equipment to do the needed operations. And, markets for the wood removed are needed to keep the operations from becoming prohibitively expensive. Currently, there is a shortage of labor and forest management equipment (Knight 2013) in much of the United States. Plus, there is little incentive to invest in such equipment because more wood is growing than is being used in the United States (figure 7.6) and in the rest of the world (Oliver et al. 2014). Moreover, cheaper sources of wood from overseas are driving down prices, against which US operators are having difficulty competing. Consequently, wood is a "buyer's market," and most stands are not being treated. In addition, much of the present forestry operations equipment is too large for small-scale operations such as removing small trees for thinnings or operating on small forest tracts (Cushing and Straka 2011).

The needed forest operations and processing infrastructure could be developed through a concerted effort to harvest and utilize more wood sustainably for wood construction using new technologies such as "cross-laminated timber" (CLT), with the residues used for wood energy. Greater use of wood in long-duration wood products that substitutes for steel and concrete could also reduce net CO_2 emissions and fossil fuel use (MGB 2012, Oliver et al.

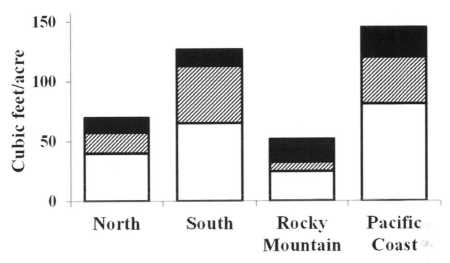

FIGURE 7.6. Forests in all regions of the United States are accumulating biomass (white = net growth) with relatively little harvest (hachure) and some mortality (black) (Smith et al. 2009).

2014). A paradigm shift toward greater logger skill and smaller, less expensive equipment that can operate more carefully could allow more stands to be managed in appropriate ways (Scott 2013). Silviculture, including the infrastructure, can be important but needs to be conducted as part of a broader effort of forest restoration rather than as a series of stand-alone mandates. The silviculture activities need the "dynamic" perspective described by Botkin (1979, 1991), rather than the now-outdated "steady state" perspective. The "steady state" perspective tended to separate management from some preferred, "natural" development that would occur without human intervention (Colwell et al. 2012). We now understand that "natural" trajectories are random and may not protect biodiversity or provide other forest functioning and values.

Similarly, the "sustained yield" economic paradigm served well (Davis et al. 2001) but proved problematic in the long term because the world is dynamic, rather than stable. Subsequent paradigms defaulted to short-term profitability, such as following price signals, as an alternative (Dowdle 1984). We need an economic paradigm in resource management that bridges the best of the long and short terms. Looking at forestry from the landscape level (Oliver 1992; Boyce 1995; Heilig 2003) facilitates this adoption process since single stands provide only parts of the overall values from forest ecosystems. It will be helpful for many small landowners to work together to provide all values. Such cooperation may also enable marketing and managing benefits through the economies of scale. Each landscape has different properties (Oliver et al. 2012), so local knowledge will be needed for appropriate management.

Finally, introduced and aggressive organisms such as forests pests and pathogens that attack trees are often extremely vigorous at first; however, over time populations of species that prey on them can develop and keep their aggressiveness in check. A gradual approach of trying to slow an aggressive species may give time for these predator populations to grow and limit the harm of the aggressive one. This gradual approach may be combined with integrated pest management (Kogan 1998), which also combats pests through a combination of monitoring and taking increasingly extreme actions only as needed. Together the two approaches may be very effective.

More stringent measures may be needed to prevent new pests from entering North America—or leaving it. Since many regions of the world are facing species losses from introduced pests, it may be prudent to stop overseas shipping of raw organic materials (Oliver 2014). This would prevent "hitchhiker" pests on these raw materials, and ensure that secondary manufacture occurs in the region where the organic material is grown. This approach in the anthropocentric age may have an additional benefit of stimulating economies throughout the world by helping them develop value-added products instead of exporting raw materials (Acemoglu and Robinson 2012).

SECTION III

ADAPTATION STRATEGIES FOR BIODIVERSITY CONSERVATION AND WATER PROTECTION

8

Planning the Future's Forests with Assisted Migration

MARY I. WILLIAMS AND R. KASTEN DUMROESE

If the climate changes faster than the adaptation or migration capability of plants (Zhu et al. 2012; Gray and Hamann 2013), foresters and other land managers will face an overwhelming challenge. Growing trees that survive may become more important than growing perfectly formed trees (Hebda 2008) and may require selection of adapted plant materials and/or assisting the migration of plant populations (Peters and Darling 1985). Agencies, land managers, and foresters are being advised to acknowledge climate change in their operations, but current client demands, policies, and uncertainty about climate change predictions and impacts constrain active measures (Tepe and Meretsky 2011). For example, the practice of restricting native plant movement to environments similar to their source has a long history in forest management (Langlet 1971); however, transfers must now factor in climate change because plant materials guided by current guidelines and zones will likely face unfavorable climate conditions by the end of this century.

The findings and conclusions in this chapter are those of the authors and do not necessarily represent the views of the United States Forest Service.

DOI: 10.5876/9781607324591.c008

To facilitate adaptation and migration, we will need to rethink the selection, nursery production, and outplanting of native trees in a dynamic context, such as modifying seed transfer guidelines in the direction of climatic change to suit target species and populations. A challenge lies in matching existing plant materials (i.e., seeds, nursery stock, or genetic material) to ecosystems of the future that have different climate conditions (Potter and Hargrove 2012). To alleviate the challenge, strategies such as *assisted migration* (also referenced as *assisted colonization* and *managed relocation*) have been proposed in adaptive management plans (e.g., USDA Forest Service 2008), but without specific guidance.

Foresters have been moving tree species and populations for a very long time. Usually, these movements are small and properly implemented by using seed transfer guidelines. Occasionally, these movements are drastic and intercontinental to support commercial forestry (e.g., exporting Monterey pine [*Pinus radiata*] from the United States to New Zealand). The concept of assisted migration, first proposed by Peters and Darling (1985), builds on this forestry legacy of moving species and populations, but deliberately includes management actions to mitigate changes in climate (figure 8.1) (Vitt et al. 2010). This does not necessarily mean moving plants far distances, but rather moving genotypes, seed sources, and tree populations to areas with predicted suitable climatic conditions with the goal of avoiding maladaptation (Williams and Dumroese 2013). How far we move plant materials to facilitate migration will depend on the target species and populations, location, projected climatic conditions, and time. For a species or population, this may require target distances across current seed-zone boundaries or beyond transfer guidelines (Ledig and Kitzmiller 1992). Target migration distance is the distance that populations could be moved to address future climate change and foster adaptation throughout a tree's lifetime (O'Neill et al. 2008). Target migration distance can be geographic (e.g., distance along an elevation gradient), climatic (e.g., change in number of frost-free days along the same elevation gradient), and/or temporal (e.g., date when the current climate of the migrated population equals the future climate of the outplanting site). Instead, evaluating species that might naturally migrate is an option. Alberta, Canada, for example, is considering ponderosa pine (*Pinus ponderosa*) and Douglas fir (*Pseudotsuga menziesii*), now absent in the province but occurring proximate to the province, as replacements for lodgepole pine (*Pinus*

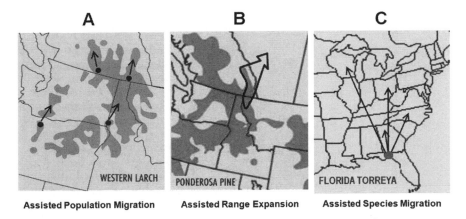

A

B

C

WESTERN LARCH   PONDEROSA PINE   FLORIDA TORREYA

Assisted Population Migration   **Assisted Range Expansion**   **Assisted Species Migration**

FIGURE 8.1. Assisted migration can occur as assisted population migration where seed sources are moved climatically or geographically within their current ranges, even across seed transfer zones—for example (A), moving western larch (*Larix occidentalis*) 200 km north within its current range. Seed sources can also be moved climatically or geographically from current ranges to suitable areas just outside to facilitate range expansion (B), such as by (hypothetically) moving seed sources of ponderosa pine (*Pinus ponderosa*) into Alberta, Canada,. In an assisted species migration (or assisted long-distance migration) effort (C), species are moved far outside current ranges to prevent extinction, such as planting Florida torreya (*Torreya taxifolia*) in states north of Florida. (Distribution ranges are shaded gray; terms from Ste-Marie et al. 2011; Winder et al. 2011; Williams and Dumroese 2013 and distribution maps from Petrides and Petrides 1998; Torreya Guardians 2016.)

contorta); lodgepole pine is predicted to decline in productivity or become extirpated under climate change (Pedlar et al. 2011).

Moving plants has been practiced for a long time in human history, but the movement of species in response to climate change is a relatively new concept (Aubin et al. 2011). As an adaptation strategy, assisted migration could be used to prevent species extinction, minimize economic loss (such as declining timber production), and sustain ecosystem services (e.g., wildlife habitat, recreation, and water and air quality) (figure 8.1) (Aubin et al. 2011). Assisted migration may be warranted if a species, such as lodgepole pine, establishes easily, provides more benefits than costs, and is at high risk of extinction and loss of the species would disrupt ecosystem services (Hoegh-Guldberg et al. 2008). Reducing fragmentation, increasing landscape connections, collecting and storing seeds, and creating suitable habitats are all viable options (depending on the species and population) to facilitate adaptation and migration. Some species may migrate in concert with climate change, thus conserving and increasing landscape connections should take precedence over other management actions. Other species may adapt to changes in climate, while yet others may have limited adaptation and migration capacities. Assisted migration needs to be implemented within an adaptive management framework, one that assesses species vulnerability to climate change, sets priorities, selects options and management targets, and emphasizes long-term monitoring and management adjustments as needed.

Frameworks, tools, and guidelines on implementation (table 8.1) (Beardmore and Winder 2011; Pedlar et al. 2011; Williams and Dumroese 2013) have been introduced to make informed decisions about climate change adaptation strategies. Programs such as the Climate Change Tree Atlas (Prasad et al. 2007), Forest Tree Genetic Risk Assessment System (ForGRAS, Devine et al. 2012), NatureServe Climate Change Vulnerability Index (NatureServe 2015), System for Assessing Species Vulnerability (SAVS, Bagne et al. 2011), and Seeds of Success program (Byrne and Olwell 2008) are available to determine a species's risk to climate change. Species most vulnerable to climate change are rare, long-lived, locally adapted, geographically and genetically isolated, and threatened by fragmentation and pathogens (Erickson et al. 2012). Listing species as suitable candidates—those with limited adaptation and migration capacity—is a practical first step, but requires a substantial amount of knowledge about the species and their current and projected habitat conditions. Provenance data exist for several commercial tree species and should be used to estimate their response to climate scenarios. The Center for Forest Provenance Data provides an online database of tree provenance data (St. Clair et al. 2013). Bioclimatic models coupled with genetic information from provenance tests and common garden studies in a GIS can be used to identify current and projected distributions (e.g., Rehfeldt and Jaquish 2010; McLane and Aitken 2012; Notaro et al. 2012). These forecasts can assist land managers in their long-term management plans, such as where to collect seeds and plants. Although modeled projections have some uncertainty in predicting climate changes and tree responses (Park and Talbot 2012), they provide an indication of how climatic conditions will change for a particular site.

The movement of species in response to climate change does not come without economic, ecological, ethical, and political issues (Schwartz et al.

TABLE 8.1. Resources related to forest management, native plant transfer guidelines, climate change, and assisted migration for the United States and Canada. Most programs are easily located by searching their names in common web browsers. All URLs were valid as of 19 March 2016.

Resource or Program	Description	Authorship
Assisted Migration Adaptation Trial https://www.for.gov.bc.ca /hre/forgen/interior/AMAT.htm	Large, long-term project to evaluate the response of 15 tree species to climate change and assisted migration	Ministry of Forest and Range, BC
Center for Forest Provenance Data http://cenforgen.forestry .oregonstate.edu/	Public users are able to submit and retrieve tree provenance and genecological data	Oregon State University and USDA Forest Service
Centre for Forest Conservation Genetics http://www.genetics .forestry.ubc.ca/cfcg/	Portal for forest genetics and climate change research conducted in British Columbia, Canada	The University of British Columbia
Climate Change Response Framework http://climate framework.org/	Collaborative framework among scientists, managers, and landowners to incorporate climate change into management	Northern Institute of Applied Climate Science
Climate Change Tree Atlas http://www.fs.fed.us/nrs/atlas/	An interactive database that maps current (2000) and potential status (2100) of eastern US tree species under different climate change scenarios	USDA Forest Service
Forest Seedling Network http:// www.forestseedlingnetwork.com	Interactive website connecting forest landowners with seedling providers and forest management services and contractors; includes seed zone maps	Forest Seedling Network
Forest Tree Genetic Risk Assessment System (ForGRAS) http://www.forestthreats .org/research/projects/project -summaries/genetic-risk -assessment-system	Tool to identify tree species risk of genetic degradation in the Pacific Northwest and Southeast	North Carolina State University and USDA Forest Service
MaxEnt (Maximum Entropy) www.cs.princeton.edu/~schapire /maxent/	Software that uses species occurrences and environmental and climate data to map potential habitat; can be used to develop seed collection areas	Phillips, Anderson, and Schapire (2006)
Native Seed Network http:// www.nativeseednetwork.org/	Interactive database of native plant and seed information and guidelines for restoration, native plant propagation, and native seed procurement by ecoregion	Institute for Applied Ecology

continued on next page

TABLE 8.1.—*continued*

Resource or Program	Description	Authorship
Seed Zone Mapper http://www.fs.fed.us/wwetac/threat_map/SeedZones_Intro.html	An interactive seed zone map of western North America; user selects areas to identify provisional and empirical seed zones for grasses, forbs, shrubs, and conifers; map displays political and agency boundaries, topography, relief, streets, threats, and resource layers	USDA Forest Service
Seedlot Selection Tool http://sst.forestry.oregonstate.edu/	An interactive mapping tool to help forest managers match seedlots with outplanting sites based on current climate or future climate change scenarios; maps current or future climates defined by temperature and precipitation	Oregon State University and USDA Forest Service
ClimWhere (formerly SeedWhere) http://gmaps.nrcan.gc.ca/climwhere/	GIS tool to assist nursery stock and seed transfer decisions for forest restoration projects in Canada and the Great Lakes region; can identify geographic similarities between seed sources and outplanting sites	Natural Resources Canada, Canadian Forest Service
System for Assessing Species Vulnerability (SAVS) www.fs.fed.us/rm/grassland-shrubland-desert/products/species-vulnerability/	Software that identifies the relative vulnerability or resilience of vertebrate species to climate change; provides a framework for integrating new information into climate change assessments	USDA Forest Service

2012). Assisted migration is a sensitive strategy because it disrupts widely held conservation objectives and paradigms (McLachlan et al. 2007). Adoption requires us to balance conservation of species against risks posed by introduced species (Schwartz 1994). Current natural resource management plans were not written within the context of climate change, let alone rapid changes in climate. The US Forest Service anticipates using assisted migration of species to suitable habitats to facilitate adaptation to climate change (USDA Forest Service 2008). But these management statements imply that assisted migration should only be implemented in cases where past research supports success (Erickson et al. 2012; Johnson et al. 2013). Assisted migration is essentially incompatible with existing US state and federal land management frameworks (Camacho 2013). For example, in current tree-improvement programs in the United States, seed transfer guidelines and zones are used to determine the safest distance that a population can be moved to avoid maladaptation (Johnson et al. 2004). For most jurisdictions in the United States, the guidelines and zones prohibit the movement of seed sources between and among zones. As they currently stand, seed transfer policies do not account for changes in climate, even though research has identified that suitable habitat for some important commercial tree species will shift north and to higher elevations during this century (Aitken et al. 2008; Rehfeldt and Jaquish 2010). The existing policies hamper any formal actions and may encourage more privately funded operations, such as the Florida torreya (*Torreya taxifolia*) project in the southeastern United States. Since 2008, it has been planted on private lands in five southern states in an effort to curtail extinction (Torreya Guardians 2016).

Even so, the debate about its implementation is largely focused on an ecological assessment of risks and benefits (see Ricciardi and Simberloff 2009; Aubin et al. 2011; Hewitt et al. 2011; Lawler and Olden 2011). Given the uncertainties inherent in climate prediction, knowing how and whether an ecosystem will be affected is difficult. We have limited knowledge about establishing native plants outside their range in anticipation of different climate conditions let alone the impact of climate change on ecosystem properties important to the survival and growth of trees (e.g., photoperiod, soil conditions, and pollinators). To further complicate matters, we know little about the long-term ecological effects of assisted migration, such as invasiveness, maladaptation, and site stability (Aubin et al. 2011). Uncertainty about future climate conditions and risks, such as genetic pollution, hybridization, impairment of ecological function and structure, introduction of pathogens, and bringing in invasive species are major constraints to consensus and implementation (Gunn et al. 2009; Aubin et al. 2011).

Economic costs and ecological risks will vary across assisted migration efforts (figure 8.1) and will likely increase with migration distance (Mueller and Hellmann 2008; Vitt et al. 2010; Pedlar et al. 2012). Establishment failure could occur if the species or population is moved before the outplanting site is climatically suitable or if the seed source is incorrectly matched with the outplanting site in a projected area (Vitt et al. 2010). Assisted migration to areas far outside a species current range would carry greater costs, management responsibilities, and ecological risks than assisted population migration and assisted range expansion (Winder et al. 2011). Principal to reforestation success is using locally adapted plant materials, so the greater the difference between seed origin and outplanting site the greater the risk in maladaptation. An increase in distance (either geographic or climatic) is usually but not always associated with loss in productivity, decrease in fitness, or mortality (Rehfeldt 1983; Campbell 1986; Lindgren and Ying 2000).

Forest tree species are highlighted most often in the assisted migration literature because of their economic value and the consequent focus on them in climate change research; however, assisted migration conducted for economic rather than conservation reasons is cited as another major barrier to implementation, meaning that economic benefit may be an insufficient justification (Hewitt et al. 2011). On the contrary, the forestry profession is well suited to evaluate, test, and employ an assisted migration strategy, given its long tradition of research, development, and application of moving genetic resources through silvicultural operations (Beaulieu and Rainville 2005; Anderson and Chmura 2009; McKenney et al. 2009; Winder et al. 2011). For commercial forestry, assisted migration could address health and productivity in the coming decades (Gray et al. 2011) because operational frameworks already exist. Forest management policy drafts to allow assisted migration and trials of assisted migration are currently underway in North America. The Assisted Migration Adaptation Trial (AMAT) is a large collection of long-term experiments undertaken by the British Columbia (BC) Ministry of Forests (Canada) and several collaborators, including the US Forest Service and timber companies; it tests assisted migration and climate warming (Marris 2009). The program evaluates the adaptive performance of fifteen tree species collected from a range of sources in BC, Washington, Oregon, and Idaho and planted on a variety of sites in BC. Important components of the trial test how sources planted in northern latitudes perform as the climate changes and evaluate endurance of northern latitude sources to warmer conditions in southern latitudes. For decades in the southeastern United States, some southern pine seed sources have been moved one seed zone north to increase growth (Schmidtling 2001). Similarly, Douglas firs have been planted around the Pacific Northwest to evaluate their growth response to climatic variation (Erickson et al. 2012). The only known assisted species migration project in the United States is a grassroots initiative to save the Florida torreya, a southeastern evergreen conifer, from extinction by planting it outside its current and historic range (McLachlan et al. 2007; Barlow 2011). The project has prompted the US Fish and Wildlife Service to consider assisted migration as a management option for this species (Torreya Guardians 2016).

Assisted migration will be best implemented where seed transfer guidelines and zones are currently in place, and most successful if based on climate conditions (McKenney et al. 2009). Provenance data, seed transfer guidelines, and seed zones can be used to facilitate the adaptation of trees being established today to future climates of tomorrow (Pedlar et al. 2012). In Canada, several provinces have modified policies or developed tools to enable assisted migration. Seed transfer guidelines for Alberta were revised to extend current guidelines northward by 2° latitude and upslope by 656 feet (200 m) (NRC 2016) and guidelines for some species were revised upslope by 656 feet (200 m) in BC (O'Neill et al. 2008). Policy in BC also allows the movement of western larch (*Larix occidentalis*) to suitable climatic locations just outside its current range (NRC 2016). To test species range limits in Quebec, some sites are being planted with a mixture of seed sources from the southern portion of the province. Canada and the United States have tools to assist forest managers and researchers in making decisions about seed transfer and in matching seedlots with outplanting sites (e.g., Optisource [Beaulieu 2009] and BioSim [Regniere and Saint-Amant 2008] in Quebec, ClimWhere (formerly SeedWhere) in Ontario [McKenney et al. 1999], and the Seedlot Selection Tool in the United States [Howe et al. 2009]). ClimWhere can map out potential seed collection or outplanting sites based on climatic similarity of chosen sites to a region of interest. The Seedlot Selection Tool is a mapping tool that matches seedlots with outplanting sites based on current or future climates for tree species such as Douglas fir and ponderosa pine.

Target migration distances must be short enough to allow survival, but long enough to foster adaptation toward the end of a rotation, or lifespan of a tree plantation (McKenney et al. 2009). Preliminary work in Canada on most commercial tree species demonstrates that target migration distances for populations would be short, occurring within current ranges (O'Neill et al. 2008; Gray et al. 2011). For some tree species, target migration distances are less than 125 miles (< 200 km) north or less than 328 feet (< 100 m) up in elevation during the next twenty to fifty years (Beaulieu and Rainville 2005; O'Neill et al. 2008; Pedlar et al. 2012; Gray and Hamann 2013). Target migration distances are needed for short- and long-term planning efforts and will require adjustments as new climate change information comes to light. Methods using transfer functions and provenance data have been developed to guide seed movement under climate change (e.g., Beaulieu and Rainville 2005; Wang et al. 2006; Crowe and Parker 2008, Thomson et al. 2010; and Ukrainetz et al. 2011). Bioclimatic models mapping current and projected seed zones have been assessed for aspen (*Populus tremuloides*) (Gray et al. 2011), lodgepole pine (Wang et al. 2006), longleaf pine (*Pinus palustris*) (Potter and Hargrove 2012), whitebark pine (*Pinus albicaulis*) (McLane and Aitken 2012), dogwood (*Cornus florida*) (Potter and Hargrove 2012), and western larch (Rehfeldt and Jaquish 2010). The lack of genetic, provenance, and performance data on which seed transfer guidelines and zones are based impede making informed decisions about assisted migration for noncommercial species. At best we can consult provisional seed zones (e.g., Seed Zone Mapper; table 8.1) developed from temperature and precipitation data and Omernik level III and IV ecoregion boundaries (Omernik 1987). Furthermore, we can shift the focus to producing plant materials that grow and survive by modifying past and current projects and implementing studies and strategies. Many existing projects, such as provenance and common garden studies, can be transformed with little modification to look at adaptation and response to climatic conditions (Matyas 1994). Information such as where the plant comes from, where it is planted on the site, and how it performs (growth, survival, reproduction, and so on) can guide forestry practices to increase the proportion of the species that survives and grows well (McKay et al. 2005; Millar et al. 2007; Hebda 2008).

Climate change poses a significant challenge for foresters and other land managers, but given its long history of selecting and growing trees, the forestry profession has the knowledge and tools to test and instigate assisted migration; we need dynamic policies that allow action. The frameworks and techniques for production and outplanting already exist, therefore researchers and practitioners can work with nurseries to design and implement adaptive measures that consider assisted migration and hopefully curtail significant social, economic, and ecological losses associated with impacts from a rapidly changing climate. The science and practice of growing trees to sustain ecosystems will greatly benefit with collaboration (McKay et al. 2005). The Adaptive Silviculture for Climate Change (Linda Nagel, project lead) is one such collaborative effort in the United States that focuses on the understanding of long-term ecosystem response to adaptation options and to help forest managers integrate climate change into silviculture planning (Northern Institute of Applied Climate Science, table 1). Framing the discussion to identify objectives and produce frameworks, such as the Climate Change Response Framework, that lead to practical and dynamic strategies is pertinent. Changing policies will require collaboration and discussion of how predicted conditions will affect forests, how managers can plan for the future, and how landowners can be encouraged to plant trees adapted to future conditions, such as warmer conditions and variable precipitation patterns (Tepe and Meretsky 2011).

Assisted migration may not be appropriate for every species or population. Whatever the chosen adaptive strategies, foresters need to be included in the dialogue with scientists and land managers in climate change planning. We have little time to act given current climate change predictions and uncertainty regarding the adaptation and migration capacities of species and populations. Establishment of healthy stands is vital now to prepare forests as changes occur. This might entail small scale experiments, such as planting fast-growing trees adapted to projected climate in the next fifteen to thirty years (Park and Talbot 2012) or randomly planting a variety of seed sources in one area and monitoring their adaptive response (similar to provenance testing) (Pedlar et al. 2011). Planting the standard species or stocks in regions highly sensitive to climate change will be unwarranted (Hebda 2008), given that reductions in fire frequency from 100 to 300 years to 30 years have the potential to quickly shift some forest systems to grasslands and woodlands (Westerling et al. 2011b). Instead, we need to shift our focus to plant species adapted to the novel conditions and/or those anticipated to migrate into these areas. Implementation of complementary actions, such as ecosystem engineering (e.g., using drastically disturbed areas as sites to test assisted migration), increasing landscape connectivity, emphasizing genetic diversity in seed source collections, targeting adaptive traits, and focusing on ecosystem function and resilience rather than a historical reference are also necessary considerations for any climate change strategy (Jones and Monaco 2009; Lawler and Olden 2011; Stanturf et al. 2014).

9

Maintaining Forest Diversity in a Changing Climate

A Geophysical Approach

MARK ANDERSON AND NELS JOHNSON

Climate change in recent decades has already begun to affect the composition of forest ecosystems in the northeastern United States. Average annual temperatures have increased by nearly 1.8°F since 1970 (Huntington et al. 2009). Tree species migration appears to be indicated by much higher seedling densities for northeastern tree species in the northern parts of their ranges compared to central and southern parts (Woodall et al. 2009). Climate-induced range shifts for other biota may also be underway. Bird species winter ranges in the eastern United States, for example, have shifted steadily north, although not as rapidly as temperature shifts (La Sorte et al. 2012). Projections of geographic ranges under various global climate model scenarios suggest most tree species will experience large range shifts in response to rising temperatures and more erratic precipitation regimes during the next century (Iverson et al. 2012). Differential migration rates and varied abilities to cope with environmental gradients and human land uses are likely to dramatically re-sort the species composition of forest communities (Rustad et al. 2012). Our ability to predict future species composition at any given location is extremely limited, calling into question the ability of current conservation networks to maintain biodiversity in the future.

Evidence suggests the strong correlation of geophysical factors with geographic patterns of biodiversity allows a new approach to designing conservation networks that will be effective even as species and communities continuously rearrange in response to climate change (Anderson and Ferree 2010). Here, we summarize such an approach for northeastern North America. We first review geophysical underpinnings of species diversity. The strong correlation of geophysical factors—number of geological classes, landform diversity, and elevation—with species diversity suggests that conserving geophysical settings is the key to conserving current and future forest biodiversity. We next review internal connectivity, which increases the ability of forest blocks to maintain species diversity and ecological processes, and regional connectivity, which facilitates regional movement in response to changing climatic conditions. By focusing on the representation of physical diversity instead of current species composition, plus accounting for internal and regional connectivity, we identify a resilient network of sites that have the potential to represent the full spectrum of forest diversity now and into the future. It should be noted that forests in less-resilient areas, such as flatlands and coastal areas, are still important for a wide variety of benefits from watershed protection, carbon storage, wildlife habitat, and timber production. Actively maintaining diversity at the site level within a resilient conservation network brings in additional factors, such as structural diversity, soil replenishment, and space to accommodate disturbance regimes, which are important for local management. We conclude with summarizing strategies to manage for diverse forests at the landscape scale within a regional forest conservation network. *Site resiliency*—that is, the ability for ecosystems to retain species diversity and basic relationships among ecological features even as environmental conditions change—is driven by geophysical settings and landscape permeability in northeastern North America. Geophysical settings are important because they are the best predictors of species diversity at a regional scale, such as temperate northeastern North America (Anderson and Ferree 2010). While climate may be a better predictor of species diversity at a continental scale, most conservation decisions are made at regional, landscape, and site scales. In addition, landscape permeability and regional connectivity are critical elements of site resiliency since species and ecological processes will need to shift to new locations as temperatures and precipitation regimes affect their viability in any given location. Projections of tree species migration indicate many of today's natural forest communities will be substantially altered by the end of this century (Prasad et al. 2007), though forests will continue to be the region's dominant ecosystem type. Assessing forest areas across the region based on their geophysical setting (e.g., geology, landform, and elevation), internal patch permeability, and regional connectivity yields a potential network of natural strongholds for future diverse forest areas. This network can be compared against the current network of secured forestlands and help focus future land protection and forest restoration efforts.

Site resilience is distinguished from species or ecosystem resilience because it refers to the capacity of a geophysical site of 40 ha to 4,000 ha (90 acres to 9,885 acres) to maintain species diversity and ecological function as the climate changes (definition modified from Gunderson 2000). This is important since neither species composition nor the ranges of variation of its processes are static in the context of climate change. Our working definition of a *resilient site* is a structurally intact geophysical setting that sustains a diversity of species and natural communities, maintains basic relationships among ecological features, and allows for adaptive change in composition and structure. Thus, if adequately conserved, resilient sites are expected to support species and communities appropriate to the geophysical setting for a longer time than will less-resilient sites.

Several factors—including geologic classes, elevation range, latitude, and area of calcareous substrate—are highly correlated with the distribution of terrestrial species diversity in the northeastern United States and adjacent Canadian Maritime Provinces (Anderson and Ferree 2010). Regressing the total number of species against these factors yields a strongly predictive relationship for terrestrial biodiversity across the northeastern United States and adjacent Canada (figure 9.1). These factors work equally well for predicting species diversity across the region even though, for example, Virginia shares only 30 percent of its biota with Prince Edward Island and the region spans 1,400 km of latitude. Moreover, the region has been in flux during the past century with many range expansions and contractions, extinctions, and species introductions. These changes appear not to affect the basic relationship between species diversity and geophysical factors. As a result, conserving the

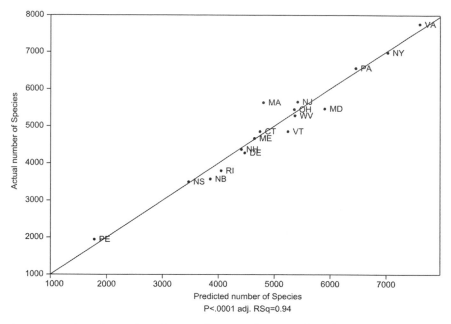

FIGURE 9.1. Actual state/province total species diversity plotted against predicted diversity based on number of geology classes, hectares of calcareous bedrock, latitude, and elevation range for northeastern North America.

full spectrum of geological classes stratified across elevation zones and latitude offers an effective approach to conserving forest diversity under current and future climates.

To capture the spectrum of geophysical settings in the northeastern United States, Anderson et al. (2014) used 405 ha (1,000 acre) hexagons to classify geology, elevation, and landform types. The region's highly diverse geologic history is manifested in over 200 bedrock types, which were grouped into nine geology classes based on shared genesis, weathering properties, chemistry, and soil textures (Robinson 1997). Likewise, elevation for the region, which ranges from sea level to 6,288 feet atop Mount Washington, New Hampshire, was divided into low-, mid-, and high-elevation classes. Finally, landform types have a major influence on species distribution and they were grouped into seventeen categories.

Anderson et al. (2014) then assigned each of the region's 156,581 hexagons—on the basis of hierarchical cluster analysis (McCune and Grace

2002) for similarity in terms of geology, elevation, and landform—to one of thirty geophysical settings. These include fifteen low-elevation settings, eight mid-elevation settings, six high-elevation settings, and one miscellaneous high-slope setting (see figure 9.2). Examples of geophysical settings include "coastal coarse sand," "low elevation fine sediment," "mid-elevation shale," and "high-elevation granitic." Data from state natural heritage programs were used to identify natural community types associated with each geophysical setting. While the species at any given site are likely to change in response to a warming climate, the ecosystem types are likely to persist. Community types that have commonly been named after predominant plant species can now be referred to by their geophysical settings. For example, a cattail (*Typha latifolia*)-marsh marigold (*Caltha palustris*) marsh becomes a freshwater-marsh ecosystem on shale at low elevation.

To assess the relative resiliency of sites associated with each geophysical setting, Anderson et al. (2014) developed scores based on landscape complexity and local permeability (or connectivity), which are summarized below.

Landscape complexity is driven by an area's topography and associated landforms and by the length of its elevation gradients. Landscape complexity creates microtopographic thermal climate options that resident species can move to, buffering them from changes in the regional climate (Willis and Bhagwat 2009) and slowing the velocity of change (Loarie et al. 2009). Under variable climatic conditions, areas of high landscape diversity are important for the long-term population persistence of plants (Randin et al. 2008), invertebrates (Weiss et al. 1988), and presumably for the more mobile species that depend on them. For example, Weiss et al. (1988) measured microtopographic thermal climates in relation to butterfly species and their host plants. They concluded that areas of high local landscape complexity—even at the scale of tens of meters—are important for long-term population persistence under variable climatic conditions. Because species locations shift to take advantage of microclimate variation and stay within their preferred temperature and moisture regimes, extinction rates predicted from coarse-scale climate models may fail to account for topographic and elevation diversity (Luoto and Heikkinen 2008; Wiens and Bachelet 2010).

A landscape complexity index was developed by tabulating the number of landforms and elevation ranges within a 40-ha circular area around

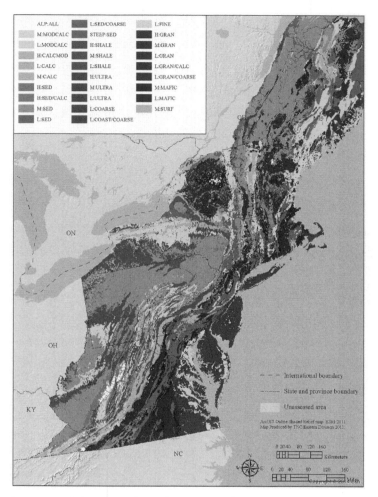

FIGURE 9.2. Thirty geophysical settings in the northeastern United
States were classified on the basis of geology class, elevation zone,
and landform type.

every 30-m cell. It was assumed that sites with a larger variety of landforms would provide more microclimate options within their local neighborhoods. The number of landforms ranged from one to eleven (there are
seventeen landform types across the region). Elevation gradients for the
cells ranged from 1 m to 795 m, which were log transformed for analysis since the gradients were heavily skewed toward narrow ranges. The

landform and elevation information were combined using a weighted sum with landform variety given twice the weight of elevation (Anderson et al. 2014). The final index was:

Landscape Diversity $= (2^*LV = 1^*ER)/3$
where LV $=$ landform variety, ER $=$ elevation range

Landscape complexity is likely to offer less resilience if the site is geographically constricted since species populations and ecological processes could be overly confined as regional climate changes. Permeability is the degree to which a given landscape supports the movement of organisms and the natural flow of ecological processes such as water or fire (definition modified from Meiklejohn et al. 2010). A highly permeable or locally connected landscape promotes resilience by facilitating local species movements and range shifts, and the reorganization of communities (Krosby et al. 2010). Maintaining a connected landscape is a widely cited strategy for building climate change resilience (Heller and Zavaleta 2009). Botkin et al. (2007) have suggested large landscape connectivity as an explanation for why there were few extinctions during the last period of comparably rapid climate change. Accordingly, our measure of permeability "local connectedness" is based on measures of landscape structure: the hardness of barriers, the connectedness of natural cover, and the arrangement of land uses. To assess landscape permeability, Anderson et al. (2014) used the resistant kernel analysis that assumes that the connectedness of two adjacent cells increases with their ecological similarity and decreases with their contrast (Compton et al. 2007). The theoretical spread for a species or process out from a focal cell is a function of the resistance values of neighboring cells and their distance from the focal cell out to a distance of up to 3 km. A focal cell score for local connectedness is equal to the amount of spread accounting for resistance divided by the theoretical amount spread if there was no resistance. Cell scores are then multiplied by 100 to create a range from 1 to 100 and converted to standard normal distributions for the region. The resistance surface was based on a classified land use map with roads and railroads embedded in the grid (NLCD 2001; Tele Atlas North America 2012). Land cover was simplified into six types including natural land (evergreen, deciduous, and mixed forest, with shrub/scrub, grassland, and woody/herbaceous wetland), water, artificial barrens, agriculture (pasture and cultivated), low-intensity development (land cover classes: low-intensity, and developed open space), and high-intensity development (land cover classes: high density, medium density, and roads). Natural land was given the lowest resistance score (10) and high-intensity developed land was given the highest weight (100). Scores for the other land classes included artificial barrens (50), agricultural lands (80), and low-intensity development (90).

The landscape complexity and landscape permeability scores were combined to develop a single resiliency score for each 30-m cell. The complexity and permeability scores were transformed into standardized normalized (Z-scores) to combine and compare resilience factors. Each factor was given the same weight in the integrated score:

Estimated Resilience = (LC1 = LC2)/2
where LC1 = local connectedness and LC2 = landscape complexity

The results show a wide range of estimated resiliency for terrestrial ecosystems across the northeastern United States and Canadian Maritime Provinces. The vast majority of areas with high terrestrial resiliency scores are forest ecosystems, although wetland and smaller patch communities are embedded within large blocks of resilient landscapes.

The last step in identifying a resilient conservation network is to locate regional linkages between large forest landscapes. Anderson et al. (2014) used the Circuitscape software tool (McRae and Shah 2009), based on electric circuit theory, to identify potential larger-scale directional movements and pinpoint the areas where they are likely to become concentrated, diffused, or rerouted, due to the structure of the landscape. As with the local connectedness analysis, underlying data for this analysis includes land-cover and road data converted to a resistance grid by assigning weights to the cell types based on their similarity to cells of natural cover. However, instead of quantifying local neighborhoods, the Circuitscape program calculates a surface of effective resistance to current moving across the whole landscape. The output of the program, an effective resistance surface, shows the behavior of directional flows. Analogous to electric current or flowing water, the physical landscape structure creates areas of high and low concentrations similar to the diffuse flow, braided channels, and concentrated channels one associates with a river system. Three basic patterns can be seen in the output, as the current flow will (1) *avoid* areas of low permeability, (2) *diffuse* in highly

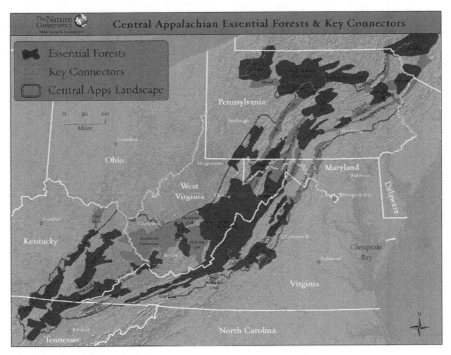

FIGURE 9.3. Central Appalachians "essential forests" and "key connectors" network based on regional resiliency and connectively analyses (Anderson et al. 2014).

intact/highly permeable areas, or (3) *concentrate* in key linkages where flow accumulates or is channeled through a pinch point. Concentration areas are recognized by their high *current density*, and the program's ability to highlight concentration areas and pinch points made it particularly useful for identifying the linkage areas that may be important to maintaining a base level of permeability across the whole region. The Nature Conservancy's Central Appalachians Program has combined the regional resiliency and connectivity results into a network of "essential forests" and "key connectors" (figure 9.3).

While geophysical factors are important for maximizing the potential for conserving forest diversity, other factors—especially biological—are important for actively maintaining or restoring biodiversity at a given site. Some factors, such as reducing stress from forest pests and pathogens, have been traditional tools in landscape forest management. Others, such as selecting species on the basis of their tolerance for anticipated temperature regimes, introduce new approaches to managing today's forests so they remain diverse and productive decades from now.

A wide variety of recommendations have been made to help managers prepare for future conditions, including the considerable uncertainty that accompanies climate projections at the landscape scale (Millar et al. 2007; Heller and Zavaleta 2009; Puettmann 2011; Cornett and White 2013). We've grouped strategies for maintaining and improving ecological resilience at the landscape scale into three categories: promote diversity, reduce existing stresses, and anticipate future conditions. These are briefly summarized below.

Three dimensions of diversity can help forests be more productive and resilient at a landscape scale. *Species diversity* has been linked to ecosystem productivity and functioning by a variety of researchers (e.g., Tilman et al. 1997; Chapin et al. 2000, Flombaum and Sala 2008; Thompson et al. 2009; MacDougall et al. 2013). Because many forest species have very specific environmental requirements and functions (niche partitioning), their loss may lead to a reduction in productivity and/or function until the niche space is occupied by another similar species. Species diversity at the stand and landscape level is also associated with resistance to ecologically destructive disturbances such as severe fire and pest and pathogen outbreaks (Thompson et al. 2009). Emphasis should be placed on increasing the native diversity of forest specialist and late-seral species associated with topographic and structural microclimates as opposed to generalist and nonnative species associated with forest fragmentation. Management should increasingly consider model results (USFS Tree Atlas; LANDIS) and emerging data that provide information on which species may benefit from a changing climate (e.g., oaks, hickories) and which may suffer (e.g., beech, some maples, spruce). *Successional diversity* is another feature that can promote ecological productivity in the face of environmental change. A range of successional or age classes at the landscape scale promotes species and structural diversity by creating a mosaic of environmental gradients with respect to light, humidity, ambient temperature, and coarse woody debris. These gradients can help perpetuate a wider variety of disturbances, species, and ecological functions that contribute to overall forest health than a landscape dominated by a single successional cohort (Franklin et al. 2007). Likewise, *structural diversity* in forest ecosystems of any age class or mix of classes can promote forest ecological health as well as provide a wide range of habitats for a variety of species. Nurse logs, for example, facilitate the regeneration of moisture-sensitive species in northern hardwood forests. The acceleration of natural successional processes, such as the use of small patch cuts to simulate gap-phase dynamics, the creation of snags, or the introduction of coarse woody debris, can improve the ability of forest ecosystems to adapt to changing conditions (Cornett and White 2013).

Most effects from a changing climate on northeastern forests will be expressed through stresses that already exist in the region. The dominant existing regional stressor—habitat fragmentation—makes forests more vulnerable to the spread of invasive species, wind damage along exposed forest edges, and altered species movement. Deer populations have had severe impacts on forest regeneration in many areas of the northeastern United States. While there is little evidence that white-tailed deer populations will increase as a result of climate change, the combined effects of excessive browse and other stressors (i.e., atmospheric deposition) could compound the regeneration challenge for many tree and understory species under a changing climate regime (Galatowitsch et al. 2009).

Many forest management programs at the stand and landscape scales already address existing stresses but climate change could change the relative threat each poses to forests. To respond to these potential threats, management should attempt to improve the forests' ability to resist pests and pathogens, work to prevent the introduction and establishment of nonnative and / or invasive pest and plant species (and remove existing populations), and manage herbivory that impacts regeneration (e.g., establishing deer fencing). Management can also reduce the risk of catastrophic fire by establishing fuel breaks, altering forest structure to minimize risk and severity of fires, or conducting prescribed burns to reduce fuel loadings. Chapin et al. (2009) suggest that management targets for existing stressors should be updated to incorporate trajectories of expected change rather than relying on historic ranges of variability. Given uncertainties about exactly how these threats will change increases the importance of establishing monitoring networks to detect unexpected changes in stress impacts and management responses to them (Joyce et al. 2009). Conn et al. (2010) provide several recommendations to reduce the total amount of stress on forest ecosystems. These include (1) strengthening state and local programs to slow forest loss and fragmentation; (2) revising forestry best management practices (BMPs) to account for expected impacts from climate change to existing and new stressors; (3) working with sustainable certification programs such as the Forest Stewardship Council (FSC) and Sustainable Forestry Initiative (SFI) to promote integration of climate adaptation measures into their performance measures; and (4) expanding state and local capacity to monitor and respond to pests, pathogens, storm damage, and fire risk. Forest management actions taken today can anticipate future conditions and reduce the ecosystem's vulnerability to expected disturbance (Bolte et al. 2009). Actions that can be taken to anticipate future impacts include selecting or promoting tree species with a wide environmental tolerance, maintaining and restoring habitat connectivity to facilitate species movement, and stand/site design to minimize edges that are vulnerable to wind storms and invasive species. Current species that have a broad natural range with regard to temperature and moisture (i.e., oaks, *Tilia, Sorbus*, white pine) can be expected to do well (Iverson et al. 1999). Species with a narrow range of ecological tolerance are likely to persist only in topographic microclimates, and attention to the recruitment dynamics within these microsites may be important to long-term persistence of these species in the forest. This may be preferable to selecting species not currently in the area that may or may not be adapted well to future conditions. Prioritizing and protecting refugia of unique habitats or populations of sensitive or rare communities should therefore be a goal of management. Ultimately, any management activities should attempt to protect the fundamental ecological functions of that system, including the protection of soil quality and nutrient cycling, and maintaining and restoring hydrological flows (figure 9.4)

Forest conservationists need methods to conserve the maximum amount of biological diversity while allowing species and communities to rearrange in response to a continually changing climate. By focusing on the representation of physical diversity instead of on the current species composition, we identify a regional network of sites that will represent the full spectrum of forest diversity now and into the future. We advocate that this geophysical approach to identifying a network of core forest areas and key connectors be used to inform and augment the traditional conservation focus on large forest reserves nested within a matrix of well-managed forest. At specific sites within a resilient forest conservation network, forest managers should continue to do many of the things we have long known are important for maintaining healthy forest ecosystems. These strategies include

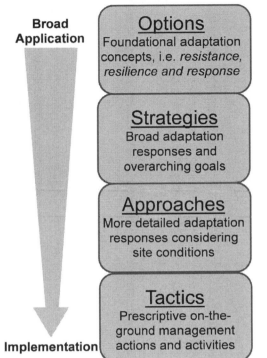

Broad Application

Options
Foundational adaptation concepts, i.e. *resistance, resilience and response*

Strategies
Broad adaptation responses and overarching goals

Approaches
More detailed adaptation responses considering site conditions

Tactics
Prescriptive on-the-ground management actions and activities

Implementation

FIGURE 9.4. Diagram of the process of the Change Response Framework. (Adapted from Swanston and Janowiak 2012.)

(1) managing for species, structural, and successional diversity; (2) reducing existing stress from invasive species, habitat fragmentation, altered fire regimes, and other factors; and (3) making forest management decisions—such as selecting species with broad tolerances—that anticipate future climatic conditions. Since there is considerable uncertainty about the rate and degree of future change, managers will need to remain flexible, experimental, and innovative. Adaptive management frameworks will be increasingly essential for sustainable forest management within a regional resilient forest conservation network.

10

Adaptation

Forests as Water Infrastructure in a Changing Climate

TODD GARTNER, HEATHER MCGRAY, JAMES MULLIGAN,
JONAS EPSTEIN, AND AYESHA DINSHAW

As plainly stated in the draft 2013 National Climate Assessment, climate change, once considered an issue for a distant future, has moved firmly into the present. Climate change can have substantial implications for the provision of clean and abundant water that is so fundamental to public health, economic development, and prosperity. In some regions of the United States, heavy precipitation has increased over the last century. At the same time, the drought in western states over the last decade represents the driest conditions in 800 years (Karl et al. 2009; Schwalm et al. 2012). Changes in timing of snowmelt and associated streamflow have already reduced summer water supplies in regions like the Northwest (US EPA 2012). All told, the costs to society of ongoing and expected water-related climate impacts are immense. They include escalated water treatment costs, lost economic activity associated with water shortages, private property and public infrastructure damage, and losses in general human health and wellbeing. Drinking water and waste

The findings and conclusions in this chapter are those of the authors and do not necessarily represent the views of the United States Forest Service.

water utilities alone are expected to incur an estimated $448 billion to $944 billion in infrastructure and operations and maintenance costs through 2050 to manage climate impacts (Association of Metropolitan Water Architects 2009). As affected communities scope strategies to secure water resources in the face of a changing climate, investments to restore and maintain healthy forests should be carefully considered for the role forests can play in buffering against expected climate impacts.

At the same time, investments should be shaped to take into account the sensitivity of forests to climate change and the new risks forests may face, among them the sustainable provision of water resources. *Natural infrastructure* provides a first line of defense for communities as the impacts of climate change intensify. Natural infrastructure is defined as a "strategically planned and managed network of natural lands, working landscapes, and other open spaces that conserves ecosystem values and functions and provides associated benefits to human populations" (Benedict and McMahon 2006). Healthy, well-managed forested watersheds, for example, can reduce peak storm flows, maintain snowpack, shield water bodies from temperature extremes, and filter sediment, nutrients, and pollutants in runoff (Gartner et al. 2013). The manner in which forests are managed also has bearing on water resources. For example, robust forest road and stream crossing designs can help to mitigate sedimentation risks associated with extreme wet weather events, and maintaining forested riparian buffers is critical for combating elevated water temperatures.

While forests alone are not a panacea to climate impacts, they provide a suite of services that can help to buffer against those impacts (Peterson et al. 2011 as cited in National Climate Assessment 2013). Some of the most important of these services are summarized in table 10.1 and detailed below. By strategically investing in the conservation, restoration, and management of ecosystems like forests, communities can build an integrated and cost-effective system of natural and built infrastructure to help adapt water provision systems to a changing climate.

Floods are expected to increase in most regions of the United States, even where average annual precipitation is projected to decline (Pan et al. 2010 as cited in National Climate Assessment 2013). The largest increases in very heavy precipitation events have occurred in the Northeast, Midwest, and Great Plains (Karl et al. 2009), damaging public infrastructure and private

TABLE 10.1. Forest Functions as a First Line of Defense against Climate Impacts

Climate Impact	Related Forest Function
Flooding and consequences of extreme precipitation	Erosion control and flow regulation
Increasing incidence of summer drought	Flow regulation and snowpack maintenance
Elevated water temperatures and lower flows	Cooling effect of forested riparian buffers

property and threatening human health and wellbeing. Meanwhile, earlier snowmelt in the Northwest, combined with more extreme precipitation events, has led to increased water flows and associated flood risk during the spring (Hidalgo et al. 2009).

In addition, as the rate of precipitation exceeds the ability of the soil to maintain an adequate infiltration rate, and as heavy precipitation increases the kinetic energy of surface water, soils will erode (National Climate Assessment 2013). Accelerated erosion causes increased sedimentation and movement of nutrients, dissolved organic carbon (DOC), pathogens, and pesticides (Delgado et al. 2011 in National Climate Assessment). For example, DOC in rivers and lakes is strongly driven by precipitation (Pace and Cole 2002; Raymond and Saiers 2010; Zhang et al. 2010) and is expected to increase in regions where precipitation is expected to increase (National Climate Assessment 2013). Elevated levels of pollutants will drive capital and variable costs of drinking water treatment, requiring investments in new and expanded treatment facilities as well as increasing levels of chemical additives. Increased sedimentation can also reduce the storage capacity in reservoirs needed for drinking water and hydropower generation, and can impact freight navigation.

Forests have multiple layers of vegetation (Dohrenwend 1977) and have particularly thick litter layers that help to slow falling rain and reduce its erosive force during heavy rain events (Stuart and Edwards 2006). Sturdy, long-lived roots also help to anchor soil against erosion (Beeson and Doyle 1995; Geyer et al. 2000). Multilayered forest canopies have more interception (Brooks et al. 2003; Briggs and Smithson 1986), greater photosynthetic area, and deeper roots than other plant communities, and so promote greater evapotranspiration and thus soil water deficits (de la Cretaz and Barten 2007). The forest litter layer promotes infiltration of water into the soil and provides a barrier that slows downslope water movement (Dudley and Stolton 2003). These characteristics, together with the very high infiltration rates of forest soils created by complex pore structures, minimize storm-flow peaks, minimize overland flow and associated erosion in intense storm events, and provide ample opportunity for nutrient uptake by plants and microbes in the soil (de la Cretaz and Barten 2007; Bormann and Likens 1979; Vitousek and Reiners 1975). In the Pacific Northwest, the forest canopy can minimize the impact of rain-on-snow events through interception. Rain falling on snow has been associated with mass-wasting of hill slopes, damage to river banks, downstream flooding, and associated damage and loss of life (US Geological Survey 2013).

Most regions of the United States are expected to increasingly experience drought in summer months. Impacts will be most pronounced in the Southeast (Zhang and Georgakakos 2011) and Southwest (Milly et al. 2008; US Department of the Interior, Bureau of Reclamation 2011a, 2011b), where longer-term reductions in water availability are expected with rising temperatures and general declines in precipitation (NCADAC 2013). These trends are occurring in confluence with growing population and demand for water in the Southeast, and increased competition for scarce water resources in the Southwest (Averyt et al. 2011). In the Northwest, changes in the timing of snow melt and associated streamflow pose challenges for water availability in the summer months. Models indicate with near certainty that reductions in summer flow (by 38% to 46% compared to 2006) will occur by 2050 for snow-dense basins (Elsner et al. 2010).

In addition to clear implications for the availability of drinking water, droughts also reduce the potential capacity for hydroelectric generation (NCADAC 2013) and can hamper other forms of energy production that consume large quantities of water such as shale and hydraulic fracturing. Drought has also created hardships for farmers and ranchers, reducing crop yields and forage available to livestock (Hedde 2012).

While forests can reduce overall water yield through interception and transpiration (Hornbeck et al. 1995), forests can also help to address summer droughts by regulating the timing of flow. Forest soils and debris can act as sponges, storing and then slowly releasing water. This process recharges groundwater supplies and maintains baseflow stream levels, although the overall effect must be measured against the "use" of water by forests.

In addition, snowmelt is most sensitive to temperature and wind speeds (van Heeswijk et al. 1996). Consequently, snowmelt is substantially higher in cleared areas than beneath forest canopies where wind speeds are lower and snow is shaded (Marks et al. 1998). Thus, forest cover can help to maintain snowpack and hedge against dry-season water supply issues in regions like the Northwest that rely on snowmelt.

Elevated stream temperatures and lower base flows can affect aquatic habitat for critical species (Spooner et al. 2011; Xenopoulos et al. 2005) and may require additional treatment by wastewater facilities to meet requirements under the Clean Water Act (US EPA 2011). It can also reduce the reliability of water withdrawals for electric power plant cooling and the efficiency of those cooling processes (Backlund et al. 2008; Gotham et al. 2012).

Rising stream temperature is also a factor, among others, in downstream lake temperature. Within the past forty years, lake temperatures have increased by an average of up to 1.5°C in over 100 lakes in Europe, North America, and Asia (IPCC 2001). Warmer surface waters can lead to blooms of harmful algae (Paerl and Huisman 2008), which are estimated to impose costs of $2.2 billion each year (Dodds et al. 2009). Higher air and water temperatures are also decreasing lake mixing, decreasing dissolved oxygen, and releasing excess nutrients, heavy metals, and other toxics into lake waters (NCADAC 2013). Increased evapotranspiration due to higher temperatures may also increase groundwater salinization in more arid regions, raising filtration and treatment costs for industrial plants, hydroelectric generators, and wastewater facilities (IPCC 2001).

Many factors affect stream temperatures: for example, stream surface turbulence, shading, stream size, and stream water travel time (Bourque and Pomeroy 2001). Shade is also critically important—direct solar radiation has been found to be the largest contributor to changes in daily temperature in streams (Johnson and Wondzell 2005). Forested riparian buffers provide shade to streamwater and have been shown to prevent temperature increases (Groom et al. 2011). Harvesting forests along streams can increase daily maximum and mean water temperatures by as much as 2°C to 10°C (Bourque and Pomeroy 2001).

These examples illustrate a key twofold point: while forests can address only some elements of expected and ongoing water-related climate impacts, investing in forests can be a timely and effective component of a broader community adaptation strategy as a "first line of defense." Given the multiple benefits associated with healthy ecosystems (e.g., wildlife, recreation, property values, carbon sequestration, and air quality), investments in natural infrastructure can be a "win-win" measure that addresses parallel community needs.

As communities consider large-scale investments to conserve, restore, or manage forests and wetlands, decision makers must understand how a changing climate may impact their water-related functions. For example, changes in precipitation and temperature can contribute to changing species composition and increasing incidence of disturbance in forests. If not carefully managed, these impacts may affect the water-related function of upstream ecosystems, potentially compromising the ability of forests to serve effectively as natural infrastructure under a changing climate. Thus, even as we argue that the forest functions enumerated above help to mitigate climate risks to water services, we also call for attention to the pathways whereby climate change impacts may compromise water-related forest functions. To date, however, a limited body of literature directly treats the impact of climate change on the provision of ecosystem services. Here we highlight two climate impacts affecting the water-related functions of forests and associated management interventions to support maintenance of those functions as the climate changes.

The increase in severe high-temperature days in combination with dry air mass events—as well as fuel changes, successional growth, invasive species, insect and disease, longer fire seasons, and more severe episodic drought—is contributing to an increase in wildfire frequency and intensity in the Intermountain West and California (Sexton 2013; NCADAC 2013; Dietze and Moorcroft 2011). Eleven of the twelve largest fires in modern US history have occurred since 2004 (Sexton 2013). These "megafires" are unprecedented in their social, economic, and environmental impacts (NCADAC 2013). Catastrophic wildfire can prime a watershed for dramatic surges in peak flows—documented to be up to 900 times greater than the unburned reference case for up to *fifteen years* after a fire, triggered by rainfall above a certain threshold (Martin 2013). These fires also disrupt the water quality–related functions of forests, and elevated postfire flows can cause massive sedimentation. Sediment exports due to wildfire are increased for up to one year following the fire; increased concentrations have been observed at well over 1,000 times the concentrations of unburned forested waterways. Similarly, postfire concentrations of nitrogen and phosphorus have been observed at over 400 times the previous concentrations in the same, unburned waterways (Smith et al. 2011). In some cases, postfire runoff can also release potentially toxic "legacy sediments" into drinking water systems.

Forest management activities like prescribed burning and mechanical thinning play a critical role in mitigating catastrophic wildfire risk. Historical fire suppression in fire-prone ecosystems like western forests led to the unnatural accumulation of fuels, a risk that is magnified by climatic trends. The behavior of fires that escape suppression is determined by available fuel, weather, and topography. The only one of these factors that can be controlled by forest managers is fuel (Thompson et al. 2012). Management interventions like prescribed burning and mechanical thinning are geared to strategically reduce the fuel load in the forest to avoid catastrophic fires—for example, by limiting canopy ignition by increasing the distance from surface fuels to flammable canopy biomass (Mitchell et al. 2009). Fuels management can also protect human communities and restore fire-adapted ecosystems to natural function. As the climate becomes increasingly variable, the impact of changing species composition on forest functions becomes more pronounced. Many species have already begun to be eliminated from areas that are dominated by human influence. A changing climate will further affect the species composition of forest ecosystems throughout the country, either causing species to migrate to cooler northern regions, or expanding vegetative ranges that sustain invasive species (Chapin et al. 2000). Invasive species can displace native organisms while modifying habitat, altering ecosystem processes, and changing the interval of fire and water utilization (National Academy of Sciences 2008). It is likely that without intervention, invasive species will come to dominate migration in many places due to the water-intensive and resource consumption habits maintained by many nonnative species. Such species are spread through climate-linked disturbances like flooding and wildfire and usually have traits that favor rapid establishment and population spread, high rates of seed production, and vegetative reproductive persistence in the soil seed bank (Watterson and Jones 2006).

Invasive species outcompete native plants and organisms while altering the ecosystem functioning of forests. A forest hydrology report completed in 2008 by the National Academy of Sciences emphasizes an extreme hydrological sensitivity to species composition. As the genetic makeup of forests shifts through competition and predation, vegetation density is often impacted—although effective wildlife management can affect changing density by altering the intensity of browsing by herbivores (Gill and Beardall 2001). Vegetation density in turn affects transpiration rates of tree species. Partial or complete removal of forest canopy can reduce transpiration and interception of rain, which can in turn increase soil moisture and water availability to plants. Increased saturation of the land reduces slope stability in the long run, while causing greater nutrient and sediment runoff and turbidity via erosion (National Academy of Sciences 2008). In some instances, the scenario might be reversed depending upon the type of tree displacement— eastern deciduous trees with higher transpiration rates and increased leafy surface area can severely deplete the water availability of forests. This suggests the importance of the delicate ecological balance of species to maintain forest functions (Brantley et al. 2013).

To combat the detrimental hydrological impacts of species composition shifts on forest functions, a holistic adaptive management approach is needed to allow forests to recover from devastating disturbances and provide critical ecosystem services despite species composition shifts. Such a management approach includes enhancements to forest biodiversity and redundancy to act as a buffer against invasive species. Functional diversity in forests is directly related to production in the ecosystem (Chapin et al. 1997); *redundancy* refers to the capacity of various forests to sustain abundant populations of the same species to ensure ecosystem functioning following an ecological disturbance. While several tree species have been lost or reduced in temperate forests, there has been relatively little or no loss of productivity in that ecosystem, which suggests compensation by other species (Thompson 2009). Both biodiversity and redundancy contribute to forest resilience by maintaining productive capacity of existing species, allowing them to better utilize and partition resources. In complex systems, many organisms provide regular ecological processes (transpiration, decomposition, respiration) compared to simpler systems, where vacant niches are likely available to nonnative organisms (Hooper et al. 2005).

When controlling for invasive species, scientists and managers must collaborate across scales and jurisdictions to identify priority areas and critical species, and to establish a system of accountability that ensures efficient use of limited resources. In the past, Adaptive Management Areas have been established in the Pacific Northwest with a focus on iterative learning, testing, and monitoring to ensure biodiversity and ecological resilience in the face of changing climate and land-use (Stankey et al. 2003). The US Department of Agriculture's *National Strategy and Implementation Plan for Invasive Species Management* identifies regulation through prevention, early detection and rapid response, control and management, and rehabilitation and restoration phases. Implementing these phases involves development of a national tracking system for invasive species, emergency response capabilities and technology, as well as shared education and outreach for proper protocols to limit the spread of nonnative organisms (USDA Forest Service 2004).

Current best practices such as those outlined above for addressing two forest management challenges—wildfire and invasive species—are important inputs to adaptation planning that could enable forests to help safeguard water provision as the climate changes. Given pervasive uncertainties regarding the future impacts of climate change on forests, however, it may not be sufficient to incrementally expand and improve application of known management techniques.

It is unclear how postfire recovery of forest ecosystems may change under warmer temperatures, new precipitation regimes, or with a shifting species mix (Anderson-Texeira et al. 2013). Such complexity makes confident predictions about the implications of climate change for specific localities and regions a substantial challenge (Dessai et al. 2009). While climate change and impact modeling continue to improve, it is unlikely that uncertainties at scales relevant to forest management will be reduced significantly in the near- to mid-term. In fact, the Intergovernmental Panel on Climate Change (IPCC) has warned that uncertainties in many instances will increase for some time to come as scientific inquiry diversifies and deepens (IPCC 2007a).

In response, a growing number of decision-makers are addressing climate change through the use of scenarios. The IPCC defines a *scenario* as "a coherent, internally consistent and plausible description of a possible future state of the world." It is not a forecast; rather, each scenario is one alternative image of how the future can unfold (IPCC 2007a). Scenario planning and analysis is the process of evaluating possible future events through the consideration of a set of plausible, though not always equally likely, scenarios. Rather than relying on predictions, scenarios enable a creative and flexible approach to preparing for an uncertain future (Means et al. 2005; Carpenter and Brock 2006; De Lattre-Gasquet 2006). Scenario planning can be conducted in many ways (Briggs 2007) and it is particularly useful for decisions that have long-term consequences, such as a forest management plan or a major infrastructure investment. Scenarios are also used in *robust decision-making* (Lempert and Collins 2007), which is increasingly being applied to urban infrastructure investments (Lempert et al. 2013). Under robust decision-making, each of a set of possible management options is tested against different future scenarios. Ultimately, a decision that fares well against a range of scenarios is chosen. In the absence of a robust option, the scenarios can also be used to identify the vulnerabilities of a potential adaptation, so that it can be modified or its risks otherwise addressed. It is important to note that robust decision-making does not weigh the scenarios with probabilities, nor does it depict the imprecise probabilities as a range. This is appropriate for the climate change context, in which probabilities typically are highly uncertain.

In the forest sector, scenarios are sometimes used for planning under the rubric of adaptive management (Cissel et al. 1999). However, practical challenges abound, and adaptive management has not attained as widespread or as thorough application as may be needed in a changing climate. An analysis of the Northwest Forest Plan (Stankey et al. 2003) highlighted how time lags confound experiments in forest management, and cited the need for greater coordination between regulators and managers under an adaptive management approach. Adaptive management also demands a willingness to acknowledge that current actions and beliefs might be wrong, and that the resources needed for iterative planning and implementation can be considerable. Despite these challenges, adaptive management will be an important strategy for ensuring that potential future climates are considered seriously in forest management, so that forests may help safeguard water benefits from climate change, rather than themselves falling victim to climate impacts. The approach needs renewed emphasis in general, new solutions to implementation challenges, and specific adjustments to consider potential climate change impacts and climate-related ecosystem thresholds.

Existing climate change adaptation efforts in the forest sector appear to be moving forward with limited attention to ecosystem services. Important recommendations for adaptation of forests focus on buffers and corridors, maintenance of large-scale ecosystem function, active management of species mixes, and improvements in monitoring. However, many of these recommendations come through a biodiversity lens, with little explicit attention to sustaining natural infrastructure functions for water (National Fish, Wildlife and Plants Climate Adaptation Partnership 2012; Heller and Zavaleta 2009).

In cases where forests are being used as part of a water infrastructure solution, adaptation planning should explicitly address infrastructural functions. This means focusing specifically on climate risks to water services, not only risks to the forest as a whole. Borrowing from emerging adaptation practice in the gray infrastructure realm, managers could consider charting a "decision map" or "flexible adaptation pathway" that links management decisions to key benchmarks for water provision over time, and enables monitoring of ecosystem services against expected levels of water demand (Fankhauser et al. 1999). Such a map or pathway charts a risk management approach that can evolve as iterative risk assessments, evaluations, and monitoring provide new information over time. London used this approach in designing its new Thames Barrier (Reeder and Ranger 2010), and New York City has used it for citywide adaptation planning (New York City Panel on Climate Change 2009).

The development of a flexible adaptation pathway requires identification of critical thresholds beyond which key system functions are compromised. For example, a particular forest ecosystem may have thresholds for climate-induced fire risk or altered species composition, beyond which the forest's ability to provide water services becomes significantly impaired. Determining which thresholds are relevant is a significant challenge, but once they have been identified, having monitoring systems in place for these thresholds is central to implementation of the adaptation pathway. Given likely changes in species composition and potential geographic movement of the overall forest system, as well as shifting water demand, critical thresholds for water provision may, in part, be distinct from critical thresholds for the ecosystem as a whole. Climate change calls on forest managers to consider whether and how monitoring systems for natural infrastructure initiatives should differ from systems for monitoring the biodiversity functions of a protected area, or from general monitoring of forest health.

In the face of a changing climate and aging water infrastructure, never has it been more important to invest in water security. Increasingly, communities are looking to strategically invest in networks of natural and working lands like forests as natural infrastructure to secure the critical functions they provide. These efforts can contribute to community resilience by securing forests as a first line of defense against water-related climate impacts. While forests can provide several critical water services *now* and as the climate continues to change, as much as 34 million acres (13.75 million ha) of forest are projected to be lost in the contiguous forty-eight states by 2060 (USDA Forest Service 2012a). Now is a critical moment to reverse this trend.

Yet these forests face a number of climate-related risks that may affect the provision of ecosystem services like clean water and flood protection. While forest management practices are well established for historical climate and ecological conditions, uncertainty will figure prominently in future approaches to management as the climate changes and ecosystems respond. To date, applications of adaptive management planning to natural infrastructure investments are instill in their infancy. Going forward, it is essential for researchers to further explore climate risks to the water services of forests, and for practitioners to incorporate an adaptive management approach in natural infrastructure investment programs.

11

Water Source Protection Funds as a Tool to Address Climate Adaptation and Resiliency in Southwestern Forests

The Southwest's fire-adapted forests are experiencing widespread changes resulting from a century of fire exclusion, climate change, and land-use decisions that have a direct impact on water sources and supplies for people who live in the region. The historical fire regime in the Southwest's extensive ponderosa pine and dry-site mixed-conifer forests was frequent, low-severity fire (Swetnam and Baisan 1996). Tree density increased significantly when humans removed fire from the ecosystem, resulting in ladder fuels and dense, continuous canopy fuels (Fulé, Covington, and Moore 1997; Allen et al. 2002). In recent decades, rising temperatures have extended the length of the fire season. Currently, wildfire intensity has increased and caused a higher percentage of moderate- and high-severity burns, a consequence of the historic accumulation of dense canopy fuels and the current condition of fires burning during periods of higher summer temperatures (Westerling et al. 2006; Williams et al. 2010). This has resulted in the two largest fires in New Mexico history in 2012 and 2011—the 297,845 acre (120,533 ha) Whitewater-Baldy Fire and the 156,593 acre (63,370 ha) Las Conchas Fire,

DOI: 10.5876/9781607324591.c011

respectively—and the largest fire in Arizona history in 2012, the 538,049 acre (217,740 ha) Wallow Fire.

Southwestern forests are critical sources of water for people and play a key role in the hydrologic cycle. Most precipitation comes as snowfall and is stored in forested mountains until spring. Snowmelt is the primary source of surface water for agriculture and municipal and industrial use (Leopold 1997). The recent large wildfires with significant areas of moderate- and high-severity burn have caused extensive and severe hydrologic damage in many watersheds across the region. The magnitude of postfire flooding can be many times greater than prefire flows (Veenhuis 2002) and in some locations has resulted in catastrophic debris flows (Cannon and Reneau 2000). Rising temperatures are predicted to further threaten water supplies and forests, not only due to longer fire seasons with more large fires (Westerling et al. 2006), but also through drought-induced forest die-off (Breshears et al. 2005), reduced snowpack, and altered stream flow (Barnett et al. 2008).

Community and political leaders responded to the 2000 Cerro Grande Fire in New Mexico with changes in national policy and local practices. A National Fire Plan was created in 2001 as a policy response to large fires such as Cerro Grande (McCarthy 2004). The National Fire Plan evolved as a result of the work of the interagency Wildland Fire Leadership Council, established in 2003, as large fires continued in western forests (US Government Accountability Office 2009). The primary issue addressed in the National Fire Plan was protecting human life, homes, and communities. Preventive efforts emphasized proactive treatments to cut and remove overgrown brush and trees around homes in natural areas; this work was to take place in what was termed the wildland-urban interface. National programs like FireWise and Community Wildfire Protection Planning were launched to increase local engagement in preparing for wildfire. The Healthy Forests Restoration Act of 2003 was passed in part to simplify the environmental review process for thinning projects (Executive Office of the President 2003). The Collaborative Forest Landscape Restoration Act of 2009 created a funding mechanism for thinning and burning at a larger scale (Schultz et al. 2012). After ten years, the National Fire Plan was replaced with the Cohesive Strategy that is currently the guiding policy for fire management and forest restoration by federal and state agencies.

Congressional appropriations for the USDA Forest Service and Department of the Interior agencies were established for treatments in a Hazardous

Fuels Reduction Program as part of the National Fire Plan (McCarthy 2004). Analysis of Congressional appropriations shows the level of funding for hazardous fuels reduction increased significantly between 2001 and 2012, growing from about $100 million to over $500 million for the Forest Service and Interior Departments combined. However, even with these major increases, funding for hazardous fuels reduction was insufficient to meet the full need for fuels reduction in western forests. Funding remained a fraction of the amount spent on fire suppression, which exceeded $1 billion in seven of the nine years from 2003 to 2012.

Early in the National Fire Plan implementation, thinning treatments in Southwestern forests averaged in the hundreds of acres per state, despite wildfires that might destroy thousands of acres in a day (McCarthy 2004). Throughout the last decade the average treatments cost has been $500 to $1,000 per acre in the Southwest. Funding is allocated to the forest or district level, and a 500 acre treatment at a cost of $250,000 to $500,000 might be all a unit can afford in a given year. The Collaborative Forest Landscape Restoration Program (CFLRP), authorized in 2009, provides up to $4 million annually for selected large landscape projects. It can finance treatments of thousands of acres and was enacted to boost the scale of restoration that can be accomplished (Schultz et al. 2012). However, the authorized appropriation for CFLRP is capped at $40 million, which is sufficient to fund twenty large landscapes around the United States. Despite the CFLRP, scientists and increasingly recognizing that the policy and funding context is making it impossible to restore large areas of fire-prone forests at a scale that can make a difference in fire behavior (Ecological Restoration Institute 2013; Stephens et al. 2013). Funding decreases for federal fuels reduction, coupled with slower economic growth, federal budget cuts, and declining state revenue prompted some to look at other possible funding mechanisms for forest restoration. Water funds are among the most successful funding mechanisms under the model of payments for ecosystem services—that is, mechanisms whereby payments are made for ecological benefits or services that are not captured in traditional market prices (Goldman-Benner et al. 2013). The Nature Conservancy in Latin America established its first water fund in 2000 in Quito, Ecuador. Today there are twelve established water funds in countries in Latin America, each providing a mechanism for water users to help pay for land management in headwaters that improves water quality and reliability.

Water storage and release is an important service provided by forests in the arid Southwest. A number of cities and towns in Colorado, Utah, Arizona, and New Mexico have created mechanisms that link the water that forests provide to downstream users with the funding needed to restore forest health—arrangements that are payments for water services (Carpe Diem West 2011). Given that forest conditions have deteriorated to the point that federal appropriations for the Hazardous Fuels Reduction Program are insufficient to meet the need in fire-prone forests, community leaders are increasingly seeking to play a role in leveraging solutions. In New Mexico, wildfire damage to water sources is prompting deeper community engagement. New Mexico is currently experiencing significant drought, higher temperatures, and increases in wildfire intensity and severity (Williams et al. 2012). With 9.4 million acres (3.8 million ha) of National Forest System lands (Western States Data 2007) in New Mexico, accounting for the majority of mid- and high-elevation forests, water managers have strong incentive to partner with forest managers on proactive solutions. The following two case studies describe the development of water funds as a tool for municipal water source protection in the fire-prone interior West. The first example is a water fund in Santa Fe, New Mexico, established in 2009. The second example is a new water fund in development for the Rio Grande and Rio Chama watersheds in New Mexico to protect water sources for Albuquerque, Rio Rancho, Los Alamos, Santa Fe, Espanola, several pueblos, and numerous rural towns and villages. Both examples are based on the model of Latin America water funds, using the manual written by Nature Conservancy staff as a guide to the design, creation, and operation of water funds (Nature Conservancy 2012).

The Cerro Grande Fire of 2000 had direct effects on Los Alamos, New Mexico, which lost 280 homes (Gabbert 2010) and was without municipal water delivery for four months while fire-damaged pipes were repaired. One year after the fire, reservoir sedimentation was 140 times higher than the previous fifty-seven years and remained significantly elevated for at least five years after the fire (Lavine et al. 2005)

In nearby Santa Fe, the city considered the risk of a similarly damaging wildfire, should one ignite in their 17,000 acre (6,900 ha) municipal watershed, contained entirely within the Santa Fe National Forest. Even though the city sustained no direct costs from Cerro Grande fire, the threat of wildfire to their two reservoirs, supplying 30 percent of municipal water, was of serious concern. Local scientists noted similarities between the overgrown forest conditions in Santa Fe's watershed and the area where the Cerro Grande fire burned, and considered it only a matter of time before Santa Fe experienced a large fire of its own. A few months after Cerro Grande was extinguished, community leaders in Santa Fe launched a concerted effort to proactively cut and remove the overgrown brush and trees, replicating historical forest conditions and reducing the amount of vegetation that could act as fuels in future wildfires.

An Environmental Impact Statement for treatments was approved in 2003 and over the next four years more than $7 million of Congressionally earmarked funding was appropriated to thin 7,000 acres (2,830 ha) of forests in the lower watersheds that are critical to supply Santa Fe's water. Controversy over the forest treatments was high at first, with local and national environmental groups expressing concern about removing trees. Concerns diminished after dozens of public meetings, several science forums, and establishment of a multiparty monitoring process to ensure community oversight. Historically, fire burned in the Santa Fe watershed every fifteen years (Derr et al. 2009), prompting forest and water managers to plan for maintenance of the thinned forest with controlled burning. The Nature Conservancy offered the "water fund" model as a potential vehicle to pay for maintenance with controlled burning and other treatments. In 2008 the City of Santa Fe Water Division formed a partnership with the Santa Fe National Forest, Santa Fe Watershed Association, and The Nature Conservancy to seek water user funding for long-term management of Santa Fe's critical water sources in the National Forest.

Data about the full economic costs of wildfire was limited in 2008, so The Nature Conservancy developed cost estimates based on the few actual costs available from other communities. Based on this, an estimated cost of $22 million to the City of Santa Fe and the Forest Service was projected from a 10,000 acre (4,050 ha) wildfire in the watershed (Derr et al. 2009). These cost estimates were important to make the case for investment in preventive treatments.

Public opinion research conducted in 2011 as part of the Santa Fe program found overwhelming voter support for the establishment of a fund to protect Santa Fe's water supply from forest fires. In a poll conducted by telephone, voters were presented with a description of the threat that a major forest fire poses to the city's water supply; steps the US Forest Service currently takes to manage this threat; and the need for a stable source of funding to help prevent fires on lands that surround the city's water supply (Metz et al. 2011). The poll found that by a nearly four-to-one margin, voters voiced support for this concept. Voters were also asked how much they would be willing to pay for a Santa Fe Water Source Protection Fund, which would protect water sources and reservoirs from damaging wildfire. More than 80 percent of voters indicated they would be willing to pay, on average, an additional sixty-five cents per month on their water bill to go toward the Santa Fe Water Source Protection Fund. Voters also were asked whether they would support an average fee of one dollar, one dollar and fifty cents, or two dollars. Even at the highest potential price point—two dollars per month—nearly two-thirds of voters who were surveyed said they would be willing to pay the fee (Metz et al. 2011).

The Santa Fe Water Source Protection Fund was approved by the City Council in 2011 as a program of watershed investment. In the final agreement, the city approved a Watershed Management Plan for sharing up to 50 percent of costs with the Forest Service for twenty years to maintain the current conditions in restored forest areas through burning, adding new fuels breaks, and restoring some additional lands. The commitment also included funding for monitoring of water quality and restoration treatment effects, and for community outreach and watershed education programs for Santa Fe youth. The approved Watershed Management Plan describes the expected management needs over twenty years and includes a financial plan that outlines the cost-sharing agreement between the city and the Forest Service (Derr et al. 2009). According to the financial arrangement, the City of Santa Fe is to pay just over $3 million over twenty years to the Forest Service to ensure protection of its water sources. The watershed treatment costs are split 50–50 between the City of Santa Fe and the Forest Service (Derr et al. 2009). Considering the additional education, water quality and monitoring costs, the expenses are shared as follows: 62 percent City of Santa Fe, 36 percent Forest Service, and 2 percent Santa Fe Watershed Association. The initial years of funding for the city and the Santa Fe Watershed Association were provided by a $1.4 million grant from the New Mexico Water Trust Board, funded by New Mexico gross receipts tax. The Water Trust Board funding enabled the city to finish paying for another water infrastructure project before using revenue from the Water Division budget to pay their half of the water source protection (Lyons, personal communication with Laura McCarthy, March 18, 2013).

The Santa Fe case study predates Denver, and was the first application of the water fund model to US public lands forests. Testing the water fund model on a small watershed with a few partners made it possible to prove the concept in just a few years. The key lessons from Santa Fe are to keep the funding mechanism simple and to develop a good monitoring and feedback mechanism to keep water fund investors up to date. Historically, Albuquerque's political leadership, business community, and water utility have put significant effort into planning for a sustainable water future. About half of Albuquerque's water today comes from the Colorado River basin via a transmountain diversion known as the San Juan-Chama Project. Planning for the importation of this water from the Colorado River basin to New Mexico began in the 1950s, at a time of growth for Albuquerque and in the middle of a ten-year drought cycle. The San Juan- Chama Project is a system of diversion structures and tunnels that moves water from the Navajo River in the San Juan River basin to the Rio Grande basin where it flows into the Chama River, a series of reservoirs, and then the Rio Grande. About 110,000 acre-feet of water are authorized for diversion, and most New Mexico cities have purchased rights to this water. Albuquerque owns the biggest share of San Juan-Chama Project water, but Santa Fe, Los Alamos, and other towns own San Juan-Chama water, as well as the Jicarilla Apache Nation and the Middle Rio Grande Conservancy District, which uses the water for irrigated agriculture (US Department of the Interior, Bureau of Reclamation 2013).

The 2011 Las Conchas Fire and 2000 Cerro Grande fire had a large impact on municipal water sources. The Las Conchas fire occurred in New Mexico's Jemez Mountains, within thirty miles of roughly half of the state's population living in Albuquerque, Rio Rancho, Los Alamos, and Santa Fe, and numerous pueblos and small towns. The fire was notable for the extent of moderate- and high-severity burn, which affected 42 percent of the area (Tillery et al. 2011). The severely burned areas in Las Conchas left nothing but ash and occasional standing dead trees and boulders. Heavy rains about six weeks after the fire started created major debris flows in four canyons draining directly to the Rio Grande. For example, rainfall of 1.5 inches on August 21/22, 2011, caused debris flows in Bland and Cochiti Canyons. The debris flows flooded the popular Dixon Apple Orchard, deposited tons of debris into the US Army Corps of Engineers' Cochiti Reservoir, and lowered dissolved oxygen content of the Rio Grande well past the point where fish and other aquatic species could survive (Dahm et al. 2013). Utility operators in Albuquerque and Santa Fe decided the water was unfit for treatment and shut down their surface water use for forty and twenty days, respectively, switching to groundwater wells at a time of peak summer usage.

The Nature Conservancy began exploring the idea of a water fund focused on protecting water sources from damage by wildfire and postfire flooding in the Rio Grande valley in 2012 with funding from Lowe's Charitable and Educational Foundation (Nature Conservancy 2014). Unlike Santa Fe, Albuquerque had not yet considered the possibility of wildfire and postfire debris flow threatening their surface water or contaminating their San Juan-Chama water. However, the Las Conchas fire provided a tangible demonstration of the problem, and city and business leaders were soon convinced that a solution must be found. The Nature Conservancy's initial presentation to the water and energy subcommittee of the Greater Albuquerque Chamber of Commerce was met with a surprisingly high level of support. Additional outreach led to endorsements of the need to find a solution for this problem from other business groups, including the New Mexico Association of Commerce and Industry, which functions like a statewide chamber of commerce, and the New Mexico Business Water Task Force, a group initially formed to advocate for the San Juan-Chama Project.

The underlying problem of dense forests and high-severity wildfire adjacent to important water supplies was relatively easy to establish; the more difficult task was to build support and establish funding for a large-landscape program of forest and watershed treatments to improve resiliency to climate change and wildfire. The Nature Conservancy convened a Rio and Forest Advisory Board in April 2013 for the specific purpose of establishing a water source protection fund for the Middle Rio Grande and its forested watersheds. The advisory board is made up of leaders from federal and state forest and water management agencies, business community leaders, university experts, and a diverse cross-section of interest groups ranging from traditional agriculture to recreation to the wood products industry. As the convener and facilitator, The Nature Conservancy has organized the advisory board into a set of task-oriented working groups.

The Conservancy's efforts are focused on creating a dedicated funding mechanism for large-scale investment in forest and watershed treatments from Albuquerque north to the Colorado border (figure 11.1). The Rio

Grande Water Fund area includes all of the forested watersheds and tributaries to the Rio Grande and Rio Chama, as well as the headwaters of the San Juan-Chama water just over the state line in Colorado.

Studies are underway to establish a clear case for a water source protection fund for the Rio Grande. The studies are necessary to guide development of the water fund in these ways:

1. Identify the watersheds that are most vulnerable to high-severity wildfire and postfire damage to set priorities for water fund expenditures;

2. Estimate watershed response to forest treatments, including water increases that may sustain forests (Grant et al. 2013) or streamflow;

3. Assess the full economic costs of the Las Conchas wildfire to inform a cost-benefit analysis; and

4. Survey municipal water users and agricultural users to determine their understanding of the threats to water security and willingness to pay for restoration treatments of at-risk forests.

The outcome of these studies and engagement of the advisory board and working groups will be to produce a comprehensive water security plan for the Rio Grande from Albuquerque north to the Colorado border. The plan will include a prioritized list and map of restoration treatments for forests and riparian areas; estimated costs and capital needs to implement the plan, including NEPA assessment for federal lands; wood product utilization and investment needs in infrastructure; and a detailed plan for water fund structure, governance, and revenue.

Early estimates by The Nature Conservancy are that the Rio Grande and forested watersheds in the area from Albuquerque north to the Colorado border includes 1.7 million acres (688,000 ha) of ponderosa pine and mixed-conifer forests (Nature Conservancy 2014). Historically, these forests experienced frequent low-severity fire. Mechanical thinning and controlled burning recommended by scientists and land managers are effective treatments to reduce fuel loads. The Nature Conservancy's estimate assumes that 40 percent of the 1.7 million acres (688,000 ha) of eligible forests would actually be treated, with a preliminary goal to treat 700,000 acres (283,300 ha) in ten to thirty years, depending on how quickly the rate of treatment can be accelerated. Current treatment levels in this area are estimated at roughly 3,000

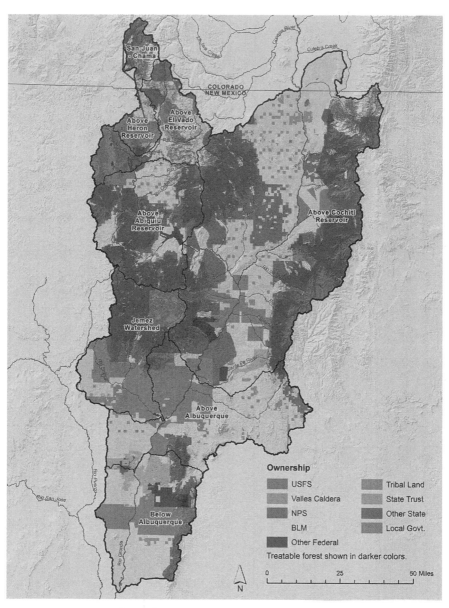

FIGURE 11.1. Proposed area for the Rio Grande water source protection fund.

acres (1,215 ha) annually, so a tenfold increase would be 30,000 acres (12,140 ha) per year, and it would take roughly twenty-three years to reach the goal. At a cost of $500 per acre, about $7 million to $15 million revenue would be needed annually, assuming current markets for low-value wood and assuming federal appropriations at current levels are available as matching funds.

Raising $7 million to $15 million nonfederal funds each year for thirty years for forest and watershed restoration will not be easy. The water fund needs to be structured in a way to receive funding from a variety of sources, including payments by municipal water users and irrigation district members, home-owner's insurance premium taxes, and corporate and voluntary donations. These options are under study now. After the investment period needed to reduce fuels substantially, a program of controlled burning and mechanical thinning with commercial by-products will need to be sustained in the long term. The annual costs to maintain forest and watershed resiliency after the initial treatments should be far less and are estimated at $1 million to $3 million. The evolution of water funds in New Mexico has progressed from a small-scale, proof of concept in Santa Fe to a large and complex Rio Grande Water Fund that includes many diverse partners and a complex landscape. The Rio Grande Water Fund is framing the issue as water security, and is gaining far more traction for forest restoration than was achieved when the issue was framed as wildfire protection.

All aspects of New Mexico life are touched by water availability and reliability. The Santa Fe and Rio Grande Water Funds are, in essence, climate change adaptation strategies, focused on garnering long-term funding to maintain resiliency in large, forested watersheds. It remains to be seen if a project as large in scale as the proposed Rio Grande Water Fund for treatments across 1.7 million acres (688,000 ha) of forest can be achieved. The concept, however, is gaining serious traction and its success or failure may be assessed within a few years.

SECTION IV

TRANSDISCIPLINARITY IN THE ANTHROPOCENE

12

Implementing Climate Change Adaptation in Forested Regions of the Western United States

JESSICA E. HALOFSKY, LINDA A. JOYCE, CONSTANCE I. MILLAR, DAVID L. PETERSON, AND JANINE M. RICE

Federal land management agencies in the United States are beginning to incorporate climate change into their management planning and operations. For example, Secretary of the Interior Order 3289, signed in 2009 and amended in 2010, suggests that potential climate change impacts necessitate changes in how the US Department of the Interior (USDOI) manages natural resources and requires its agencies to incorporate climate change in planning, prioritization, and decision-making (USDOI 2009). One of four strategic goals in the US Department of Agriculture (USDA) strategic plan for fiscal years 2010–2015 (USDA 2010) is to ensure that national forests and private working lands are conserved, restored, and made more resilient to climate change; a 2011 Departmental Regulation (DR-1070-001) (USDA 2011), required the USDA and each agency within to prepare a climate change adaptation plan. More recently at the executive level, President Obama issued an Executive Order, "Preparing the United States for the Impacts of

The findings and conclusions in this chapter are those of the authors and do not necessarily represent the views of the United States Forest Service.

Climate Change" (Executive Order 13653). The Executive Order requires the heads of the Departments of Defense, the Interior, and Agriculture, the Environmental Protection Agency, the National Oceanic and Atmospheric Administration, the Army Corps of Engineers, and other agencies to work with the Chair of Council on Environmental Quality and the Director of the Office of Management and Budget to "complete an inventory and assessment of proposed and completed changes to their land- and water-related policies, programs, and regulations necessary to make the Nation's watersheds, natural resources, and ecosystems, and the communities and economies that depend on them, more resilient in the face of a changing climate." The assessments are required to include a timeline and plan for making changes to policies, programs, and regulations.

Agency directives have spurred a flurry of climate change–related activity in federal land management agencies over the last few years. From that activity, science-management partnerships have emerged as effective catalysts for development of vulnerability assessments and land management adaptation plans at the strategic and tactical level (Cross et al. 2013; Gaines et al. 2012; Littell et al. 2012; McCarthy 2012; Peterson et al. 2011; Swanston and Janowiak 2012). Science-management partnerships typically involve iterative sharing of climate and climate effects information by scientists, and of local climate, ecological, and management information by managers. This iterative information exchange aids identification of vulnerabilities to climate change at the local scale and sets the stage for development of adaptation strategies and tactics, often developed through facilitated workshops (for a detailed description of the steps involved in developing a vulnerability assessment see Joyce and Millar 2014).

US Forest Service scientists and land managers are tasked with reducing the effects of climate change on ecosystem function and services (USDA FS 2008, 2011a). Partnerships among scientists in the Forest Service Research and Development branch, managers in the National Forest System, and other agencies and universities have played a major role in advancing climate change adaptation in the agency. Development of science-management partnerships is a performance measure in the USDA FS Climate Change Performance Scorecard (USDA FS 2011b), which rates national forests on how well they are responding to climate change. Here, we describe the process and outcome of several recent science-management partnerships led by the

USDA FS in the western United States, identify key elements, and discuss future application to other regions.

As a part of the WestWide Climate Initiative (Peterson et al. 2011), a science-management partnership was initiated among a research scientist from the USDA FS Rocky Mountain Research Station office in Fort Collins, Colorado, the regional ecologist from the USDA National Forest System, Rocky Mountain Region's office in Lakewood, Colorado, and the resource staff officer from the Shoshone National Forest supervisor's office in Cody, Wyoming. The partnership was initiated to determine the potential effects of climate change on Shoshone National Forest and to develop tools to help national forest land managers adapt their management to climate change. Over time, involvement from the regional office and Shoshone National Forest expanded to include experts in wildlife, water, ecology, and planning. The scientists at Western Water Assessment at the Cooperative Institute for Research in Environmental Sciences, University of Colorado, Boulder, became important partners. In addition, scientists from the US Geological Survey (USGS) and several universities participated in partnership activities. Periodic briefings were held at the Shoshone National Forest and in the regional office to keep interested staff updated on activities.

Initial discussions identified the need to synthesize the literature on climate change specific to the Shoshone National Forest and surrounding area. A report, "Climate Change on the Shoshone National Forest, Wyoming: A Synthesis of Past Climate, Climate Projections, and Ecosystem Implications," was jointly written by Rocky Mountain Research Station and National Forest staff to synthesize current scientific information about prehistoric, recent, and future climate and how future warming may affect natural resources on Shoshone National Forest (Rice et al. 2012). A focused review of the potential impacts of climate change on quaking aspen (*Populus tremuloides* Michx.) was also conducted in cooperation with other scientists in the WestWide Climate Initiative, because aspen is a high priority for management across the western United States (Morelli and Carr 2011).

Staff on the Shoshone National Forest expressed a desire to interact with scientists on specific topics and to have sufficient time for discussion of each topic. Thus, the Natural Resource and Climate Change Workshop was held in Cody, Wyoming, in 2011. The workshop was attended by over fifty participants from Shoshone National Forest, other federal and state agencies, and private-sector organizations. Topics included climate change, snowpack, and vegetation models. Seven local experts in the fields of climate and climate change effects, water resources, snow and glaciers, ecosystem modeling, Yellowstone cutthroat trout (*Oncorhynchus clarkii bouvieri*), and recreation and tourism offered information about climate and potential effects in the Shoshone area (USDA FS 2011b).

During the workshop and afterwards, discussions among scientists, and regional and Shoshone National Forest staff led to the identification of key resources for further consideration. These high-priority resources included water availability, Yellowstone cutthroat trout, and quaking aspen, and partnership scientists and managers conducted a vulnerability assessment for each resource. The vulnerability assessment for water availability was developed to provide information about the effects of climate change on water resources in the Shoshone National Forest region, an important source of water for human uses. For the Yellowstone cutthroat trout, a customized vulnerability assessment tool was developed using indicators for climate change exposure as well as inherent landscape, anthropogenic, and ecological factors of sensitivity and adaptive capacity. This tool provides information to guide adaptation strategy development and conservation and monitoring planning.

Aspen in Shoshone National Forest currently occupies a fraction of its potential habitat based on climate, topography, and soils, which suggests that its distribution is constrained by other factors. The question of where aspen may exist in the future could not be completely addressed in the assessment, although the assessment pointed to the potential for an expansion of aspen because of projected changes in climate. The effects of other factors, such as conifer competition, fire regime, insects and pathogens, and wildlife browsing likely cause spatial and temporal variability of aspen distribution and abundance that may continue to be dynamic under climate change. The results of the vulnerability assessments conducted through the partnership have informed conservation project planning for Yellowstone cutthroat trout, helped to inform the selection of hydrologic monitoring locations, and to provide vulnerability information for the future management of water availability in grazing allotments. Rice et al. (2012) was extensively used in the recent Shoshone National Forest planning process.

The first formal project to address climate change in a national forest emerged in 2008 from a science-management partnership initiated by the

USDA FS Pacific Northwest Research Station, Olympic National Forest, and Olympic National Park (Halofsky et al. 2011). Building on a long history of cooperation between the national forest and national park located on the Olympic Peninsula, this project engaged scientists and resource managers in a sequence of educational workshops, development of a vulnerability assessment, and compilation of adaptation options.

Early discussions among scientists and natural resource managers identified the need to increase awareness of climate change among federal natural resources managers on the Olympic Peninsula and assess the vulnerability of natural resources at Olympic National Forest and Olympic National Park. Four separate workshops were convened on the topics of hydrology and roads, fisheries, vegetation, and wildlife, with participants from different federal and state agencies, tribes, and other groups attending each workshop. At each workshop, scientists from the Forest Service and University of Washington provided state-of-science summaries on the effects of climate change on natural resources, and scientists and managers worked together to identify the most important impacts on the Olympic Peninsula. Smaller workshops were then convened with a core group of scientists and managers to review the vulnerability assessment and develop adaptation strategies and tactics for each of the four resource areas. All information was subsequently peer reviewed and published as documentation for management and decision-making (Halofsky et al. 2011).

Building on knowledge gained from working with Olympic National Forest and Olympic National Park, the North Cascadia Adaptation Partnership (NCAP) was subsequently initiated in north-central Washington in 2011. The partnership covers 2.4 million ha across the west and east sides of the Cascade Range and includes Mount Baker-Snoqualmie National Forest, Okanogan-Wenatchee National Forest, North Cascades National Park Complex, and Mount Rainier National Park (Raymond et al. 2013; Raymond et al. 2014). This diverse, mountainous region contains temperate rainforest, subalpine and alpine ecosystems, extensive dry forests subject to frequent fire, and shrub-steppe ecosystems. It also contains 17,000 km of roads and is adjacent to densely populated areas of western Washington.

The NCAP project started with one-day educational workshops at each of the four management units, which included resource managers, line officers, administrative personnel, and various stakeholders. Then, four two-day workshops were convened for all management units combined, with one workshop focused on each of the following topics: hydrology and access, fisheries, vegetation and ecological disturbance, and wildlife. The first day of each workshop focused on developing summaries of resource sensitivities as components of the vulnerability assessment, with scientists leading the discussion and managers contributing data and information on specific locations. The second day of the workshop focused on adaptation to sensitivities identified for each of the four resource areas, with managers providing adaptation strategies useful for planning and adaptation tactics useful for on-the-ground applications. Information discussed and written at workshops was compiled in peer-reviewed documentation that will be used by the national forests and national parks (Raymond et al. 2013; Raymond et al. 2014).

The NCAP project was premised on an "all lands" approach that considered public, private, and tribal lands other than Forest Service and National Park Service lands. Individuals from about forty different federal and state agencies, tribes, and conservation groups participated in the workshops and assisted with identification of resource issues and adaptation options. The NCAP catalyzed an ongoing dialogue on climate change in the North Cascades region that has persisted beyond the formal phase of the project. For example, additional workshops have been convened on how climate change will affect extreme flood events that have the potential to damage roads on the west side of the Cascades. Workshops have also been convened to focus on how climate change will affect fisheries and estuarine systems in the Skagit River basin, a major watershed in the NCAP project area.

In 2009, as a part of the WestWide Climate Initiative (Peterson et al. 2011), a science- management partnership was established between research scientists at the USDA FS Pacific Southwest Research Station and managers at Inyo National Forest and Devils Postpile National Monument in the Sierra Nevada, California, to develop tools and information to help forest and park managers adapt their management and planning to climate change. At the start of the project, Inyo National Forest was beginning to revise its land management plan, and Devils Postpile National Monument was launching a general management plan to identify long-term desired conditions for the monument and guide park managers as they decide how to best protect monument resources and manage visitation. After initial meetings to determine the direction of the partnership, the team determined that facilitated sharing of knowledge about climate change and its effects through targeted workshops and assessment reports (developed by scientists) would help managers integrate climate change into planning and management. Inyo National Forest staff had several specific requests of scientists in the partnership: (1) compile a summary of climate trends and adaptation options relevant to the eastern Sierra Nevada, (2) develop a regional bibliography of information on climate change, (3) establish a technical advisory board that includes climate scientists conversant in eastern Sierra Nevada regional issues, (4) prepare a report and field survey for a potentially novel climate threat to quaking aspen in the eastern Sierra Nevada, (5) participate in the public engagement process before the land management plan revision process, and (6) conduct facilitated climate applications workshops. A desired outcome of the partnership was for Inyo National Forest to implement resource treatments developed by partnership discussions and products, and to incorporate climate considerations in the land management plan revision.

Priorities identified by staff at Devils Postpile National Monument, whose lands are surrounded by Inyo National Forest, included a need for high-resolution climate monitoring and information on the potential role of the monument as a cold-air pool that could serve as a climate refugium for some species in a warmer climate. The latter led to a request for scientists to develop an analysis of cold-air pooling in the upper watershed of the monument. Partnership activities at the monument had a strong focus on science, including a combined field and classroom workshop, summary of ongoing research, and synopsis of research and monitoring efforts needed to guide future adaptation efforts. A scientific technical committee was also convened to consult on general management plan development and advise on implementation of adaptation treatments.

The science management workshop conducted at Inyo National Forest, "Evaluating Change in the Eastern Sierra," was attended by a mix of federal, university, and other scientists, resource specialists, and concerned citizens. Scientists presented information on climate projections at the global, regional, and local scale, and discussed effects on other resources, such as vegetation (Morelli et al. 2011). Implications for the Inyo National Forest were then discussed. For Devils Postpile National Monument, a science management workshop was held with scientists from the USDA FS and USGS and managers from the National Park Service. The workshop included presentations on climate and hydrologic projections relevant to Devils Postpile National Monument as well as physical and ecological vulnerabilities and potential effects on visitors and infrastructure. Presentations were followed by a general discussion on implications for managing Devils Postpile National Monument as a refugium in an uncertain future.

In addition to education and training through facilitated workshops, outcomes of the science-management partnership in the eastern Sierra Nevada included several reports and tools. For Inyo National Forest, scientists developed a report reviewing aspen response to climate and describing an aspen screening tool (Morelli and Carr 2011). The climate project screening tool (Morelli et al. 2012) was developed to provide a screening process to assess if climate change would affect resources involved in management projects in line for implementation. A report summarizing some of the latest data on climate change projections and effects relevant for eastern California was developed for use by land managers in the Sierra Nevada (Morelli et al. 2011). In anticipation of the potential for Devils Postpile National Monument to serve as a climate change refuge, owing to its position at the bottom of a canyon with cold-air drainage, a network of temperature sensors in multiple-elevation transects and a climate monitoring station were recently installed to measure temperature patterns. To date, all adaptation projects in national forests and adjacent lands have used similar approaches and accomplished similar outcomes despite the fact that they were conducted in different geographic locations with varied natural resources issues and with different groups of managers (Peterson et al. 2011). First, each project was developed on the foundation of a strong and enduring science-management partnership (Littell et al. 2012) that a Forest Service research station initiated. Building these partnerships, which typically included other agencies (especially the National Park Service) and stakeholders (table 12.1), required substantial time and energy to establish personal relationships and build trust. Having individuals to serve as liaisons between climate scientists and managers was critical, and the partnerships went well beyond simply providing climate data on a website or in a database for managers to access. The partnerships have persisted through time, even beyond the end of the original project, because of the effort that went into establishing relationships and providing information that can be directly applied to management.

Second, each project included an educational component in which natural resource personnel, line officers, and in some cases, administrative staff

TABLE 12.1. Units involved, focus topics, and products for five science-management partnerships conducted with US Forest Service Research and Development across the United States.

Partnership name	Geographic region	Primary partnering units	Focus topics	Published tools and reports
Inyo National Forest and Devils Postpile National Monument Case Study	Pacific Southwest	US Forest Service Pacific Southwest Research Station, Inyo National Forest, and Devils Postpile National Monument	Quaking aspen, cold air pooling	Morelli and Carr 2011; Morelli et al. 2011, 2012
North Cascadia Adaptation Partnership	Pacific Northwest	US Forest Service Pacific Northwest Research Station, Mt. Baker-Snoqualmie National Forest, Okanogan-Wenatchee National Forest, North Cascades National Park Complex, and Mount Rainier National Park	Hydrology and access, fisheries, vegetation and ecological disturbance, and wildlife	Raymond et al. 2013; Raymond et al. 2014
Northwoods Climate Change Response Framework Project	Lake States	US Forest Service Northern Institute of Applied Climate Science, Chequamegon-Nicolet National Forest	Forest ecosystems, carbon stocks	Swanston et al. 2011; Swanston and Janowiak 2012
Olympic National Forest Case Study	Pacific Northwest	US Forest Service Pacific Northwest Research Station, Olympic National Forest, Olympic National Park, University of Washington Climate Impacts Group	Hydrology and roads, fisheries, vegetation, wildlife	Halofsky et al. 2011
Shoshone National Forest Case Study	Interior West	US Forest Service Rocky Mountain Research Station, Shoshone National Forest, National Forest System Rocky Mountain Region	Water availability, Yellowstone cutthroat trout, and quaking aspen	Morelli and Carr 2011; Rice et al. 2012

attended sessions in which they learned about the latest science on climate change and climate change effects, and shared their experiences with climate-related resource issues (Halofsky et al. 2011). This baseline of knowledge is critical for identifying key climate change vulnerabilities, developing adaptation plans, enhancing monitoring efforts, and generally incorporating climate change in planning and management.

Third, each project focused a great deal of effort on producing a peer-reviewed assessment of the vulnerability of natural resources to climate change (table 12.1), which helped to identify resources most at risk. These assessments, typically led by Forest Service scientists in collaboration with other agencies and universities, were state-of-the-science syntheses that focused on the topics considered by resource managers to be the most important (e.g., hydrology, fisheries, vegetation). Considerable effort was focused on downscaling and customizing information, often large scale and general in nature, to specific landscapes and resource management issues.

Fourth, each project based the development of adaptation options directly on the vulnerability assessment and known principles of climate change adaptation (Joyce et al. 2008, 2009; Peterson et al. 2011). Scientists provided information on resource sensitivity to climate change for different scenarios, and resource managers responded with solutions for mitigating resource risk (table 12.2). These responses typically included an overarching adaptation strategy (conceptual, general) and a subset of adaptation tactics (specific, on the ground) for each strategy (Peterson et al. 2011; Raymond et al. 2013).

Commitment to regular, clear communication was a key to the success of all projects. Scientists spent many days on the ground in national forest landscapes and in offices where resource managers work. These conversations and experiences were critical for getting iterative feedback on the vulnerability assessment, management issues, and potential applications of climate change information. There is no substitute for scientists (typically with more discretionary time) working directly with resource managers (typically with minimal discretionary time) to ensure that the vulnerability assessment and adaptation options are relevant to local planning and management.

Completion of a vulnerability assessment does not guarantee that adaptation measures will be implemented. Decision-makers and the general public (including local communities) need to be engaged in the vulnerability assessment or as part of the adaptation process to facilitate *collective learning*. Collective learning, that is, information emerging from experience and/ or human interaction during which people's different goals, values, knowledge, and point of view are made explicit and questioned to accommodate

TABLE 12.2. Examples of climate change sensitivities and related adaptation strategies and tactics. In science-management partnerships, sensitivities are typically communicated by scientists, and adaptation strategies and tactics are developed by land managers based on sensitivities.

Sensitivity	Adaptation strategy or tactic
Increased opportunity for invasive species establishment	Implement early detection/rapid response for exotic species treatment
Potential for mortality events and regeneration failures, particularly after large disturbances	Develop a gene conservation plan for ex situ collections for long-term storage
	Identify areas important for in situ gene conservation
	Maintain a tree seed inventory with high-quality seed for a range of species
	Increase production of native plant materials for post-flood and postfire plantings
Increased forest drought stress and decreased forest productivity at lower elevations	Increase thinning activities
	Use prescribed burns and wildland fire to reduce stand densities and drought stress
Increased winter and spring flooding	Implement more conservative design elements (more intensive treatments such as larger diameter culverts, closer spacing between ditch relief culverts and waterbars)
	Increase maintenance frequency of drainage features

conflicts, is the basis for identifying the collective action to tackle a shared problem (Yuen et al. 2013c).

Resource managers and leadership in national forests and other lands where projects were conducted consistently cite the value of the projects in providing a new context for resource management and in enhancing "climate smart" thinking. However, implementation of information derived from climate change vulnerability assessments in national forest and national park resource assessments and monitoring is uncommon. Inclusion of climate change adaptation strategies and tactics in resource planning and project plans is just starting, even though current practices are often highly compatible with deliberate actions that enhance the ability of forests to adapt to climate change. More time may be needed for the climate change context of resource management to be incorporated as a standard component of agency operations.

At the national level, the federal agencies have a strong focus on advancing climate change issues. At the local scale, many management units would like to develop vulnerability assessments and adaptation plans. However, in the absence of a mandate to do so, the process of developing projects similar to those described above will continue to be slow. The USDA FS Climate Change Performance Scorecard requires development of climate change vulnerability assessments and adaptation plans, but the mandate is largely unfunded. Efforts to accelerate climate change implementation in national forests come during a period of steep budget decreases, making it difficult to implement planned projects and initiate new projects. At the present time, relatively few national forests have undertaken significant steps toward completing vulnerability assessments and adaptation plans, and the status of adaptation planning in other agency units is similar (Bierbaum et al. 2013).

The slow pace of federal agencies in emulating the processes and applications described above (Peterson et al. 2011) can be increased by mainstreaming (or operationalizing) climate change as a part of standard operations in the National Forest System and other federal lands. This transition has been enabled by various strategic documents in the Forest Service and other agencies. Concepts such as ecosystem-based management and ecological restoration that were originally plagued by skepticism and uncertainty evolved into operational paradigms. So too must climate change become incorporated in thought, actions, and management guidance; climate change needs to become a standard component in strategic planning, project planning, monitoring, and implementation. This will likely come with increased awareness of climate change, understanding of the potential effects of climate change, and development and awareness of effective responses to decrease resource vulnerabilities.

Scientific knowledge about the effects of climatic variability and change on natural resources is for the most part not a limiting factor in moving forward with climate change activities in national forests and other federal lands. However, effective transfer of climate-related knowledge from the scientific to the management community is lacking, and thus so is the application of the information in natural resource management. Future efforts can therefore focus on the synthesis of relevant scientific information for specific landscapes (vulnerability assessment), effective transfer of that information to the management community, and then development of responses that reduce negative effects on resources (adaptation planning). This can be expedited in agencies like the Forest Service and National Park

Service by institutionalizing science-management partnerships to facilitate climate change adaptation and associated processes. An ideal partnership in the Forest Service includes scientists from research stations, resource managers in national forests, and subject-matter experts from regional offices, along with scientists and managers from other agencies, universities, and organizations. National Oceanic and Atmospheric Administration Regional Integrated Sciences and Assessments (RISA) program scientists were involved in the adaptation partnership developed with Olympic National Forest and Olympic National Park (Halofsky et al. 2011), in the NCAP effort (Raymond et al. 2013; Raymond et al. 2014), and in the Shoshone National Forest effort; scientists from RISA centers and USDOI Climate Science Centers could be key partners in future efforts. If participants in these partnerships work on multiple projects, they will accrue knowledge that will make each subsequent project more effective and efficient. In addition, vulnerability assessments and adaptation plans can be developed for clusters of national forests and parks (and, potentially, adjacent federal, tribal, and other lands) with similar biogeographic characteristics and management objectives, resulting in time and budgetary efficiencies. Different clusters of management units may be appropriate for different resources. Tools that help the user focus on their specific resources, projects, and decision space will likely best assist them in developing adaptation options. A wide variety of tools have been developed to help structure the assessment of vulnerability and can range from qualitative frameworks, such as the climate project screening tool (Morelli et al. 2012) and decision-support flowcharts, to quantitative climate and bioclimatic projections, such as the Climate Wizard (Girvetz et al. 2009) and Tree Atlas (Prasad et al. 2007). Selection of a tool should support the attainment of goals and objectives and produce actionable information. The user should also be aware of the tool's capacities and limitations, its inherent geographic and biological scope, its capacity to include climate projections and handle uncertainty, and the expertise required to use it effectively (Beardmore and Winder 2011; Wilsey et al. 2013).

Existing resource management tools, particularly ones that assess environmental risks, may overcome the need to learn new tools and offer an opportunity to incorporate climate change into the existing management practices ("mainstreaming climate change considerations"). For example, the science-management partnership on the Olympic National Forest recognized that the current technique used to prioritize road maintenance could be enhanced with climate change information on increased risk of landslides and high-intensity rainfall, using an existing tool to evaluate increased risk associated with climate change (Halofsky et al. 2011). Modifying existing tools may also facilitate the comparison of climate change considerations with other nonclimatic stressors or considerations.

We are optimistic that climate change can be mainstreamed in the policies and management of the Forest Service and other federal agencies by the end of this decade. This can be expedited by considering climate change as one of many risks to which natural resources are subjected (Iverson et al. 2012), and by considering adaptation as a form of risk management. This approach has been recently described for water resources, fire, carbon, forest vegetation, and wildlife (Peterson et al. 2014; Vose et al. 2012), and will be fully incorporated in future US National Climate Assessments and assessments by the Intergovernmental Panel on Climate Change (Yohe and Leichenko 2010). We anticipate that evaluating climate change risks concurrently with other risks to resources will become standard practice over time.

13

Challenges and Opportunities for Landscape-Scale Management in a Shifting Climate

The Importance of Nested Adaptation Responses across Geospatial and Temporal Scales

Gary M. Tabor, Anne A. Carlson, and R. Travis Belote

Large landscape conservation has emerged over the past three decades as a science-based response to increasing large-scale habitat fragmentation and degradation by advancing the concepts of ecological integrity, ecological connectivity, wildlife corridors, and comprehensive landscape matrix conservation. More recently, large landscape conservation approaches have been embraced as a strategy to facilitate the adaptation of biodiversity to the impacts of climate change. In one sense, large landscape conservation is the evolution of the "beyond parks" conservation approach (Minteer and Miller 2011) in which species and ecological processes cannot be satisfactorily sustained within most circumscribed protected landscape parcels.

Conserving nature's parts and processes requires working at a landscape, ecosystem, or even bioregional scale. Hansen and DeFries (2007) demonstrate how even the vast spatial scales of our largest national parks are insufficient to fully support many ecological processes or prevent cross-boundary effects of surrounding human-dominated landscapes. Size does matter in ecology because of the scale of processes and impacts, and, in general, the larger the

DOI: 10.5876/9781607324591.c013

scale of focus, the better chance of conserving critical ecological processes, such as hydrologic function, natural disturbance regimes, species life cycles, and functional trophic interactions (Lindenmayer et al. 2008). Conservation at such large scales increases the complexity of decision-making as collaboration and consensus among diverse stakeholders, with diverse values, is required. These processes not only sustain nature but provide vital ecological services that support human livelihoods.

The Greater Yellowstone Coordinating Committee was established in 1964 to foster ecosystem-scale collaboration among government agencies in the region, the same year that the Wilderness Act was passed. Further research of species movement ecology in later years led to the design of even larger conservation efforts such as the Yellowstone to Yukon Conservation Initiative, which recognized the inter-ecosystem movement needs of the region's medium- and large-sized mammals, migratory birds, and cold-water fish within the Rocky Mountain Cordillera (Tabor 1996; Locke and Tabor 2005). Since its inception in 1993, the Yellowstone-to-Yukon effort—through its network of 200 or so public and private organizations—has protected roughly 23 million acres (9 million ha) of existing public lands through enhanced designations and roughly 1 million acres (400,000 hectares) of private lands through conservation easements and acquisitions. This includes one of the largest private land deals in the United States: the wholesale purchase of Plum Creek timberlands within the railroad legacy checkerboard landscape, including nearly 50,000 acres (20,234 ha) of the Swan and Blackfoot Valleys in the Crown of the Continent Ecosystem.

There has been an exponential growth of large landscape efforts in the past ten years, which, for the most part, reflects a growing conservation interest in maintaining ecological connectivity and wildlife corridors as an approach to address habitat fragmentation and heightened concerns about climate change impacts on species and habitats (McKinney et al. 2010; Regional Plan Association 2012; McKinney and Johnson 2013). Large landscape efforts promote resilience to large-scale stressors such as climate change, provide a range of potential climate refugia, and support species that can respond to changing environmental conditions with the opportunity to shift their geographic distribution. In reality, the story is more complex. Species interactions are likely to change as individual species respond differentially to climate stressors, and present day trophic structures may give way to novel species interactions and ecosystems in the future. Moreover, not all species have the ability to shift their distributions to keep pace with the relatively rapid rate of climate change, and current understanding of the extent to which genetic plasticity may allow or prevent species from responding to climatic shifts in their current habitat is poor.

Within the Yellowstone-to-Yukon region at the international boundary between Canada and the United States is the Crown of the Continent Ecosystem. This 18 million acre ecosystem surrounds Waterton Lakes and Glacier International Peace Park, the first international peace park, which was established in 1932. This landscape also bears the physical evidence of climate change as all twenty-five remaining glaciers in Glacier National Park are predicted to disappear within the next two decades after surviving for more than 7,000 continuous years (Hall and Fagre 2003). Triple Divide Peak within Glacier National Park connects three major continental river basins—the Columbia, the Missouri, and the Saskatchewan. The Crown of the Continent not only serves as a focal point for landscape impacts of climate change, it also serves as a focal point for US and Canadian landscape collaboration and innovation. Within the southwestern portion of this ecosystem, a new large-scale restoration effort is being prototyped.

The Southwestern Crown of the Continent (SWCC) Collaborative Forest Landscape Restoration Program (CFLRP) project in Montana is 1.48 million acres (600,000 ha) of forested, mountainous habitat in three adjacent US Forest Service (FS) Ranger Districts (Lincoln, Seeley Lake, and Swan) and includes portions of three of Montana's National Forests (the Helena, Lolo, and Flathead National Forests, respectively). The CFLRP began in August 2009 upon passage of the Public Lands Omnibus Bill. This Congressional Act established an annual budget of $40 million to finance ten collaborative, large landscape projects on Forest Service land across the United States. Through intensive, long-term work to improve forest health and resilience in an era of shifting climate, CFLRP projects are intended to sustain ecological, economic, and social benefits in rural communities that have traditionally relied on natural resources locally for their livelihoods, drinking water, and recreational opportunities. The Act also strongly encourages a shift to adaptive management in these landscapes by requiring all twenty-three CFLRP projects to develop and implement a large-scale monitoring program; a baseline inventory of natural resource conditions, coupled with short- and long-term evaluations of the effectiveness of restoration projects, is expected to create critically important information-feedback loops for managers in an increasingly uncertain future (Hutto and Belote 2013; Larson et al. 2013b).

Through its selection for funding in 2010, the SWCC in Montana combined several existing local collaboratives into a new diverse coalition comprising US Forest Service agency staff, university faculty, conservation organizations, and citizen groups. The SWCC is sited within the larger 18 million acre (7.28 million ha) Crown of the Continent, renowned for its unusually high degree of ecological integrity. No known extinctions of plant or animal species have occurred since Lewis and Clark's travels through the region 200 years ago (Prato and Fagre 2007). In addition to the prime habitat provided for grizzly bear, elk, wolverine, deer, gray wolf, Canada lynx, forest birds, and waterfowl within the forested mountain landscapes, the cold, clear streams of the Crown are home to a variety of native salmonid species.

Nonetheless, major restoration needs exist. Noxious weeds and exotic fish species have invaded terrestrial and aquatic ecosystems across the landscape, thousands of miles of old logging roads fragment key wildlife habitat and lead to increased sedimentation in blue ribbon trout streams through erosion, and mining activities from an era gone by necessitate focused and expensive cleanup efforts in several places. Decades of fire suppression—a management response to catastrophic wildfires in Montana and Idaho during the "Big Burn" of 1910—have dramatically altered the ecology of western forested ecosystems and resulted in unnaturally high accumulations of forest fuels (Arno and Fiedler 2005; Egan 2009). While many of these restoration needs identified above are common to Western landscapes of the United States and Canada, CFLRP project partners within the Southwestern Crown of the Continent face further management opportunities and challenges associated with the Montana Legacy Project, a historic conservation deal in which 273,000 acres (110,479 ha) of Plum Creek Timber Company–owned land was sold to a consortium of conservation organizations (led by The Nature Conservancy and the Trust for Public Land) before being transferred into public ownership through the US Forest Service. The checkerboard ownership pattern associated with the Montana Legacy Project began a century ago when the lands were initially purchased by the Northern Pacific Railway, but remains visible from space today given major differences in the management of these and adjacent lands through time. The absorption of these former commercial timberlands into the public domain highlights the significant, and often dynamic, challenges of developing conservation projects across jurisdictionally fragmented lands. Addressing restoration and climate adaptation challenges in the region requires explicitly dealing with the challenging issue of scale. Ecological processes operate across spatial scales where large-scale patterns (e.g., climate regimes) govern small-scale processes (e.g., seedling recruitment), while small-scale patterns (e.g., stand-level structure of forest patches) also scale up to large-scale processes (e.g., fire behavior and resulting emergent properties of landscape composition and arrangement, Hessburg et al. 2013). Policy and management responses to coupled ecological pattern-processes span vast spatial scales as well (table 13.1; Ban et al. 2013). Understanding cross-scale patterns and mechanisms of linkages across spatial scales will be critical to sustain ecological systems in the age of climate change. Working across scales requires an appreciation for different processes—ecological and social—that operates at scales spanning orders of magnitude (table 13.1).

The Southwestern Crown of the Continent effort provides a concrete example of the outright necessity of collaborating and coordinating across nested scales to sustain ecological functions across geospatial scales, as the SWCC continues to prototype much of the science and implementation for climate adaptation that is being scaled up throughout the entire ecosystem. Consider, for example, the largest landscape-scale collaboratives in the region—Yellowstone to Yukon Conservation Initiative, the Great Northern Landscape Conservation Cooperative, and various Crown of the Continent coalitions (see figure 13.2). These groups generate much of the regional, scientific vision for sustained ecological functions across twenty-five latitudinal degrees of topographically complex, mountainous ecosystems and work collaboratively to establish and share data about the impacts of climate change and other stressors. Attributing phenomena to climate change impacts may be detectable only at regional scales (e.g., van Mantgem et al. 2009). Thus, regional monitoring programs—and their associated costs—may require science consortia (e.g., Climate Science Centers) or national programs (e.g., National Phenology Network). Also, opportunities to attract and match funding are often highest at very large scales and are most challenging at the smallest scales. Appreciation and funding for this vital function is often lacking.

TABLE 13.1. Examples of nested scales where key patterns and processes occur in ecological and sociopolitical realms. Understanding impacts of global changes at each scale and mechanisms that operate across scales is needed to sustain ecological services and conserve biodiversity in the Anthropocene.

Spatial Scale	Area (hectares)	Ecological Process Example	Sociopolitical example
Global	51,000,000,000	Water, carbon, energy cycling; global climate variability	G8 Global Summits on climate change and carbon emissions; geopolitical treaties and trade agreements
Bioregional	100,000,000	Long-distance animal migrations, river basin hydrology, continental-scale climatic influences	River basin compacts, Canadian Boral agreements, Landscape Conservation Cooperatives, Yellowstone to Yukon
Regional	1,000,000	Regional populations and genotypes of species	Forest Service Planning and Planning Rule
Landscape	100,000	Habitat composition; contagious landscape processes (fire, insects, spread of invasive species); large animal (e.g., grizzly bear) home ranges	Collaborative Forest Landscape Restoration Program
Watershed	1,000	Hydrologic function; home range for small animals	Watershed Condition Class Framework
Stand level	100	Local habitat for animal foraging and nesting; maintenance of tree diversity and local restoration and local disturbance dynamics; seed dispersal	Local Forest Service districts and local restoration committees
Local level	0.1	Regeneration niche; interactions between individuals (e.g., competition mutualisms)	Contractor decisions and work; monitoring common stand exams; local restoration committee field trips

At the other end of the spectrum are those projects and groups operating at smaller geospatial scales than the SWCC (e.g., local restoration committees, the Montana Legacy Project, Forest Service Districts, and individual timber contractors). Despite the significant amounts of time and effort required to coordinate with each group, these collaborative efforts have turned out to be absolutely critical given that management control is highest at more local scales. From stand-level treatments to district-level project planning, this is the scale at which an extremely detailed knowledge base of the threats and opportunities for treatments exists, and at which managers subsequently

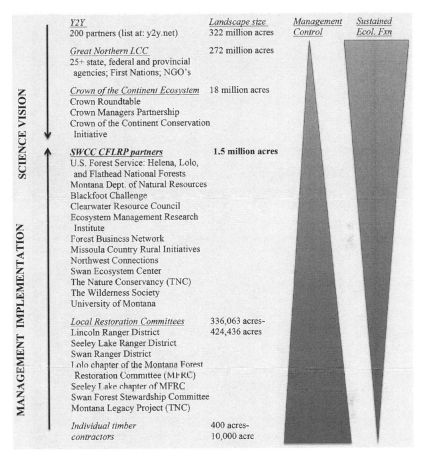

SCIENCE VISION / MANAGEMENT IMPLEMENTATION		*Landscape size*	*Management Control*	*Sustained Ecol. Fxn*

Y2Y
200 partners (list at: y2y.net) — *Landscape size* 322 million acres

Great Northern LCC
25+ state, federal and provincial
agencies; First Nations; NGO's — 272 million acres

Crown of the Continent Ecosystem — 18 million acres
Crown Roundtable
Crown Managers Partnership
Crown of the Continent Conservation
Initiative

SWCC CFLRP partners — **1.5 million acres**
U.S. Forest Service: Helena, Lolo,
and Flathead National Forests
Montana Dept. of Natural Resources
Blackfoot Challenge
Clearwater Resource Council
Ecosystem Management Research
Institute
Forest Business Network
Missoula Country Rural Initiatives
Northwest Connections
Swan Ecosystem Center
The Nature Conservancy (TNC)
The Wilderness Society
University of Montana

Local Restoration Committees — 336,063 acres-
Lincoln Ranger District — 424,436 acres
Seeley Lake Ranger District
Swan Ranger District
Lolo chapter of the Montana Forest
Restoration Committee (MFRC)
Seeley Lake chapter of MFRC
Swan Forest Stewardship Committee
Montana Legacy Project (TNC)

Individual timber — 400 acres-
contractors — 10,000 acre

FIGURE 13.1. Levels of partnership and coordination required across nested geospatial and temporal scales for one large landscape project in Montana. The SWCC CFLRP provides a real-world example of the types of capacity and coordination required to successfully manage large landscapes in the Anthropocene, given that management control is highest at small geospatial scales, while sustained ecological function and connectivity are most effectively addressed at extremely large geospatial scales.

implement treatments on the land or intervene to manage wildlife populations that range across public and private lands.

In the SWCC, treatments are still designed as traditional Forest Service–led projects conducted within one of three Forest Service Districts. Project boundaries are typically about 250 acres to 2,500 acres (100 ha to 10,000 ha)

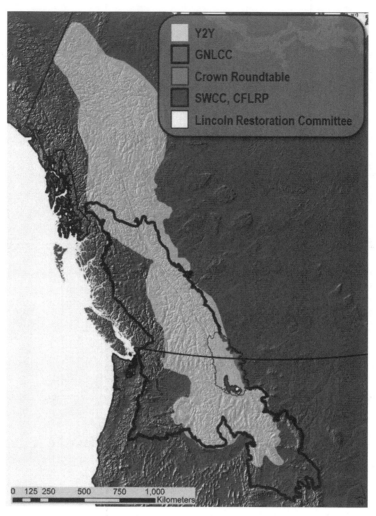

FIGURE 13.2. Map illustrating the geospatial scales associated across vertical nested conservation scales. Yellowstone to Yukon (Y2Y) connects conservation planning across bioregions. The Great Northern Landscape Conservation Cooperative (GNLCC) networks federal and state agencies and nonprofits to handle complex conservation challenges across large regions. The Roundtable of the Crown of the Continent coordinates all stakeholders in the region. The Southwestern Crown of the Continent works to coordinate collaborative design and monitoring of landscape forest restoration. Nested within the SWCC, the local Lincoln Restoration Committee helps design on-the-ground projects, including use of experimental design of treatments as a way of implementing a science-based portfolio approach to restoration forest planning.

in size, though not all areas are treated in the larger project boundaries. Treatments are still typically applied to stands ranging in size from about 5 acres to 250 acres (20 ha to 100 ha), and projects are still designed by agency specialists. However, increasingly collaborative input has become more influential in designing projects and their treatments. While projects are still designed with "stands" as the primary unit of treatment, placing treatments in the context of landscape processes increasingly occurs. Landscape modeling tools have provided an idea of landscape characteristics within a historical range of variability as well as predicted landscape-level effect of treatments on landscape fire behavior and resulting landscape composition and arrangement. More work is needed to connect scales from less than 250 acres (100 ha) patterns in stands to processes (e.g., fire, wildlife movement) operating at much larger (e.g., 2,500 acre, or 10,000 ha) scales, but the collaborative effort continually revisits the question about landscape function, rather than mere stand-level structure and composition.

Consideration of nested spatial scales may help move forest management beyond stands, but climate change also requires a consideration of temporal scales beyond traditional harvest rotation and schedules. A consideration of the "lifespans of treatments" now being implemented similarly should be factored into economic and ecological decisions, including plans for adjusting management decisions following monitoring and evaluation of data. Re-entry into stands and landscapes may be required to sustain initial restoration and adaptation investments. Implementation of adaptive management strategies usually implicitly considers time, because future decisions should be adjusted as new information and understanding become available now. Global changes require that actions and policies implemented today consider an uncertain future marked by altered climatic regimes and shifting species ranges, while anticipating ecological surprises (Williams and Jackson 2007). Perhaps most important, it is becoming apparent that some issues may be best addressed at very large scales (e.g., long-term planning, development of scientific data sets or tools for assessment of connectivity, monitoring of climate change impacts, funding) while other matters may be best coordinated at much more local scales (e.g., prioritization of on-the-ground projects, decisions about which scientific data sets and tools to use in informing project development, etc.). As described above, conservation biologists have understood for decades that protected areas with boundaries may not sustain biodiversity because (1) global changes impact "protected" areas, and (2) populations of animals and plants need more room to move and maintain genetic diversity. Addressing the second issue by working across land management jurisdictions remains one of the most challenging elements of landscape conservation. Lands adjacent to conservation reserves may enhance core regions for sustaining biodiversity or serve as regions of connectivity, especially as climate change shifts the geographic distribution of habitat. Ecologically compatible land use approaches across patterns of land ownership have been labeled *matrix conservation* (Noss 1983). In other words, matrix conservation considers protected areas to be embedded in a landscape matrix of land uses. UNESCO's Man and Biosphere Program recognized this issue beginning in the early 1970s by advancing the implementation of landscape-scale conservation with ecologically intact core areas surrounded by gradients of increasing human use buffer zones. Noss and Cooperrider (1994) and Soulé and Terborgh (1999) improved on this design by advancing the concept of ecological connectivity or corridors between core-protected areas, thus creating an interconnected ecological network of protected areas.

Ecological connectivity has become a major element of large-scale landscape conservation and is defined as the degree to which the landscape facilitates movement processes across habitat patches on multiple spatiotemporal scales (Taylor et al. 1993). Over individual lifespans, daily and seasonal movements among patches ensure access to required resources (Dingle 1996); over generations, dispersal maintains metapopulation structure and provides rescue effects from population extinction (Harrison 1994); over multiple generations, long-range dispersal sustains genetic diversity and the ability to respond to long-term trends, including climate change. Connectivity is now a major element in many revised State Wildlife Action Plans, the Western Governors' Association Wildlife Corridors Initiative, the US Forest Service's Forest Planning Rules, and the new national Fish, Wildlife, and Plants Climate Adaptation Strategy. Connectivity is an ecological characteristic of landscapes, but achieving connectivity requires that conservation scientists and practitioners work across political boundaries. Connecting people to connect landscapes is the only approach that can sustain conservation outcomes through the vagaries of political and fiscal cycles. Conservation across jurisdictions requires time-consuming, facilitated collaboration processes, however, to bring key conservation stakeholder interests together.

For instance, the Yellowstone to Yukon Conservation Initiative began as a bottom-up nongovernmental organizational effort to connect conservation efforts with similar goals across an ecologically defined and relatively intact region. Today, there are nearly a dozen Crown of the Continent–wide ecosystem-scale initiatives that span the US-Canada border and bring various stakeholder groups together from tribal nations, government, private land owners, businesses, watershed groups, local communities, universities, environmental educators, and the nonprofit conservation community.

A landscape-scale network of all the ecosystem-wide initiatives, known as the Roundtable of the Crown of the Continent, was established in 2007. The Roundtable has created an informal governance structure based on a charter of common principles and shared goals that establishes a framework for multijurisdictional landscape conservation and land management collaboration; its purpose is to facilitate multijurisdictional, large scale, climate adaptation implementation across all major land ownership communities across the entire ecosystem.

Even at the smaller nested scale of the SWCC, cross-jurisdictional work is required. Ecological (e.g., fire and animal movement) and social (e.g., fire management, recreation) processes operate across diverse ownership boundaries in the region (figure 13.1). Communication and collaboration among diverse jurisdictions from federal agencies to state lands to local landowners can be a challenge, but also offer great opportunity. Partnerships between groups, facilitated by local conservation groups, create the kind of information exchange needed for land stewards of various affiliations to respond to ecological impacts as climate changes (figure 13.1; see also Wyborn and Bixler 2013 for another regional example of partnerships across scales). Without cross-jurisdictional partners, social responses to conservation challenges and threats across spatial scales would be stymied.

Uncertain impacts of climate change require new approaches and strategies. A nested portfolio approach using elements of experimental design continues to build trust, sets up resilient landscapes by focusing on diversity and heterogeneity at various spatial scales, and may be a way of hedging against uncertainty (see below; Millar et al. 2007). The value of this approach is that it (1) is science based and will allow management adjustments to be conducted with strong inference and understanding; (2) spreads risk by not doing the same thing everywhere; (3) honors various perspectives and empowers collaborative stakeholders; (4) confronts uncertainty head on through the use of multiple treatments or experimentations; and (5) embraces uncertainty through humility. In the SWCC, we have designed two projects with a rigorous approach to experimental design (Larson et al. 2013b).

Using a robust experimental design, several projects of the SWCC will be implemented by turning diverse management perspectives into replicated treatments (Larson et al. 2013b). For instance, the best method for restoring and sustaining forested values in lodgepole pine (*Pinus contorta*) landscapes where mountain pine beetle has caused significant mortality is a controversial topic. Lodgepole pine forests are considered to have been maintained historically by stand-replacing fires. It remains an active area of research and controversy whether mountain pine beetle, climate change, fire exclusion, or their convergence have altered landscape structure and composition, putting ecological and social values at risk of regime-shift-inducing fires. Competing science and social perspectives have suggested that lodgepole stands and landscapes are either a very low or a very high priority for active management to restore structure and function. In situations of high scientific and social uncertainty, the SWCC and local restoration committee have begun designing a subset of projects as replicated experiments where various management options are viewed as experimental treatments.

Experimental approaches using stands and even small watersheds to replicate various treatments and monitor ecological responses helped move ecology from a descriptive to an experimental science (Bormann and Likens 1979). In addition, nesting experimental applications of a portfolio of approaches can be accomplished across spatial scales ranging from 0.1 ha to entire landscapes, while simultaneously accommodating the legal framework associated with different land designations (e.g., wilderness areas, roadless areas, etc.). This approach is consistent with a portfolio approach to managing climate risk (sensu Aplet and Gallo 2012). Such an approach would consider designated wilderness areas *observation zones*, where managers can accept and learn from climate-induced impacts. *Restoration zones* are areas where managers resist climate-induced changes by working to restore resilience to degraded lands in the face of climate change. Existing lands administered by federal, state, and local agencies outside of wilderness where nonclimate stressors impact ecosystems would be good candidates for assignment of restoration zones. Finally, *innovation zones* would allow managers to facilitate transition to novel ecosystems given expectations that these ecosystems will undergo large scale, climate-induced regime shifts (Aplet and Gallo 2012). The SWCC CFLRP project, for example, offers the opportunity to incorporate two of these three portfolio approaches at the landscape scale: the Bob Marshall Wilderness is an observation zone, in which managers are legally required (by virtue of the wilderness land designation) to manage this area minimally, while the SWCC CFLRP project area is a restoration zone, in which substantial intervention by managers could help reverse environmental degradation associated with a range of historic stressors and land use, thus sustaining key ecological values into the future.

Nesting experimental treatments of stands within treated watersheds and landscapes could help create a resilient landscape by implementing diverse approaches across scales while simultaneously creating a landscape set up to contribute to our understanding of best approaches in the Anthropocene. While not yet intentionally implemented by the SWCC, CFLRP projects offer a rare opportunity for pairing treated watersheds and landscapes with untreated controls. Untreated lands—where nature is left untrammeled— have come under increasing fire in recent years (e.g. Kareiva and Marvier 2012, but see Kareiva 2014) as pernicious threats of global change (altered climate, invasive species, altered nutrient loadings, acidification, etc.) have impacted ecosystems once regarded as pristine. A hands-off approach to ecosystem management was once held as the preeminent conservation strategy of conservation reserves. In the Anthropocene, it may be important for managers to intervene at the expense of untrammeled lands in exchange for the benefit of sustaining ecological patterns and processes upon which we depend. Does this new era render those reserves where nature is left untrammeled passé?

Wilderness and protected lands still constitute a viable conservation strategy in an age of shifting climate, as unmanaged wild lands serve many ecological and social purposes in rapidly changing conditions. Wilderness lands provide a benchmark by which to assess managed lands and various management strategies implemented in the nested portfolio approach described above. In fact, untreated control landscapes of about 250,000 acres (100,000 ha) may be regarded as part of the experimental portfolio approach to climate-adaptation project design. Uninterrupted or reestablished fire regimes and top-predator trophic interactions exist primarily within large unmanaged wild lands, and the presence of large predators on the land is strongly correlated with significantly higher levels of biodiversity in ecosystems around the world (Stolzenburg 2009; Terborgh and Estes 2010). We believe that the Anthropocene wilderness baby should not be thrown out with the Holocene bathwater.

In the SWCC, the unlogged forests in the Bob Marshall Wilderness where fire regimes have been reestablished in recent years provide a compelling case study of how untrammeled (or untreated) "control" lands can provide insights into appropriate restoration strategies in a managed landscape. Fire has returned to ponderosa pine, western larch, and mixed-conifer forests of gentle terraces above the South Fork of the Flathead River. Effects of fire in terms of mortality, recruitment, and composition of new trees, fuel loadings, spatial arrangement of tree clumps and gaps, and woody debris loads are currently being studied. These data indicate that some forest types may be more resilient to reestablished fire than once perceived (Larson et al. 2013a), while also providing insights into appropriate restoration treatments that could mimic nature's patterns.

Sustaining nature's parts and processes in the Anthropocene requires maintaining biological diversity across life's hierarchy of organization. Growing numbers of studies link ecological function across scales of biodiversity from genetic diversity (e.g., Crutsinger et al. 2006) to heterogeneity in the spatial arrangement of organisms (Larson and Churchill 2012) to landscape heterogeneity within and among ecosystems (Turner et al. 2013). Therefore, to sustain the processes upon which humans depend requires maintaining sufficient biological diversity across scales and levels of biological organization.

Biophysical diversity sets the stage to give rise to ecosystem and species diversity (Beier and Brost 2010), which occurs at various spatial scales (from diverse climatic regimes and landforms within and among continents to local edaphic and topographic effects). Local and landscape processes, such as species interactions and disturbance, further govern habitat and species diversity across more local scales. Understanding the patterns and processes that give rise to and sustain species diversity across spatial scales has been a cornerstone of ecology for over a century and remains an important research theme of the science. Shifting climatic regimes, altered atmospheric chemistry, and introduced species may profoundly influence patterns of biodiversity distributions and ecosystem function. Basic understanding of the mechanisms that govern distributions and abundances of species and patterns of biodiversity should still provide important insights into best conservation approaches to sustain biological diversity—in all its forms—in the Anthropocene.

Large landscape conservation is an emerging approach to address large-scale impacts to the ecological integrity of the planet. Conservation within a landscape context sustains ecological processes across an array of land jurisdictions and helps to align diverse land management approaches so that ecosystem benefits and services are optimized. All land has ecological potential depending on how it is managed. Restoration practice is a key element of resilient land management. Wilderness and protected areas enhance the resilience potential of lands, especially in the face of climate change.

"How much is enough?" is a question that has vexed conservationists since the beginning of the modern conservation era. This question has little meaning in the Anthropocene as the planet edges toward an ecological regime shift. Ecological processes that sustain nature and humanity are dependent on functional ecology and the species interactions it consists of. The planet is now the scale of consideration and planning, and the solution agenda needs to mirror the global impact of humanity.

Large landscape conservation requires local societal efforts to reach toward management scales that are novel and often challenging. While vision may guide these large-scale efforts, social glue is required to maintain and cement them over time. New approaches to conservation need to be prototyped. In the Southwestern Crown of the Continent Ecosystem, a 1.5 million acre landscape within the much larger Yellowstone-to-Yukon region, we are working with our many partners to prototype such an approach. While relatively young in its inception, the Southwestern Crown of the Continent Collaborative is testing the following elements of large landscape conservation. First, large landscape conservation is an approach nested within larger and smaller scales of science and implementation. Vertical integration of scales of action is needed and requires intensive work to connect individuals, institutions, and resources to perform this function. All land has ecological potential, even though all land has mixed ownership. For instance, the Southwestern Crown of the Continent Collaborative is represented at larger scales of action through the Roundtable of the Crown of the Continent, the much larger US Great Northern Landscape Cooperative, and the even larger Yellowstone to Yukon Conservation Initiative. At the same time, the SWCC embodies smaller-scale initiatives such as the Blackfoot Challenge, three US Forest Service districts, and various local communities.

Second, large landscape conservation is a fusion of the spatial and temporal aspects of ecology and those of human society. Multijurisdictional-facilitated processes are the new norm for conservation. Collaborative approaches to science and management that include stakeholder engagement and participation are essential. Professional conservation approaches need to embody the "servant leadership" approach that empowers stakeholders as conservation practitioner partners. Mechanisms that foster societal trust are essential to the success of these efforts.

Third, large landscape conservation will have a broad array of governance designs ranging from formal to informal approaches. The work in the Crown of the Continent suggests the role of network governance among stakeholder groups. The Southwestern Crown of the Continent Collaborative has developed a multi-stakeholder project implementation roundtable structure. A larger roundtable structure exists in the Crown of the Continent to bring all ecosystem-wide efforts and stakeholders together. While a common set of principles and an organizing charter serve as collaborative touchstones for this coordination, these roundtable efforts are an example of network governance.

And finally, large landscape conservation science and monitoring integrates formal and informal information processes from rigorous experimental methods to traditional ecological knowledge. Interdisciplinary science is an essential element of this work. Science and monitoring should embrace a networked science approach where science, monitoring, metadata, and local information are handled in transparent and accessible fashions. This includes enlisting all stakeholders in the practice of science and monitoring.

14

Climate Change Effects on Forests, Water Resources, and Communities of the Delaware River Basin

WILL PRICE AND SUSAN BEECHER

The mainstem Delaware is the longest undammed river east of the Mississippi, flowing freely for 330 miles from southern New York (2,362 square miles or 18.5% of the basin's total land area), through eastern Pennsylvania (6,422 square miles or 50.3%), New Jersey (2,969 square miles, or 23.3%), and Delaware (1,004 square miles, or 7.9%) to the Atlantic Ocean. The Delaware River's 13,539 square mile watershed drains only about 0.4 percent of the United States land area yet supplies drinking water to 5 percent of the US population—some 16 million people in four states. The Delaware River basin (DRB) also supports the largest freshwater port in the world within the 782 square mile Delaware Bay. Three reaches of the Delaware River, about three-quarters of the nontidal river, are included in the National Wild and Scenic Rivers System (Kauffman 2011).

In 2010, over 8.2 million residents lived in the basin, including 654,000 people in Delaware, 2,300 in Maryland, 1,964,000 in New Jersey, 131,000 in New York, and 5,469,000 in Pennsylvania. An additional 8 million people in New York City and northern New Jersey receive their drinking water through

DOI: 10.5876/9781607324591.c014

interbasin transfers from upper Delaware River reservoirs. Between 2000 and 2010, the population in the DRB increased by 6.1 percent. Over the last decade, a number of counties in the basin showed double-digit population increases and Philadelphia gained population for the first time in centuries (Kauffman 2011). Flows from upper tributaries are partially controlled by dams, holding back 238 billion gallons in two reservoirs supplying the city of New York, for an average daily detention and diversion of 665 million gallons. The remaining 90 percent (land area) of the watershed downstream of the NYC reservoirs is controlled by the weather, and serves the needs of another 7 million people.

Approximately 64 percent of the approximately 8.6 billion gallons withdrawn daily from the Delaware River is used for thermoelectric cooling. Typically less than 5 percent of cooling water is actually consumed; the rest is discharged back into the basin. Less than 20 percent is used for drinking water—an average of approximately 665 million gallons per day is allowed for diversions to the city of New York and to New Jersey, and the rest (860 million gallons per day) is diverted for residents throughout the region, many of whom live outside the basin's boundaries. In recent years, per capita water consumption has declined with the increased efficiency in all sectors and the declining productivity associated with the economic downturn. Declines in water consumption are offset to some extent by increased deployment of closed-cycle cooling (CCC) for electricity generation. This type of cooling reduces total water withdrawn but increases water consumption, as more water is lost from closed-cycle systems. Despite improved efficiency, a projected increase in population within the DRB is predicted to result in a gradual increase in total water consumption for drinking water, industrial processes, and energy generation (PDE 2013).

A recent study estimates the annual regional economic value of water supply at $25 billion (Kauffman 2011). Estimates in this study do not include the embodied water content and energy in goods and services exported to the world from the New York, Philadelphia, and Wilmington corridor—a region containing the first- and sixth-most-populous metropolitan areas in the United States.

That the entire main stem of the river is free flowing and is principally managed by the storage and release of water in the very upper reaches of the watershed, with implications for the annual water availability to New

York City, makes the system vulnerable to weather extremes. Periods of low flow can have several interrelated effects. Drought conditions became the first test of the federal compact between New York, New Jersey, Delaware, and Pennsylvania, and resulted in a 1978 "Good Faith Agreement" that was developed and implemented by the Delaware River Basin Commission and designed to protect aquatic life by maintaining minimum flows through reservoir releases (DRBC 1982; Albert 1987). Drought conditions and extended periods of low flow pose many other problems for water use in the basin, not the least of which is allocation for drinking water and energy generation. Energy generation—the largest category of water withdrawal—is limited by temperature requirements set for the river downstream of facilities. The Schyulkill Restoration Fund in the Schyulkill River watershed (a major river flowing into the Delaware River) was created through an agreement that also includes augmentation of "pass-by flows" for a nuclear generation station, which had at times exceeded temperature limits.

Perhaps one of the biggest problems posed by low flows is the upstream movement of the "salt-line." The Delaware River has the largest freshwater tidal prism in the world, as the Delaware Bay's depth and "cone-shaped geometry" allow for more than 100 miles of tidal exchange upriver. Freshwater flows normally impede tidal advance. However, low flows can allow high chloride levels (> 250 ppm) to reach drinking water and cooling intakes within the tidally influenced portion. Floods are also a concern in the DRB, especially the potential for lower-basin flooding resulting from a combination of coastal storm surge and floodwaters from upstream—a scenario that was narrowly missed during Hurricane Sandy in 2012, which still devastated lower portions of the basin. Prior serious flooding in 2004 and 2006 led to the development of a Flexible Flow Management Program (FFMP), which balances drinking water needs for New York City and the rest of the basin, energy generation, ecological requirements, and upstream movement of the salt-line (Gong et al. 2010).

Changing climatic conditions are already being felt in the upper Delaware region. Annual mean temperature and annual mean precipitation in the upper basin have increased significantly over the past 100 years. The trend over the past 30 years for temperature and precipitation is more than three and five times the 100-year trend, respectively. The number of days per year with heavy precipitation shows a significant upward trend. Future projections generally show the basin getting progressively warmer and wetter through-out the twenty-first century (Najjar et al. 2012).

Higher average temperatures, increased magnitude and frequency of heavy precipitation events, a longer growing season, warmer winters with more precipitation falling as rain, and changing hydrologic conditions all put multiple sectors at risk, including forests, water resources, agriculture, and human health.

Information on how these projected trends could affect both the quality and quantity of the water of the Delaware River is only beginning to emerge as global climate change models are downscaled to the region, and interpreted for the watershed. As described above, global models predict that the region encompassing the DRB will experience more precipitation, and greater variability and intensity of events (Najjar et al. 2012; McCabe and Ayers 1989). The magnitude of changes in streamflow is less clear. Milly and others have modeled the relative change in runoff over the next century under a number of scenarios, generating an "ensemble mean" of percentage increase in streamflow, which for portions of the northern mid-Atlantic states and New England constituting the DRB is projected to increase between 5 and 10 percent above 1900–1970 levels (Milly et al. 2005). Considering storage capacity in the upper basin, this alone may not be a problem, were it predictable from season to season and year to year. However, increased unpredictability of seasonal storms and the difficulty of predicting the pathway of large single events can cause problems. Some models also suggest that the net increase could be accompanied by more severe droughts, earlier snowmelt, and more intense precipitation events in late fall through spring—an increase in droughts and floods (Najjar et al. 2012). Similarly, the differences in projections for streamflow generated by the Hadley and CCC scenarios for the neighboring Susquehanna River basin illustrate the challenges in understanding just how to prepare for climate change. Both show increasing and earlier streamflow in the late fall and early winter, but differ in their predictions for the spring (24% increase for Hadley, and 4% decrease for CCC) (Neff et al. 2000).

Other model results more explicitly include the effect of increasing temperatures on forests, which shifts the story to some extent, and perhaps adds to the "dampening effect" that forests can have on floods. Huntington (2003) shows that in the east, forests are an important determinant of predicted streamflow based on temperature-related changes in evapotranspiration (ET). For every 1°C increase in mean annual temperature (MAT), ET increased 2.85cm, suggesting that with a predicted 3°C increase in MAT over this century; the annual reduction in streamflow in a New England forested watershed could reach 11–13 percent (Huntington 2003). Their results are annual averages, reflecting longer growing seasons, less snowmelt during spring green-up, and other seasonal dynamics. In the western United States, vegetative water demand is also predicted to further decrease water availability (Westerling et al. 2002). A suite of water quality changes will also likely occur due to climate change, and will vary depending on conditions. Predicted increases in precipitation could expand stream networks and the volume of shallow subsurface flow in forested areas of the watershed, mobilizing more nutrients and delivering them along with increased sediment to stream channels (Murdoch et al. 2000). In urban and exurban areas the increase in volume would mean more nonpoint source pollutants. At the same time, warming temperatures would increase microbial processing of nitrogen in forest soils and accelerate metabolic processes in-stream, resulting in reduced nitrate loading in source water, and dilution and increased assimilative capacity of inputs from all sources (Murdoch et al. 2000, 1991). The other possibilities for potential water quality impacts are too numerous to describe, and many are speculative and depend on what happens in forests, along streams, and in developed parts of the watershed.

How modeled trends balance out in the DRB has not been determined. Longer growing seasons extending into later months with more rain could result in more, less, or similar amounts of runoff, depending on the role of ET. Less forest cover would increase the runoff the combined result of less ET and infiltration—increasing the possibility of flooding, especially during hurricane season. Warmer summers with less rain and increased vegetative water stress in upper portions of the basin surely seem a recipe for more severe droughts. Water quality will be tightly correlated with the flow regime—how, when, and where nutrients and sediments are delivered to the system. The magnitude of all of these changes is uncertain. Calling these "amplified water-related extremes," Kundzewicz et al. (2002) reviewed the causes of major floods and droughts around the world, stating that "mechanisms of climate change and variability are intimately interwoven with more direct anthropogenic pressures." They go on to emphasize that global increases in the intensity of flood events are confounded, and often exacerbated, by changes that have already occurred in river basins.

What happens to forests will influence how a changing climate affects water quality and quantity. Coined the "forest water controversy," scientists and politicians have debated how forests influence water quantity and quality since the emergence of hydrology and forestry as fields of study (Andréassian 2004). The debate is not entirely settled in that studies still reveal conditions in which different forest types, physiographic characteristics, and weather patterns produce unexpected outcomes. To some extent much of what has been learned in particular watersheds now changes, especially for watersheds managed based on models using historical conditions.

In the DRB the effect of longer growing seasons, warmer temperatures, and seasonal changes in precipitation must be considered along with concurrent anthropogenic forces that will also affect forest ecosystems. Along with climactic changes, forests of the DRB are being lost at a rate of 100 acres (40 ha) each week. Forests of the region were largely denuded for agricultural and industrial uses during the 1700s and 1800s, leading to extreme floods. With regrowth through the second half of the twentieth century, in 2006 small watersheds (HUC12) in the upper portions of the DRB were more than 75 percent forested, but this is predicted to change. There are also indicators of long-term forest unsustainability: forests are generally even-aged and maturing; dominated by larger, sawtimber-sized trees; lacking in diversity; not fully stocked; and predominantly privately owned by an aging demographic. Additional nonclimate forest stressors include parcelization and fragmentation driven by population increases and changes in land ownership and land use. An array of diseases, insects, and invasive species are present in forests throughout the region. Regeneration is negatively impacted by white-tailed deer populations and forest harvesting practices such as "high-grading" and diameter limit cuts (Pennsylvania Department of Conservation and Natural Resources, Bureau of Forestry 2010; New York State Department of Environmental Conservation 2010; New Jersey Department of Environmental Protection 2010). Nonclimate stressors on water resources include population growth; associated land use and impervious cover changes; competing demands for water; and flow management practices that result in flow fluctuations, thermal stress to fish, and other ecological impacts. Natural gas drilling, not currently a factor in the upper

Delaware region due to a moratorium on drilling while the Delaware River Basin Commission develops regulations to address potential risks, could become a stressor to water quality and quantity in the near future.

As described by Kundzewicz et al. (2002), the effects of climate change are critically dependent on how anthropogenic pressures shape future conditions. In some cases, ongoing reductions in forest cover coupled with engineered storage (e.g. urbanized watersheds) may alleviate water scarcity in the face of increased droughtiness. The tradeoff is that with fewer forests the quality of the water declines and flows are more responsive to precipitation events. For regions in which climate change poses risks related to the increasing intensity and frequency of storms, it is conceivable that the dampening effect provided by increased infiltration in forests soils and ET later into wet seasons is perhaps one of the greatest services forests can offer. Forests in the DRB, as in many temperate systems, are the top consumptive use of water in the basin (Sloto and Buxton 2005), and are therefore a major factor in managing flows. The impact of climate change on the DRB's forests will regulate the change in evapotranspiration and rates of infiltration, and as a consequence the severity of floods and droughts.

To date, there has been no attempt to assess the combined influence of forest loss, ecosystem change, and climate change on water resources in the DRB. More troubling is that models developed for managing water allocation, determining permitted uses, and understanding flood probability have not yet accounted for significant land use and climate change effects. In addition, most studies of climate change and water availability have not taken into account the effects of competition, response, and adaptation to changes, factors that are critical in a basin like the DRB with its numerous and diverse forms of use. Hurd et al. (2004) used Water Allocation and Impact Models (Water-AIM) to "simulate the effects of modeled runoff changes under various climate scenarios." Their models allow for analysis of changes in pricing, patterns of water use, reservoir storage, and associated economic welfare, based on changes in supply driven by different climate change scenarios. Overall, their modeling predicts that negative economic impacts are mostly borne by nonconsumptive water uses that are dependent on instream flow (e.g. thermoelectric) and agricultural users. For the DRB these types of uses would also include ski areas, golf courses, and other amenities that are critical to the economy of more rural portions of the basin.

Given the certainty of changes in climate and forest cover, but the uncertainty of the magnitude of the difference this will make for water quality and quantity, a precautionary and conservative approach is perhaps the best watershed management strategy. Such a strategy should include conservation of forests that, based on available data, are most important for preserving water quality and affecting flows. A precautionary approach should also include climate adaptation planning that promotes conservation of forests and ecosystem resiliency, while preparing communities that will be affected by changes in quality and quantity of water resources. Two cases studies illustrate efforts in the DRB to pursue these two strategies. One, the Common Waters Fund (CWF) seeks to engage downstream water users in the protection and maintenance of forests that are most important for water quality and regulating flows, before they are lost or degraded. Second is the development of a climate adaptation plan for the upper basin region, which provides a roadmap for taking multiple actions that reduce the impacts and vulnerability of upstream and downstream communities and the forests and the river on which they depend. The CWF is one of more than seventy source-water protection funds or payment programs established around the country to maintain the quality and quantity of water resources on behalf of downstream beneficiaries (Bennett et al. 2013). The programs differ by origin and structure, many of which were launched by cities such as New York, Denver, and Seattle—all of which are investing in forest conservation. For example the New York City program, which also involves the DRB, emerged as an alternative to installation of additional filtration capacity by investing instead in forestland acquisition, easements, and stewardship—or the development and implementation of conservation-minded forest management plans (Pires 2004). Few of the models around the country have attempted to create a water fund/program in which the upstream protection priorities are predominately privately owned and span multiple political jurisdictions (i.e., states, counties, and municipalities). Fewer still have attempted to engage as many different kinds of downstream beneficiaries. However, many of the large watersheds of the eastern United States face similar challenges.

CWF was developed by public agencies, conservation groups, and individuals that had formed a partnership called Common Waters with the support of private foundations, the US Department of Agriculture (USDA) Forest Service, USDA Natural Resource Conservation Service (NRCS), and the US National Park Service. As a pilot, the CWF initiative seeks to protect source water through investments by downstream users (e.g. water purveyors, electricity generators, and water-intensive manufacturing) to manage future water resource risks. As an alternative, investments could also be mobilized through policies enacted on behalf of all stakeholders. The CWF demonstrates approaches that can help meet the challenge of managing risks and protecting water resources at the watershed scale: (1) developing an integrated program with the buy-in and capacity to work across a large geography with multiple political jurisdictions; (2) incorporating all readily available peer-reviewed scientific information to set consensus watershed protection priorities; and (3) engaging a diversity of water users who share a common resource.

The partnership that created the CWF formed as a collaborative for sharing information and pursuing joint initiatives that would help protect the Delaware River and forests of the region, which are considered essential to the economy and quality of life for a 3 million acre area encompassing portions of New Jersey, New York, and Pennsylvania. It was modeled after the Chicago Wilderness, and eventually developed a formal mission and voluntary self-governance structure that permitted the active engagement of a broad spectrum of interests (Helford 2000). Members included public land management agencies at the state and federal level, whose participation in the development of the CWF program for landowners helped ensure it would meet federal and state requirements (i.e., state stewardship and tax incentive programs for New Jersey, New York, and Pennsylvania, and participation requirements for the USDA NRCS Environmental Quality Incentives Program). Meeting these requirements meant that watershed protection projects involving stewardship planning and conservation practices could be implemented anywhere in the watershed, and would be familiar to partners working with landowners. Members also included land trusts engaged in working with landowners on the donation and sale of conservation easements. These organizations helped design CWF program requirements for permanent protection projects in priority areas, for which CWF paid transaction expenses. The collective capacity and expertise represented by the partnership was essential to the pace and scale of implementation of CWF, which at the end of the pilot period had enrolled approximately 50,000 acres (20,200 ha).

A precautionary approach implies that areas most important for maintenance of water quality and quantity are protected using the best information and means possible. For the CWF, this meant offering protection options amenable to private landowners at the time of enrollment (e.g. permanent easements, ten-year watershed stewardship plans, and/or conservation practices).

The CWF's precautionary approach also meant establishing priorities throughout the upper portion of the DRB based on the best available peer-reviewed science. Priorities were established by creating water resource priority tiers (0 to 4) that combined several data sets developed by Common Waters partners. These included the Natural Land Trust's *SmartConservation*™ data set (Cheetham and Billett 2003); The Nature Conservancy's priority conservation blocks; the USDA Forest Service's *Index of Forest Importance to Surface Drinking Water* (Weidner and Todd 2011); and data sets associated with the *Delaware River Basin Conservation Areas and Recommended Strategies* report (2011). In some cases CWF represented the first attempt to use these priorities for land protection. The use of combined data sets not only ensured that the highest priorities were targeted, but that there was broader agreement that CWF projects addressed goals held by participating organizations. Concurrent with creating the CWF program and initial investment in protection projects, Common Waters engaged different types of water users, mostly located in the lower portion of the basin where the majority of the electricity and drinking water demand is located. Delaware River surface water is delivered to more than 16 million people by more than 100 water purveyors, whose water withdrawal is regulated by the Delaware River Basin Commission (DRBC). As described above, more than one-half (65%) of the average daily withdrawal (8,650 MGD) is nonconsumptive use for cooling in energy generation—whose withdrawal and discharge is also regulated by the DRBC. Drinking water purveyors include public (municipal) utilities, and publicly and privately held corporations. Electricity generators are mainly publicly held corporations. There are also major beverage producers and bottling facilities, pharmaceutical companies, and manufacturers with headquarters and/or facilities located in the DRB. All depend on DRB surface waters, or in some way face risks posed by changes in quality, floods, and droughts. Of these different kinds of water users Common Waters met with the largest consumptive users and representatives of each kind of use/industry, for a total of twenty-six organizations. The purpose of the meetings was to learn how users perceive their own business risks related to water resources, assess readiness to consider investing in source water protection, and determine the information that would be necessary to justify investments. As of 2013, two companies have made some investment in CWF, in support of science activities that would help better predict and assign economic value to changes in the water quality and water quantity. Better information linking climate change and forest loss with hydrology and chemical quality will be essential for identifying and valuing the proportional benefits of source water protection in the DRB.

Common Waters joined with the Model Forest Policy Program and a network of rural forested communities working collaboratively across the nation to develop a Climate Adaptation Plan specific to the tristate upper Delaware region. The plan examines how environmental changes associated with climate could affect forests, waters, people, and economies of the region, and recommends strategies for adapting to these changes. The planning area included portions of Monroe, Pike, and Wayne Counties in Pennsylvania; Sussex and Warren Counties in New Jersey; and Delaware and Sullivan Counties in New York (Beecher et al. 2013).

To assess the potential impacts of climate change on the upper Delaware region and identify strategies by which communities might adapt and prepare, the planning group conducted an assessment and risk analysis for each sector: forests, water resources, and economics. A master list of current and potential climate risks was developed and consequences associated with those risks were ranked. The probability of each risk occurring and the ability of communities to respond were also part of the overall risk value assigned. The broad goals—education, local government policy and planning, landowner support, financial investment, mitigating business impacts, and flow management—identified to address key risks are summarized here. To reduce risks and build support for implementing solutions, generating dialogue and information exchange about climate risks was identified as a top priority for the upper Delaware region. While many of the region's residents have a general understanding of climate change as a future global problem, they might not make the connection with impacts happening in their communities now or, if they do, they don't know what can be done about it. Raising the awareness level about climate risks in the region will have the added benefit of building understanding about what it will take to reduce greenhouse gas emissions (mitigation). This is important since the ability to adapt will likely be limited if the pace of climate change continues on its present course.

In considering the findings of the risk assessments, analysis, and prioritization, it is clear that risks to the region could be reduced significantly through implementing land use policies that maintain existing forest cover, reduce forest fragmentation, maintain impervious cover at reasonable levels (e.g. < 10%), and take full advantage of the ecosystem services provided by floodplains and riparian corridors. Local governments have primary responsibility for the land use decisions that can ultimately make communities less vulnerable and more economically resilient to environmental changes. Although it is a challenge to coordinate land use policy in a region that includes three states, seven counties, and hundreds of municipalities, it has great potential for far-reaching climate resiliency benefits. To prepare for these changes, local governments can develop floodplain management policies that reduce flood risks and the substantial costs of emergency response, infrastructure damages, and property losses. Local governments can also incorporate what they know about climate change into updates of emergency plans, hazard mitigation plans, transportation plans, stormwater management plans, comprehensive plans, and other local planning efforts. Culvert sizing and bridge design standards should be examined and updated to account for changing precipitation patterns. Funding mechanisms should be identified to address the backlog of high-hazard dam maintenance and repairs, as these structures are vulnerable to increases in precipitation intensity and present a safety threat to downstream people and properties.

Management practices that improve the health and diversity of forests in the region are important to reducing forest and water stressors. With so many of the forests in the upper Delaware region under private ownership, landowners and the professional foresters that work with them are essential to enhancing forest resilience during an expected long period of climate change. Land trusts and a network of hunting and fishing clubs are also key partners in forest health initiatives, such as managing insects and invasive plants or supporting science-based deer population management that balances populations with sustainable forests and quality timber management. Collaborating with these groups and identifying funding to support management practice implementation are key strategies. Tax assessment policies that incorporate the value of ecosystem services provided by forest lands are another important tool to help landowners keep forests as forests.

Forests in the upper Delaware River watershed are essential to maintaining the extraordinary water quality of the Delaware River. The forests that keep water clean for the residents of the New York City metropolitan area are maintained by the NYC Department of Environmental Protection, a public, tax-dollar-funded authority. But the millions of people who live downstream and also depend on Delaware River water (Philadelphia, Easton, and Trenton) have no such centralized oversight of the forests on which their water quality depends. The CWF aims to fill this gap by funding stewardship and conservation by the private forest landowners in the upper Delaware region on whose forests the water quality of all downstream users depends. A permanent funding stream would include contributions from downstream users who enjoy the extraordinary water quality of the Delaware River and are willing to invest in its protection. Strategies that address climate change by conserving forest and water resources are also crucial to the region's economic vitality, quality of life, and natural and cultural heritage. Sustainable development does not represent a tradeoff between business and the environment but rather an opportunity to strengthen the synergies between them. The Climate Adaptation Plan recognizes the significant economic importance to the region of entrepreneurism, agriculture, tourism, and outdoor recreation, and the risks to these sectors—and to small businesses in general—of climate-driven extreme weather, hydrologic changes, and seasonal disruptions. Strategies that help manage impacts while identifying and capitalizing on new economic opportunities presented by a changing climate will be important to businesses in the region now and in the future.

There are many entities vying for upper Delaware water resources and few regional stakeholders directly involved in decisions about how that water gets allocated and managed. Given the hydrologic changes associated with increasing temperatures and the finite storage capacity in upper basin reservoirs, it is essential that flow management policies factor in climate change to ensure sufficient water quantity for human and ecological needs. There is much at risk with nonclimate- and climate-related stressors, but the upper Delaware region has the natural assets that can help reduce those risks: a high percentage of forest cover; private landowners with a stewardship ethic; clean water and healthy ecosystems; and institutional and organizational frameworks in place that could facilitate regional adaptation strategies. Translating the Climate Adaptation Plan to action represents an opportunity for the people and governing bodies of the region to prepare for a "new normal" set of environmental conditions while maintaining the health of the natural systems that sustain the quality of life and support the region's economic base.

Predicted changes in climate combined with anthropogenic pressures have implications for forests, water resources, and regional economies in the DRB. The Common Waters partnership has piloted approaches to avoid and adapt to these changes. The two case studies presented here—a source-water protection program (CWF) for landowners and a Climate Adaptation Plan for the upper DRB—represent strategies that could be models for watersheds elsewhere with highly diverse types of use and complex regulatory systems.

Evolving Institutional and Policy Frameworks

15

Policy Challenges for Wildlife Management in a Changing Climate

MARK L. SHAFFER

As one of the major biome types on Earth, forests are of fundamental importance to wildlife. (Note: the term *wildlife*, when used alone, functions as a shorthand for all species, plant and animal, living in an undomesticated state.) From the standpoint of species diversity, the most diverse terrestrial habitats on Earth are the great tropical rainforests. From the standpoint of sheer standing biomass, the great temperate rainforest of the Pacific Northwest may be unequalled. In terms of charismatic megafauna, which for most people are the face of wildlife, many signature North American species (e.g., deer, bear, elk, wolf, moose, cougar) are principally forest species. A large fraction of the lands American society has chosen to devote to conservation are forested lands.

Reflecting the importance of these environments, American society has developed a substantial institutional and policy framework for the management of its forests and wildlife (the US Forest Service and the National Forest

The findings and conclusions in this chapter are those of the author and do not necessarily represent the views of the United States Fish and Wildlife Service.

Management Act, the US Fish and Wildlife Service and the Endangered Species Act, state forestry and wildlife agencies and laws, etc.). This institutional and policy framework was developed in a period of relative biological stasis. The question now is whether this existing framework is adequate during a period of great biological change and, if not, what adjustments might be in order.

Climate is fundamental to biological systems. The interaction of temperature and precipitation is the major determinant of the distribution of biomes (e.g., forest, grassland, desert, etc.), which control the distribution of species dependent on those systems. Drastic changes in climate are thought to have been proximal, if not an ultimate, driver of past mass extinction events such as those that occurred at the end of the Permian and Cretaceous epochs (Twitchett 2006; Feulner 2009).

The earth is again entering a period of rapid climate change. According to the last National Climate Assessment (Melillo et al. 2014), measurements and observations show that, among other things, average air and ocean temperatures are increasing globally, the rainfall of the heaviest storms is increasing, extreme events such as heat waves and drought are becoming more frequent and intense, sea level is rising, and Arctic sea ice is shrinking.

Not surprisingly, many species are showing signs of changes in their distribution and the timing of major life history events (e.g., migration, nesting, emerging, blooming, etc.) consistent with a warming climate (Parmesan 2006). Some of these observed changes are signaling that additional climate change is likely to affect the ability of some of our conservation institutions and/or laws to achieve their stated objectives. For example, the namesake species of Joshua Tree National Park may no longer grow in that area in the coming decades (Cole et al. 2011). Moose, one of the signature species of Minnesota's North Woods—and a prime game species—are in a sharp decline that is thought to be related to increasing temperatures (Cusick 2012). The great western forest fires of the 2010s or so—fueled in part by temperature-mediated insect infestations—may, in some cases, result in a change from forest to shrubland and grassland ecosystems (Williams et al. 2010). Recognizing the emerging challenges of climate change for US wildlife resources, Congress in 2009 requested that the White House Council on Environmental Quality (CEQ) and the Department of the Interior (DOI) develop a national strategy to "assist fish, wildlife, plants, and related ecological processes in becoming more resilient, adapting to, and surviving the impacts of climate change" (CEQ/USDOI 2009). As DOI's wildlife bureau, the US Fish and Wildlife Service (FWS) took the lead in structuring a process to fulfill this request. Because of the complementary nature of US wildlife law, the FWS invited the National Oceanic and Atmospheric Administration (NOAA) and state wildlife agencies to co-lead the effort. Ultimately, a steering committee of representatives from fifteen federal agencies, five state fish and wildlife agency directors, and leaders of two intertribal natural resource commissions oversaw development of the National Fish, Wildlife, and Plants Climate Adaptation Strategy (NFWPCAS) (National Fish, Wildlife and Plants Climate Adaptation Partnership 2012).

The NFWPCAS is an unprecedented effort by all levels of government that have authority or responsibility for wildlife in the United States to work collaboratively to identify what needs be done in a period of rapid climate change. It was developed by teams of managers, researchers, and policy experts drawn from federal, state, and tribal agencies organized around major ecosystem types. The Strategy identifies seven major goals that must be achieved to give wildlife the best chance of surviving the projected impacts of current and anticipated future climate change (table 15.1). Numerous strategies (23) and actions (100+) are identified that are essential for achieving these goals.

All of the seven major goals identified in the NFWPCAS are things that the wildlife management community already does (e.g., conserve habitat, manage species and habitats, enhance management capacity, etc.). What will be new, and what the NFWPCAS illustrates, is that these things will need to be done in new ways, or in new places, or at new times, or in new combinations for conservation to be effective. In other words, conservation in a period of climate change will be equipped with the same types of tools, but they may need to be used in new ways. In some cases (as in policy adjustments), the existing tools themselves may need modification or even replacement.

The NFWPCAS includes one recommended strategy and seven actions that are policy focused. The major policy-focused strategy (3.3) is to "review existing federal, state, and tribal legal, regulatory and policy frameworks that provide the jurisdictional framework for conservation of fish, wildlife, and plants to identify opportunities to improve, where appropriate, their usefulness to address climate change impacts." This recommended strategy is further broken down into seven specific actions that focus on the following:

TABLE 15.1. Goals of the NFWPCAS

Goal 1	Conserve habitat to support healthy fish, wildlife, and plant populations and ecosystem functions in a changing climate.
Goal 2	Manage species and habitats to protect ecosystem functions and provide sustainable cultural, subsistence, recreational, and commercial use in a changing climate.
Goal 3	Enhance capacity for effective management in a changing climate.
Goal 4	Support adaptive management in a changing climate through integrated observation and monitoring and use of decision-support tools.
Goal 5	Increase knowledge and information on impacts and responses of fish, wildlife, and plants in a changing climate.
Goal 6	Increase awareness and motivate action to safeguard fish, wildlife, and plants in a changing climate.
Goal 7	Reduce nonclimate stressors to help fish, wildlife, plants, and ecosystems adapt to a changing climate.

1. Incorporating the value of ecosystem services into habitat protection and restoration

2. Developing or enhancing market-based incentives to support restoration of habitats and ecosystem services

3. Improving compensatory mitigation requirements

4. Improving floodplain mapping, flood insurance, and flood mitigation

5. Identifying existing legal, regulatory, or policy provisions that provide climate change adaptation benefits

6. Providing appropriate flexibility under the Endangered Species Act to address climate change impacts on listed species

7. Addressing sea level rise

Many other strategies and actions, although not focused specifically on policy, raise policy issues. For example, Action 2.1.8 is to "utilize the principles of ecosystem based management and green infrastructure." Depending on the specific context for using these principles, new policy might be required.

As it stands, the NFWPCAS comprises a very long to-do list of the many things the wildlife managers need to undertake to fully come to grips with the challenge of climate change to their mission and the resources for which they have authority and responsibility. Rather than rehash this extensive list, it may prove more informative to consider the underlying themes in the

TABLE 15.2. Guiding Principles of the NFWPCAS

1	Build a national framework for cooperative response.
2	Foster communication and collaboration across government and non-government entities.
3	Engage the public.
4	Adopt a landscape/seascape based approach that integrates best available science and adaptive management.
5	Integrate strategies for natural resources adaptation with those of other sectors.
6	Focus attention and investment on natural resources of the United States and its Territories.
7	Identify critical scientific and management needs.
8	Identify opportunities to integrate climate adaptation and mitigation efforts.
9	Act now.

Strategy and to identify a few of the major challenges for wildlife conservation policy going forward.

Prior to the formal launch of the NFWPCAS development process, FWS held several Conservation Leadership Forums to convene representatives from other agencies, other levels of government, and the academic and non-governmental communities to consider the climate change challenge and to develop appropriate responses. From those meetings emerged nine guiding principles that were used in development of the NFWPCAS (Table 15.2). These guiding principles are reflected in so many of the NFWPCAS goals, strategies and actions that they suggest four broad themes for wildlife adaptation efforts:

Be Inclusive and Collaborative. Climate change is so pervasive, and its impacts potentially so far reaching, that no single agency, no single level of government, indeed no single sector will be able to mount an effective response on its own. All affected agencies and interests need to be at the table working collaboratively to be effective.

Think, Plan, and Act at the Right Scale. The days are over of believing that a single set of best management practices universally applied will automatically lead to a biologically functional landscape. Different agencies and organizations work at different scales. Entities that operate at the local scale need to do so in the context of the broader physical, biological, and institutional landscape of which they are a part. And entities that operate at the national or regional scale need to be mindful of the needs, realities, and differences of the many landscapes in which they operate.

Integrate across Sectors. A corollary of being inclusive within the conservation sector is also to be inclusive of other sectors. Much of what governs the fate of wildlife is not the actions or inactions of the wildlife management community, but actions by other sectors that affect the natural world (e.g., agriculture, transportation, energy development, construction, etc.). Starting an adaptation planning process by including everyone and everything may be too large a burden for any sector to bear, but once each sector has a working understanding of its needs relative to adaptation, it needs to reach out to the other sectors relevant to its interests to identify commonalities, synergies, conflicts, and resolutions.

Engage, Communicate, and Act. The effects of climate change on species are beginning to be readily apparent. Because projections of future conditions and impacts come with great uncertainty, it is tempting to wait until more is known and the models are better, so that there is less uncertainty before we act. Unfortunately, like many large systems, Earth's climate has great inertia, and once change is entrained it will not be quickly or easily restrained. There is unequivocal evidence that the climate is changing, that the underlying cause is the growing accumulation of greenhouse gases (GHGs) in the atmosphere resulting from human activity, and that there is no plausible institutional or policy framework in place to restrain additional GHG emissions that will increase the impacts on wildlife. Species are already responding; it is time for the wildlife management community to engage, communicate, and act on what we do know, even if the rates and patterns of change and the future status of species and communities remain uncertain.

Achieving the goals of the NFWPCAS will in many instances require having the right policies. As noted above, the NFWPCAS has one major strategy and a number of actions focused on or related to having the correct conservation policies. Whether existing or new, these policies will need to be developed and employed in the face of several emerging realities about wildlife conservation in a period of climate change:

No Precedent. Depending on how it is defined, wildlife management is a few hundred to a few thousand years old. The best global circulation models are now projecting that if GHGs continue to accumulate at current rates, average global temperatures will by 2100 reach levels that have not occurred for millions of years (Houghton et al. 2001). Wildlife management, either as primitive practice or modern profession, has not seen such a period of change in its history. There is no precedent, no body of knowledge derived from experience to underpin wildlife conservation policy for a period of rapid climate change. Nor can we replicate the earth to take an experimental approach to discover the best way forward. Policy can be defined as a rule for decision-making. Many of the decisions the conservation community will have to make in the coming decades will have to be made in unprecedented circumstances. It will be a time of trial and error and wildlife conservation policies will need to be cast in flexible terms to acknowledge and adjust to that uncomfortable reality.

Unknown Destination. The world is not moving from the current climate to a new climate; it is leaving the current climate, with no fixed destination. Novel climates will emerge, presenting species with new combinations of temperature and precipitation that they may not have experienced before in their individual evolutionary histories (Williams and Jackson 2007). This makes adaptation planning and policy formulation even more challenging. Plans and policies need target conditions around which to be formulated. Even if the effects of each 1°C increase are modest by themselves, their impacts will likely prove ecologically cumulative. Having multiple degrees of temperature increase means having multiple or at least iterative plans and policies that are, perhaps, very different. Wildlife conservation plans and policies will need to be revisited regularly in light of the emerging trends in GHG accumulation and the resulting level of projected climate change.

Species Shift, Communities Change. There are many unknowns with regard to the response of living systems to climate change. One thing that is known with some certainty is that in past periods of climate change species responded individually and not as tightly integrated communities. In other words, each species shifted its range in its own way at its own rate and, therefore, the co-occurring assemblages of species that are recognized as natural communities changed in composition (Hunter et al. 1988). This has profound implications for wildlife conservation planning and policy in an era of rapid climate change.

Many of our existing conservation plans use natural communities as coarse filters for conserving wildlife diversity (i.e., as proxies for habitat). The logic is that by identifying the range of communities and then conserving some of each, their constituent species will be maintained (Hunter et al. 1988). This coarse filter approach is often complemented by the use of a fine filter that is focused on the needs of certain individual species that may be of particular importance for one or more reasons (i.e., ecological, economic, social, cultural, etc.). The individualistic response of species to climate change is already becoming apparent. Some species ranges are beginning to shift (e.g., the Joshua tree). If natural communities are defined as all the species in a given area interacting together, then is the North Woods without moose still the North Woods?

The potential replacement of forests with shrublands and grasslands after the recent major fires in the American Southwest is perhaps an extreme example of community change (Allen, chapter 4, this volume). The larger message for the wildlife conservation and management community is that we are entering an era when effective conservation will hinge more than ever on understanding the needs of individual species. With more than 1,500 native taxa already listed as threatened or endangered in the United States and likely many more to come due to the impacts of climate change, this will be a major challenge. Even without considering climate change, a major study of the conservation status of US species (Stein et al. 2000) suggested that up to a third of our native species in major taxonomic groups (e.g., vertebrates, flowering plants, etc.) is at risk of extinction in the coming decades due to existing threats. A subsequent global analysis suggested that up to 35 percent of species could be at risk due to climate change (Thomas et al. 2004). Although no cross-comparison of the two studies seems to have been done, the conservative conclusion at the moment is that anywhere from 33 to 68 percent of native species could be at risk. That is perhaps an order of magnitude greater than the 1,500 species currently listed in the United States.

Wildlife conservation policies will need to recognize the essential independence of species in terms of their response to climate change. The sheer magnitude of species that may need to be managed suggests that conservation policies will also have to try and differentiate the relative importance of species for a variety of considerations (environmental, ecological, utilitarian, etc.)

Speed Kills. Perhaps the most challenging feature of the current period of climate change from an evolutionary perspective is its projected speed. The adaptive capacity of species is currently one of the great unknowns in projecting the future of wildlife in a changing climate. Some species may prove to have greater adaptive capacity than is currently anticipated, but the fossil record suggests that evolution is a relatively slow process even in geologic terms. Relying on a slow process in a period of rapid change may leave many species unable to keep up.

There are four basic responses of species to climate change: acclimation, relocation, adaptation (in the evolutionary sense), or extinction. *Acclimation* is a function of a species phenotypic plasticity and genetic variation to address short-term changes at any point in time. Although an assessment of the genetic diversity of a species may provide some insight into its ability to acclimate in the short run and to adapt in the long run, it is no guarantee of success in novel circumstances. *Relocation* is a function of behavior and selection pressure. For relocation to be a successful survival strategy, a species needs suitable habitat to which it might relocate, and the ability to reach that habitat.

Given the speed at which climate is projected to change, short-term acclimation and relocation are likely to be the principal mechanisms by which current species might endure this period of change over the short term. Thus, the most promising interventions for maximizing the retention of species diversity will be to provide a range of habitats and some level of biological connectivity across the landscape. Given the importance of population size to demographic survival and genetic diversity, the amount of each habitat type conserved will also prove important. The modern landscape is so fragmented from a biological standpoint that the managed relocation of species may prove necessary as a component of "functional connectivity" going forward. Determining what all of this means in operational terms will prove to be the heart of wildlife management's challenge for the next century. Wildlife conservation policies need to emphasize the retention of significant amounts of the variety of habitats across the landscape and their functional connectivity, including the possibility of managed relocation.

Friend or Foe. Invasive species are one of the major challenges to wildlife conservation (see Evans, chapter 2, this volume). As of 2000, they were ranked number two as a cause of species listings under the US Endangered Species Act (Stein et al. 2000) and their impact may have grown since that time. The conservation community is predisposed to see a species new to an area as a threat and to move to contain or eliminate it. With relocation as one of the principal responses of species to a changing climate, more and more species will be showing up in areas they have not previously inhabited. Will they be invasive? Should they be viewed as exotic? Or, are they the climate pioneers? Current policies on invasive species were not formulated with this situation in mind. A blind reaction to something new as a threat might actually work against one of the principal means by which species will attempt to adjust to climate change. Wildlife conservation policies will need to develop criteria by which to differentiate climate change pioneers from invasive species.

Climate Change Is Only One of the Problems. There are few, if any, natural communities that have not been impacted to some degree by nonclimate stressors. Habitat loss, fragmentation, and degradation have already taken a toll on the status of many US species (Stein et al. 2000). As mentioned earlier, the impacts of invasive species have also been substantial for many native species. Pollution, especially in the form of pesticides and chemicals that disrupt the endocrine systems of vertebrates are also a problem in some cases (Colborn et al. 1996). It is expected that climate change will not only be a threat in its own right through direct challenges to the thermal and moisture tolerances of species, but also by exacerbating these existing stressors. Consequently, one of the major goals of the NFWPCAS is reducing these existing stressors, the theory being that species will then be better able to cope with the additional pressures of climate change. This recommendation is a common theme in the wildlife adaptation literature (Mawdsley, O'Malley, and Ojima 2009; Heller and Zavaleta 2009). Wildlife conservation policies will need to be based on an inclusive and integrated consideration of species vulnerabilities and not simply their climate-related vulnerabilities.

Realism. Wildlife management is, in some sense, a misnomer. It is really about managing human behavior that affects wildlife rather than managing wildlife itself. The human activity of harvesting is managed so as to leave wild populations that can replenish themselves. Management, however, cannot make a species reproduce. In some circumstances, to retain the conditions that support a species, certain human activities that can alter land use might need to be foregone But management cannot make a species use a certain habitat or stay in a certain area. So it is with adaptation to climate change. Management cannot make a species adapt to climate change, but it can influence the human activities that will make such adaptation more or less likely. Human activity with regard to the use of fossil fuels has now reached a level that is entraining a directional shift in Earth's climate, and wildlife is responding. As a primary driver of biological systems, climate will always trump management. Management, at best, will be a tugboat that can only nudge a much larger ship in a hopefully useful direction. Going forward, wildlife conservation policies will need to be based on a clear-eyed assessment of their potential leverage to reach a desired outcome.

The NFWPCAS is a congressionally mandated, collaboratively executed attempt by US wildlife managers to come to grips with the challenge of climate change. It identifies seven major goals and numerous strategies and actions that need to be pursued to give wildlife the best chance of coping with the increasing impacts of climate change. Many of its recommendations are focused on, or are related to, the need to review current wildlife conservation policies in light of climate change. As the community begins that work in earnest, it confronts serious challenges related to the unique aspects of climate change, including (1) its uniqueness in human history; (2) the individuality of species responses in periods of change; (3) the speed of the changes projected to come; (4) the challenge of differentiating between invasive species and climate-change pioneers; (5) the interaction of climate change with existing stressors; and (6) perhaps most important, the disparity in the power of management interventions in the face of the scope and scale of climate's inertia and its impact on living things. This last point underscores the fundamental observation that climate change adaptation efforts cannot succeed without the curtailment of CO_2 emissions at some level. There will be no adaptation without mitigation.

16

Evolving Institutional and Policy Frameworks to Support Adaptation Strategies

DAVID CLEAVES AND R. PATRICK BIXLER

In a quote attributed to Gifford Pinchot, conservation is "the application of common sense to the common problems for the common good." To reach this ideal in the future, we will have to adapt the approaches we take in resource management. The new normal is that there is no normal and the process of implementing an adaptation process on national forests and other forested lands is just as important as the particular adaptation measures or adjustments that land managers select. The challenge will be more about how we go about the adaptation process than about particular adaptation measures or adjustments. Defining common problems and the common good must be a multi-stakeholder process that will require dynamic engagement with a focus on learning.

This is no easy task, and as human influences become even more pervasive on ecological systems, the role of leadership in aligning policies and institutions and helping change human behavior will become more critical. We need

The findings and conclusions in this chapter are those of the authors and do not necessarily represent the views of the United States Forest Service.

DOI: 10.5876/9781607324591.c016

a working theory to help us navigate the climate-driven challenges ahead. Without an institutionalized guiding framework, we can fail to recognize events as lessons and we may fail to characterize and measure progress, set goals, and reshape pathways of change. The theory of adaptation must promote flexibility and itself be adaptable, not a source of new dogma. It should help codify the advances that are occurring as managers and landowners experiment and struggle with changes already impacting their forested lands. It should include the seeds of its own adjustments, allowing the working theory to catch up with practice and innovation as well as scientific advances; if not, we risk defaulting to a passive or reactive model of adaptation, defining uncertainties always as downside risks and scrambling to minimize losses without seeing opportunities and lessons that change sends our way.

In a way, *adaptation* is a misleading term because it implies a reaction to the consequences of climate change once it has occurred. However, adaptation clearly has to be anticipatory and preventative. Interacting social and ecological systems are already in motion, we have to strive to understand them and apply active adaptive management to make adjustments. A working theory of climate adaptation should embody the following five principles: (1) avoid waiting for "complete" science that never arrives, (2) boost learning by blending science and experience, (3) create "controlled" disturbances to reduce irreversible costs and losses of inevitable disturbance episodes, (4) respond to stressor complexes—climate and nonclimate— rather than single hazards, and (5) buy lead time and reduce panic responses. We have few actual "climate change" decisions to make. We have thousands of forest management decisions that influence and are influenced by the changing climate. Our decision processes can be made more adaptive to these forces. Adaptive governance must orient management strategies toward system resilience and climate "smartness." To make climate-smart decisions, land management agencies must work to match analysis detail to the level of climate-sensitivity (Dilling and Lemos 2011); test expected outcomes of alternatives in multiple, plausible futures (scenarios; see Gartner et al., chapter 10, this volume); use information about relative vulnerability to stressors (climate and nonclimate together) in designing and choosing alternatives and in ranking actions; and challenge traditional assumptions about future change. These elements of climate-smart decision-making must be integrated with local knowledge. Our theory must also guide us in helping prepare citizens to deal with change and share in adaptation. This includes engaging citizens in knowledge-creation and sharing, and in recognizing the nuances of how science can and should directly inform policy. In dealing with climate stressors we are operating in a postnormal world where uncertainty and conflict are the norm (Funtowicz and Ravetz 1993). In this postnormal world, we must recognize that the linear conception of policy formation is flawed; more science is not guaranteed to lead to less uncertainty and political action. Facts need to be understood in the context of competing scientific understandings of climate change, a built-in uncertainty that is amplified by the various political, cultural, and institutional arenas within which science is carried out. Producing logically indisputable proofs about the natural world is inherently problematic. Therefore, we must seek broader sources of experience, information, and wisdom.

Engaging citizens in the use and application of *local ecological knowledge* (LEK), defined as knowledge, practices, and beliefs regarding ecological relationships that are gained through extensive personal observation of and interaction with local ecosystems, and shared among local resource users (Charnley et al. 2008), can help inform climate-smart decisions and active adaptive management. Incorporating such local knowledge helps co-create public expectations that are more realistic for the new realities of change. Grounding adaptation in local experience and knowledge can help guide public expectations and mitigate uncertainty.

Landscape-scale conservation is quickly becoming the "place" to test a working theory of adaptation, however, as an overarching approach to adaptation begs to be defined, refined, and pressed into service (see Tabor et al., chapter 13, this volume). Landscape conservation focuses on large spatial extents (i.e., interacting mosaics of ecosystems), spatial heterogeneity, and the influence of humans on landscape composition, structure, and function (Lindenmayer et al. 2008). Important, too, it replaces an equilibrium understanding of ecosystems with a more nuanced, nonequilibrium model (Holling 1973), taking into account the inherent dynamism and complexity of ecosystems across multiple spatial and temporal scales. Landscape-scale conservation relies on concepts and skills of collaboration, sustainable resource management, climate adaptation, and risk management. By focusing on large-scale resource management it fundamentally crosses many traditional administrative, jurisdictional, and natural resource boundaries (Sternlieb et al. 2013). It uses the scalability of the landscape itself to employ a range of risk management mechanisms that include, but are not necessarily limited to, spreading out exposures to moderate systemic risks; balancing diversity with the scales of operations needed for economic activity, regeneration success, and habitat connectivity; planned redundancy and preservation of multiple adaptive options; reserves from which to cushion shocks, restart regeneration, resist invasion; and building cushions for experimentation with emerging novel systems.

Our working theory will have to fully integrate landscape conservation as an orchestration mechanism to deploy collaboration, analysis, and social engineering to manage multiple, interacting hazards. In landscape-scale approaches to adaptation, the social, economic, and institutional elements can become parts of the "baseline" and the system's adaptive capacity.

Different institutional arrangements in landscape-scale collaborations influence important abilities to think, innovate, predict, anticipate, collaborate, and self-regulate. A wide variety of policies—environmental, land use, economic and taxation, estate, and others at federal, state, local, and organizational levels—all interact to influence the decisions of landowners and managers. They shape problem frames, goals, information availability, options, analysis requirements, risk postures, and other elements of adaptation decision processes. They also introduce their own sources of uncertainty and barriers to the processes of adaptation. These various policies may not be aligned to support adaptation, sending mixed or conflicting signals to land managers. They may be aligned all too well in the wrong direction, limiting flexibility for adaptive responses, presenting structural barriers and imposing transaction costs that discourage responses to new information or experience. We must adapt our policy mix to encourage adaptive behavior, blending existing and new policies to support a future of adaptation.

Sorely needed is a cohesive policy package and forest action framework organized around active adaptive management. It should integrate climate change mitigation goals (management of the forest carbon) with adaptive responses to climate impacts. Active forest conservation, restoration, and management are critical interventions in preserving and improving the ability of forests to uptake carbon, adapt to a changing climate, and provide associated ecosystem services such as water, wood products, wildlife habitat, biodiversity, and recreational opportunities.

A policy framework to support proactive adaptation in forest systems could be built on three archetypal actions—retention, restoration, and reforestation. *Retention* involves keeping forest as forests in the face of disturbance and land use pressures. *Restoration* involves repair and recovery of health to key system functions. *Reforestation* includes bringing or returning forest systems to unforested or forests degraded by abuse or disturbance.

Policies and initiatives to support these actions should focus on developing markets for ecosystem services, wood products, and carbon sequestration; facilitating public/private partnerships; establishing principles (rather than rules) for adaptive response; and setting priorities for treatment based on science-based observation and analysis. Markets enable action through economic activity by providing better information and assistance, reducing transaction costs and gridlock, and reallocating cost and risk-bearing. Partnerships tap into new sources of investment and human resources and assure diversity of perspective. Policies should support a diverse array of different types of partnerships, including research/management, public/private, interagency, landscape coalition, and supply chain partnerships. Changes to the policy mix should be formed more around principles than around new rules that tend to become rigid and expensive to monitor and enforce. These principles would reduce the influence of the "precautionary principle" and its "if in doubt, don't" interpretation in favor of a more realistic "cautionary action" principle. The focus would be on monitoring and analysis at appropriate scales across a wide range of actions.

In the Anthropocene, the "no-action" option should not be the universal standard. We should reframe problems to allow collaborative retreat from the old battlegrounds of "action vs. no-action." We need instead to create policies to give future decision-makers the capacity to adapt across a range of interventions. We must focus on how to wisely implement proactive management, create incentives for looking ahead, and wrestle with surprise and unintended consequence.

Some of the best thinking on policy and institutional change has been captured in the Resources for the Future report series, "Reforming Institutions and Managing Extremes: US Policy Approaches for Adapting to a Changing Climate" (Morris et al. 2011). The authors described how effectiveness for the future could be enhanced with the following:

1. Provide specific guidance for federal rulemaking.

2. Create connections and synergy with other policy areas.

3. Address inefficiencies in current federal legislative and regulatory policy.

4. Supply information and data to enable policy-makers to better understand risk and uncertainty.

5. Embed flexibility and responsiveness into management structures.

6. Address equity and social justice concerns.

Do institutions have resilience? Do they create resilience? Are some institutions too resilient or perhaps too rigid for the good of the systems they were designed to steward? Adaptive capacity is the ability of an individual, organization, or a species to adjust to changes, to moderate potential damages, to take advantage of opportunities, and to cope with consequences. In other words, adaptive capacity is the ability to manage or influence resilience. Institutional resiliency is the ability to internally organize and adjust not only to uncertainty but also to other manifestations of a changing climate: increased complexity and conflict.

Adaptive capacity derives from assets and resources (such as knowledge, networks, human capital) and governance mechanisms that enable the mobilization of resources to transform and adapt. Intangible attributes and behaviors are also critical capacities, including learning to live with change and uncertainty, nurturing diversity for resilience, combining different types of knowledge for learning, creating opportunity for self-organization toward sustainability, and alertness to patterns of change, especially those that challenge the underlying assumptions that drive current strategies and programs (Berkes et al. 2003).

It may be useful, if perhaps painful, to reflect on elements of our institutional approaches and structures, and how they are performing as the demands for adaptation grow. Authoritarian and top-down approaches are giving way to more self-organized arrangements that tap into local leadership and attachments to place. The newer institutional arrangements are more like evolving institutional ecosystems, and the roles of government agencies in these evolving structures are changing from authority and intervention to providing services and enabling self-organized solutions. It may be helpful in this period of institutional readjustment to consider what Elinor Ostrom (1993) and other social scientists have referred to as principles of institutional design. Ostrom's list of principles are organized around user participation in setting boundaries for use, equalizing costs and benefits, making collective choices about operating policies, monitoring and enforcement through grad-uated sanctions, resolving conflicts, and nesting work efforts. These insights might help us better deliver government services to support self-organized adaptation and resilience building.

These new institutional forms include (1) large-scale, place-based, citizen-led collaborations; (2) forms of ownership and management such as land trusts, community forests, nongovernmental owners, private timber invest-ment and management, and real estate investment trusts (TIMOs and REITs); and (3) government configurations focusing on delivering action-able science such as USGS's Climate Science Centers, USDA's Climate Adaptation and Mitigation Hubs, NOAA's Regional Integrated Science and Assessment Centers (RISAs); or convening stakeholders and science providers toward adaptation action, the most prominent being DOI-Fish and Wildlife Service's Landscape Conservation Cooperative (LCC) system. These new players are nestling into regional- and landscape-level adaptation efforts interacting with traditional institutional players. Existing institutions can be part of this evolution by removing lingering barriers to collabora-tive adaptation efforts. Adapting to change is an energy-intensive business. Landowners and managers cannot afford to waste energy sorting through confusing arrays of information, programs, and processes. More informa-tion—however well intentioned—can still create high transaction costs for people making adaptive adjustments. Organizations that provide adapta-tion services—information, technical, financial, and others—need to work toward "one-stop shopping" by organizing their multiple programs into packages that can be easily used and customized to local needs. The limita-tions of "silo" functional structures are becoming more evident. Not only do dwindling financial resources make it less feasible to maintain internal "empires," the inefficiencies of communicating across boundaries and the needs for rapid integration and flexibility by managers are combining to pressure organizations to dissolve functional boundaries. One of the most pernicious boundary-based barriers to adaptation is found in budget struc-tures. The ability to blend different sources of funds to accomplish adapta-tion objectives is becoming more critical, despite pressures to account for every dollar in its narrow program category. Adaptation-friendly budget structures include the Forest Service's new Integrated Resource Restoration (IRR) fund, which combines seven separate program budget lines, and the

Collaborative Forest Restoration Program (CFLRP), which funds large-scale projects with multiple budget codes.

Adaptation involves the need to deliberately consider multiple risks, develop options for managing them, and wrestle with the difficult decisions of who should pay. Institutions need to develop skill sets for diagnosing patterns of risk and intervening in the most cost-effective ways. Many individual stressors are actually linked through system functions and processes as well as their common ties to the changing climate, so it is becoming more important to understand and manage systemic or connected risks. Tradeoffs and costs to be incurred at the scale of these risk complexes may demand different decision skills and tools than we relied on to independently manage each stressor. It may also require institutional adjustments that support more sophisticated and explicit ways of approaching complex risks. Risk behavior is already woven into our institutional fabric. Many government and private sector institutions are founded to transfer risk-bearing from one party to another. But how do these institutional arrangements act as a barrier to adaption by shielding us from the consequences of our actions? How long will these institutions (e.g. subsidized insurance) hold up under the changing patterns of intense events? Risk-based thinking should drive us to reconsider our own behavior in the context of a changing risk context.

Approaches and technologies for creating, sharing, and applying knowledge are being transformed. New social and institutional structures for exchanging information have developed so rapidly that assumptions about how people access and use information to make adaptation decisions may be outdated. Communities of practice such as The Nature Conservancy's fire learning network may become the knowledge management institutions of the future. More scientific organizations are using "crowd-sourcing" methods that expand their reach to diverse investigations and that may involve citizens in providing data. These represent new sources of knowledge that are less dependent on "go-to" agencies and "official information" and are more amenable to blending knowledge of different types, sources, and vintages. They combine collaborative learning with the powers of social media to give place-based meaning to information as it emerges. We may have to find new ways to nourish these networks with actionable science and lessons gleaned from adaptive management.

Our worries about the effectiveness of technology transfer and the health of the science / management interface are now part of a bigger question of how institutions participate in the relationships and networks that manage knowledge. Can the research community provide tools and platforms with which managers can investigate their own hunches and blend their tacit knowledge with broader scientific findings? How can we better involve practitioners and citizens in the development of the science base? These and other new questions have emerged. Adaptive actions and programs will be increasingly scrutinized for effectiveness and efficiency. They will have to compete rigorously with other uses of public and private capital. Measures of resilience and adaptive capacity are now being brought into some agency budget discussions, a good start. We must be able to articulate, quantify, and realize returns on investment. Adaptation investments must (1) frame future positioning and ecosystem outcomes as returns on investment, (2) distinguish among inputs, outputs, outcomes, and range of future options in "value chains," (3) estimate the true costs of conservation practices, including the benefits and costs of collaboration, and (4) establish performance measures that are meaningful to the individuals and organizations who would work together to make the adaptation successful.

The challenges of measurement and program evaluation will no doubt stimulate a lot a creative thinking about the business of adaptation in the new few years. We must figure out how to incorporate attributes of adaptive capacity such as flexibility, social license, preservation of options, learning, and scalability along with measures of ecosystem outcomes into program goals and performance evaluation.

The US Forest Service since 2011 has been using a balanced scorecard approach to measure progress in incorporating climate change into sustainable forest management programs and practices. The Forest Service Climate Change Scorecard is composed of performance hurdles and guidance in four dimensions: (1) organizational capacity; (2) partnerships, engagement, and education; (3) adaptation; and (4) mitigation and sustainable consumption. Each of the 155 National Forests and 20 National Grasslands complete the scorecard report annually. A national network of 130 collateral duty climate change coordinators evaluate the utility and the insights provided and exchange lessons learned to improve the state of climate response practice. Three years of measurement and narrative reports are providing a clearer and more useful picture of what is needed to make adaptation to climate successful.

As waves of baby boomer retirements and agency downsizing occur in tandem, organizations are undergoing important changes in their workforce—capacity reductions, losses in experience, and rapid repopulation of leadership ranks. This is a great opportunity to adjust leadership development and rewards to support adaptive decision-making. We need transformational leadership that can help an increasingly diverse citizenry through ill-structured problems and uncertainty and to take actions that improve learning. This leadership will help people confront their own expectations and wrestle with situations where the changing climate and their own responses can lead to unexpected losses and gains. It is a form of leadership practiced by all employees, not an exclusive set of titular leaders in hierarchical structures. What knowledge, skills, and attitudes should we promote in this new "gene pool" of leadership? Leaders of adaptation will have to be experts in managing organizational and social change. They will have to turn institutions more sharply than they were designed to turn and challenge organizational, political, and other barriers to flexibility. Communication and engagement skills will become more critical in helping stakeholders become partners, wrestle with issues of risk transfer, and adopt new behaviors for living "up close and personal" with extreme events. Communication will need to evolve beyond media talking points into true engagement toward better understanding of how people respond to risk and how to avoid "pseudo-certainty" in a rapidly changing world.

Leaders of the future will also have to understand the science of decision-making as well as the science of ecosystems. They must deal with human judgment in the myriad functions and decision processes of land management and be able to adjust decision process to improve learning and respond to new information. We are the benefactors of major advances in behavioral economics and decision science and of an evolution in decision practices in many fields. New models are emerging that involve clear shifts in decision processes to choosing robust rather than optimal solutions; from single decision-maker to consensus choices; from reliance on published science to wider varieties of evidence; from solving problems to coping with conditions; and from information-starvation to information overload.

Extreme events can create social and organizational chaos, divert resources, attract political scrutiny and involve reputational risk. These events are no longer rare. Our landscapes are being shaped by climate- and human-driven disturbance and it seems that we should learn how to recover from or use the disturbance event to create more resilient conditions rather than just clean up the damage. We can continue to view extreme events as "natural" disasters even though we know that damages emanate from exposure caused by human choices. But we can also view them as punctuation marks in the bigger narrative of unrelenting change. They can be teachable moments and political opportunities to nudge the process of organizational learning and change.

We will do well to emphasize opportunism, flexibility, and recovery after these events. Can we use these "unscheduled" disturbances to guide larger-scale adaptive transitions? How nimble are we in jumping expected cycles of ecological succession into new ecosystem states? We may have to rethink our translations of management "control" theory from business and the factory floor to the management of increasingly dynamic ecosystems. We need new landscape-scale science as well as new theories of management under turbulence to guide our quest for resilience.

Theodore Roosevelt once said, "in utilizing and conserving the natural resources of the Nation, the one characteristic more essential than any other is foresight." Adaptive capacity includes developing foresight to understand key uncertainties and identify emerging issues, deal better with surprise, anticipate unintended consequences, decrease reaction time to rapid change, clarify multiple external perspectives about trends and plausible futures, and shape preferred futures and future pathways (Bengston et al. 2012).

Leaders must create the organizational space and appreciation for exploring alternative futures. This includes insight from projections, models, scenarios, futuring exercises, and expert judgment, while maintaining balance between these sources of information and history, experimentation, and other forms of evidence. Foresight does not mean obsessing over forecasts. The current yearning for finer-resolution climate model projections is understandable, but must be tempered lest it grow into illusions of precision, or excuses for inaction, or anchoring choices to individual forecasts. A keen understanding of the craft of foresight development and the caveats of using forecast information and managing cognitive biases must be built into our leadership skill bank.

We search for new implications and recommendations in each new study or report from the field. Our emphasis on uncertainty, new discoveries, and new tools may at times divert us from fully using what we already know about systems and their adaptive mechanisms. It may be time to relearn from some of the "old" science in light of the adaptation challenges ahead. There may be fewer secrets to be found than there are principles and basic understanding to be applied to new situations. Are there roots of resiliency hiding in plain sight in these classic studies and science findings? We may have to reinterpret what we think we know with an adaptation "going forward" perspective. Leadership can sanction and direct this reflection with appropriate questions. What problems were the scientists who created this knowledge responding to? What did they observe about climate-forest interactions that could guide our expectations about possible futures? What about this information might be relevant to forest conservation in the Anthropocene?

17

Forest Conservation in the Anthropocene

Policy Recommendations

V. Alaric Sample, Christopher Topik, and Paige Lewis

America's forests are undergoing changes unlike any seen before in human history. With each passing year, precedents are being set for the extent and impacts of wildfires. Record areas of forests stand dead or dying, not just from exotic insects and diseases, but also from species that have been native to these forests for eons. More subtle but potentially more profound changes are taking place each day as native plant and animal species quietly disappear from their historical home ranges, perhaps to reappear at the frontiers of some other more poleward ecosystem. Scientists and forest managers puzzle over new arrivals, trying to decide whether to define them as invasive species to be eradicated, or as climate refugees to be nurtured as they continue their exodus toward destinations unknown.

In the midst of this unprecedented change, the stewards of America's forests, public and private, must decide how they will act if they are to sustain the forests themselves and the array of economic, environmental, and societal values and services forests provide—water, wildlife, biodiversity, wood, renewable energy, carbon sequestration. As they do their best to anticipate a

DOI: 10.5876/9781607324591.c017

future in which the lessons of the past offer little guidance, they must assess the risks associated with several alternative courses of action, and then manage those risks through intensified monitoring and continuous readjustments aimed at preserving as many options as possible for future resource managers. The Anthropocene, in short, is a game changer. Conserving biological diversity in the world's forests, for example, is a particular challenge as plant and animal species follow the climate-driven movement of the ecosystems and habitats in which they evolved (Williams and Dumroese, chapter 8, this volume; Oliver, chapter 7, this volume; also Hannah 2012; Hannah et al. 2002; Lovejoy and Hannah 2005). Ecological communities disassemble as species capable of migrating do so, and those that are not remain behind. Those that can migrate must now traverse landscapes that in earlier epochs were not filled with highways, cities, farms, and other manifestations of a rapidly expanding human population. Designated parks, refuges, reserves, and other traditional approaches to protecting habitat are still important (Caro et al., chapter 6, this volume, Caro et al. 2011), but may be less effective when the species themselves are on the move (Kareiva et al. 2011). This is prompting biologists, resource management professionals, and policy-makers to consider new approaches to conservation planning (Anderson and Ferree 2010) and to strategies focused on large landscapes—vast areas that stretch from Yellowstone National Park to the Yukon, or from the southern Appalachians to Labrador (Anderson and Ferree 2010). These immense landscapes encompass cities, towns, and agricultural working lands, as well as a mosaic of public and private forests that are all managed for different purposes. For these landscape-scale conservation strategies to be environmentally, economically, and socially sustainable—and politically possible—new governance models must be developed to facilitate an unprecedented level of communication, coordination, cooperation, and collaboration (see Tabor et al., chapter 13, this volume; McCarthy, chapter 11, this volume, and Halofsky et al., chapter 12, this volume).

Much more than wildlife habitat conservation is at stake. For thousands of communities, forests are pivotal to maintaining adequate supplies of clean water to meet municipal, agricultural, and industrial needs. Forests in the headwaters and riparian areas of the nation's rivers and streams are low-cost, high-return guarantors of water quality, water supply, and favorable timing of seasonal flows. These are critical considerations in the western

United States and parts of the South where higher temperatures, prolonged droughts, and shifts in precipitation patterns are already causing economic and social disruptions (Milly et al. 2008). Other regions in the eastern United States anticipate a continuing increase in precipitation in the form of extreme storm events, accentuating the essential role that intact forests serve in storm water control and flood mitigation.

The low-cost and largely self-maintaining "green infrastructure" that forests provide is vulnerable to direct and indirect effects of climate change. The direct effects of drought, elevated temperatures, and changing precipitation patterns can be seen in reduced tree growth, lower survival rates in tree seedlings and young growth, and reforestation failures in the wake of natural disturbances. The loss of certain more climate-sensitive tree species within a forest can change the overall species composition or mix, eliminating food sources and habitat for native wildlife species.

Forests that are already under a high degree of environmental stress from these direct effects of climate change are more vulnerable to its indirect effects. Forests in many parts of the world are experiencing extraordinary levels of mortality from insects and disease. Incidents involving even endemic or native pests and pathogens, which would normally kill only a small fraction of the trees in a forest, are in some regions causing near 100 percent mortality over thousands of square miles (Allen et al. 2002). The resulting volume of dead and dying trees invites wildfires that themselves are unprecedented in size and severity (Allen, chapter 4, this volume; Westerling et al., chapter 3, this volume; Brown et al. 2004). Following events such as these, the harsher climate makes reforestation and ecological restoration that much more difficult and prone to failure, often leading to increased soil erosion, stream sedimentation, impacts on terrestrial and aquatic habitat, and damages to water supplies, storm water control, and flood mitigation. These climate change effects are largely outside the experience and expertise of today's forest managers. Even the knowledge base for forest management practices that has been built up over the past two centuries is itself based on forest science developed almost entirely within a period of relative climate stability. Successfully meeting the challenges of forest conservation and sustainable management in the Anthropocene will require the development of new forest science, policies, and practices based on these conclusions:

1. A more integrated approach is needed for understanding, preparing for, and adapting to the effects of climate change on natural resources. Scientists and natural resource specialists in wildlife habitat management, biodiversity conservation, water resource protection, and other disciplines are all working to develop effective approaches, but there is still a tendency to focus within rather than across disciplines. Strategies developed independently to optimize one set of objectives may dictate management activities that run counter to strategies oriented to other objectives.

2. Wildfire management and policy is central to adaptation strategies across all resources.

3. The development and effective implementation of policies to limit the ecological, economic, and social impacts of wildfires are an essential consideration of any climate change adaptation strategy.

A more dynamic policy framework is needed as a basis for natural resource management that can adapt to climate change. The most important lesson is not that the existing policy framework should be replaced, but that policies themselves must be dynamic enough to accommodate rapidly changing environmental conditions. Statutes and regulations that provide a broad enabling framework will be more effective than prescriptive laws and rules that reflect highly changeable theories and approaches. The current strategy for forest management adaptation to climate change is an amalgam of different strategies being developed largely independent of one another. Considerable scientific research and management resources have been devoted to developing new adaptive strategies for biodiversity conservation, as geographic shifting of habitat zones raises questions about the long-term effectiveness of traditional protected-area approaches. The extraordinary increase in the size, frequency, and extent of wildfires in forest watersheds has prompted urgent development of strategies to protect municipal water supplies and water quality (McCarthy, chapter 11, this volume). Climate change is exacerbating the wildfire problem, as forests are becoming warmer, dryer, and subject to more extreme weather events and longer fire seasons. Acres burned by wildfires during 2012 were the third most of any year since 1960, with 9.3 million acres (3.76 million ha) burned, and the US Forest Service estimates that 20 million acres (8.9 million ha) will burn annually by 2050 (Finley 2013). This increase in the number of "megafires," and the immense volume of greenhouse gases emitted during the fire event itself and the often-lengthy recovery period afterwards, have become new and significant factors in climate change mitigation policy. Because of the impact of megafires, wildfire policy itself is undergoing a thorough reexamination.

Correcting the downward trajectory in forest conditions and reinforcing their resiliency to the effects of climate change is a daunting challenge, requiring ecosystem restoration on an estimated 152 million acres (61.5 million ha) of federal, state, tribal, and private forest land in the United States (USDA Forest Service 2012b). Ecosystem restoration in this context is focused on functions and processes, and strengthening the capacity to recover from significant, large-scale natural disturbances, not on attempting to restore forests to some earlier evolutionary state. Substantial improvements will be needed in the current institutional, legal, and policy framework for the management of forests and their associated values and services, especially at the federal level (Cleaves and Bixler, chapter 16, this volume; Shaffer, chapter 15, this volume). Policies and practices must also increase community engagement for local involvement in forest management and seeing the work implemented. Nowhere is this truer than in wildfire policy, not only because of the direct impacts of fire discussed above, but also because of the burgeoning costs of emergency wildfire suppression that drain away public funding available for the management and protection of other resources.

The societal, environmental, and fiscal costs of fire in the nation's forests continue their precipitous climb. Federal expenditures for emergency wildland fire suppression during 2012 alone were $1.9 billion, in addition to the nearly $1.5 billion required to maintain, staff, and equip federal fire programs. The direct and indirect cost of wildfire management currently consumes more than 50 percent of the Forest Service budget, leaving an ever-smaller pool of funds to support hazardous fuels reduction, timber management, wildlife habitat improvement, recreational access, watershed protection, and the variety of other important services that the American people value and expect.

The full economic and social impacts of this extraordinary increase in wildfires are far greater than these figures indicate, but thus far have been difficult to quantify. A study by the International Association of Fire Chiefs (2013) estimated that direct public expenditures for emergency wildfire suppression average around $4.7 billion annually—$2.5 billion from federal agencies, $1.2

billion from state agencies, and about $1 billion from local governments. But even this is only a fraction of the total economic and social costs. An analysis of six recent wildfires by the Western Forestry Leadership Coalition (2010) showed that fire suppression expenditures may be as little as 3 to 5 percent of the fires' total economic impact. Yet current federal policy and funding priorities are focused on strategies to limit direct emergency wildfire suppression costs. A more comprehensive approach based on reducing the overall environmental, economic, and social impacts of wildfires is needed, as a basis for policies that strike an optimal balance of spending on emergency wildfire suppression, and on wildfire prevention through forest restoration.

The following recommendations are aimed at providing some practical methods to enhance forest and fire management in the United States to create more resilient forests and forest-dependent communities in the Anthropocene.

RECOMMENDATION 1

Strengthen the institutional framework for long-term investment in forest restoration and sustainable management

Reduce hazardous fuels

It is essential that the federal government increase investments to reduce fire risk in a manner that make forests more resilient and resistant to fire and other stressors. This should be based on a broadly supported long-term strategy so that steady progress can be made toward overarching goals for resource protection and sustainability. Strategic, proactive hazardous fuels treatments have proven to be a safe and cost-effective way to reduce risks to communities and forests by removing overgrown brush and trees, leaving forests in a more natural condition, resilient to wildfires. A recent meta-analysis of thirty-two studies confirmed that when implemented strategically, fuels treatments make a crucial difference in the size, spread, and severity of wildfires (Martinson and Omi 2013). These treatments can improve the safety and effectiveness of firefighters and provide protection for a community or essential watershed that might otherwise see extensive loss.

Federal investments in maintaining the capacity and skills for hazardous fuels treatments have been shown to reduce property losses while also providing jobs and other economic benefits to rural communities

(Ecological Restoration Institute 2013). An economic assessment of forest restoration in eastern Oregon by the Federal Forest Advisory Committee (2012) revealed "an investment in forest health restoration has the potential to save millions of dollars in state and federal funds by avoiding costs associated with fire suppression, social service programs, and unemployment benefits." It is estimated that for every $1 million invested in hazardous fuels treatments, approximately sixteen full-time-equivalent jobs are created or maintained, representing more than a half-million dollars in wages and over $2 million in overall economic activity (Nielsen-Pincus and Moseley 2010). Nevertheless, recent federal budgets have made debilitating cuts to funding the Hazardous Fuels Reduction programs at the Forest Service and the Department of the Interior.

Strengthen results-based cooperation on forest restoration through initiatives such as the Collaborative Forest Landscape Restoration (CFLR) Program

The active involvement of local communities and stakeholders plays an essential role in the management of public lands, but the challenges of forest restoration will require an unprecedented level of cooperation among federal land managers, stakeholders, and organizations that provide the local economic infrastructure for carrying out resource protection and restoration activities. The CFLR Program is an innovative mechanism aimed at enhancing community involvement in forest restoration and management. It is being used to test a variety of approaches, bringing science and local needs together in collaborative visions for future forest management.

Through these projects, the CFLR Program is demonstrating that collaboratively developed forest restoration plans can be implemented at a large scale with benefits for people and forests. From fiscal year 2010 through fiscal year 2012, the cumulative outputs generated by the projects totaled 94.1 million cubic feet of timber; 7,949 jobs created or maintained; $290 million in labor income; 383,000 acres (155,00 ha) of hazardous fuels reduction to protect communities; 229,000 acres (92,600 ha) of thinning to reduce fire risk and improve forest resiliency; and 6,000 miles of improved road conditions to reduce sediment in waterways (CFLR Steering Committee 2012). Equally important is the long-term commitment these projects have fostered to community sustainability and forest resilience (Butler 2013).

Maintain capacity for multiresource management and protection through increased administrative and budgetary efficiencies

Given the scope of the wildfire management challenge on federal lands, it is likely that other resource programs will continue to be underfunded relative to actual needs for resource protection and stewardship. One way of addressing this challenge is to consider more efficient and better-integrated approaches to budgeting and accomplishing multiresource management on federal forests. The US Forest Service and Bureau of Land Management (BLM) have in the past considered budget reforms aimed at facilitating a more integrated approach to implementing land and resource management plans developed under the National Forest Management Act and the Federal Land Management and Policy Act (Sample 1990). These early pilot studies of consolidated budgeting, planning, and accomplishment reporting demonstrated significant cost savings, increased performance, and improved accountability in many instances.

Finding budgetary and administrative efficiencies will be an essential component in the Forest Service's and the BLM's strategies for wildfire management and broader adaptation to climate change. Among the key lessons learned from earlier efforts are that congressional and administration support is essential. When modifications in the existing budget structure result in changes to existing processes for budget development and appropriations, they have met resistance, particularly at the President's Office of Management and Budget (OMB) and among members of Congress with an interest in specific programs (Sample and Tipple 2001). The increasing proportion of the Forest Service and BLM budgets being directed to emergency wildfire suppression, in addition to the more general budget reductions, means that other resource programs are struggling to accomplish management objectives. The efficiencies discovered in previous attempts at budget reform suggest that it is time to consider this again. The Forest Service is currently experimenting with Integrated Resource Restoration (IRR), a budgetary tool that attempts to increase efficiency by blending funding sources for a variety of forest, watershed, and wildlife habitat programs. The IRR is being employed in three regions on a pilot basis (Northern, Southwest, and Intermountain). Congressional and administration support will continue to be essential for this pilot to be successful.

RECOMMENDATION 2

Create and fund a new federal fire-suppression funding mechanism to free up resources for proactive management

Policy action is needed to guarantee adequate resources for wildland fire first responders, but in a way that allows investments in the up-front risk-reduction programs discussed above. While federal fire preparedness and suppression resources will need to be maintained at an effective level to protect life, property, and natural resources, these funds must be provided through a mechanism that does not compromise the viability of forest management activities that can reduce risks to life and property and mitigate the demand for future emergency response. The current system of funding fire preparedness and suppression at the expense of hazardous fuels and other key programs threatens to undermine—and eventually overtake—the vital management and conservation purposes for which the Forest Service and Department of the Interior bureaus were established.

The dramatic increase in the number of homes near federal lands that are prone to damaging fire has added significantly to the cost of fire suppression. In the past, paying for this cost often entailed "borrowing" funding from other programs. Fire borrowing severely impacts even the most basic level of resource management planning, reducing non-fire-related agency personnel, and undermining efforts to retain skilled contractors in local communities to carry out land management and stewardship activities (GAO 2004). Congress subsequently passed the Federal Land Assistance, Management, and Enhancement (FLAME) Act of 2009 (43 USC § 1748a) as part of a bipartisan effort to change the funding mechanism for wildfire suppression by establishing two emergency wildfire accounts funded above annual suppression. These FLAME accounts were intended to serve as a safeguard against fire borrowing. Unfortunately, the implementation of the FLAME Act has not proceeded as intended. Due to several factors, hundreds of millions of dollars from the agencies' non-suppression programs were transferred into emergency response accounts (Taylor 2013).

A new, separate federal funding source should be established so that funding for vital fire suppression activities is distinct from ongoing land management requirements. One option is the establishment of a "Wildland Fire Suppression Disaster Prevention Fund" that could be used to support federal fire-suppression actions during emergencies, just as the Disaster Relief Fund is used to help communities recover after disasters. Fire suppression is different from other natural disasters, since the federal response is needed most acutely during the actual event. Such support should complement prevention and risk-reduction activities and postfire recovery and restoration actions. This fund could be established through the congressional appropriations process and supported using declarations in subsequent annual appropriations bills. It would also be appropriate to enhance state participation in such a fund. In addition, Congress could increase the ability of the Federal Emergency Management Agency to provide states impacted by wildfire with additional resources for fuel hazard mitigation. Broadening and diversifying the investments in proactive management and mitigation is far more cost effective than continuing to focus tremendous resources on emergency response.

RECOMMENDATION 3

Accelerate implementation of collaborative authorities

Stewardship contracts and agreements are among the most valuable tools the US Forest Service and BLM have to carry out ecosystem restoration actions, including hazardous fuels treatments, on federal forests. Congress first granted this statutory authority on a pilot basis to allow the Forest Service to carry out critically important land stewardship and resource protection activities, many of which had been carried out previously through National Forest timber sales. The effectiveness of these pilot studies led Congress to expand the program to authorize the Forest Service and BLM to utilize stewardship contracts anywhere on the federal forests under their management (16 U.S.C. § 2104). Permanent statutory authority for stewardship contracts and agreements was provided within the Agricultural Act of 2014 (Public Law 113-79; 2.7.14).

Over the past decade, stewardship contracting has proven to be an innovative and flexible tool that enables the Forest Service and BLM to implement projects that restore and maintain healthy forest ecosystems, foster collaboration, and provide local employment through sustainable community economic development (Pinchot Institute 2012). Stewardship contracts are the only administrative tool that can provide certainty to local contractors for up to ten years, a critically important consideration for small businesses securing financing to purchase equipment, expand facilities, or increase their skilled workforce to carry out the land management activities specified in the contract.

The following specific steps are needed to achieve the two main objectives of expediting agency-level policy direction on stewardship contracting to resource managers in the field, and immediately initiating the agency-level process for enhancing the implementation of stewardship contracting in the field:

- Provide updated guidance to agency field staff related to the permanent authorization of stewardship contracting and how these legislative authorities can be used to accelerate the pace and scale of restoration of federal lands. The Forest Service and BLM operate under different policy frameworks, but that should not prohibit interagency coordination.

- Develop a forum for interested stakeholders to remain current. Provide guidebooks to help with industry, tribal, and citizen outreach on the use of stewardship contracts and agreements as a key tool for enhancing partnerships among stakeholders and expanding the on-the-ground work that the federal agencies can accomplish. Evaluate opportunities to use the recently expanded Good Neighbor authority to work with steward-ship contracts and agreements (Public Law 113-79).

- Consider the recommendations from the FY 2012 Stewardship Contracting Programmatic Monitoring report (www.pinchot.org/gp/Stewardship_Contracting), and the recommendations from the Stewardship Contracting Roundtable and regional partners.

RECOMMENDATION 4

Increase the capacity of states and communities to become fire adapted

Given the potential for devastating increases in loss of resource values and public expense, a diverse range of agencies and organizations has begun promoting the concept of *fire-adapted communities*, defined as knowledgeable and engaged communities in which the awareness and actions of residents regarding infrastructure, buildings, landscaping, and the surrounding ecosystem lessen the need for extensive protection actions and enables such communities to safely accept fire as a part of the landscape.

Programs such as State and Volunteer Fire Assistance and Forest Health Protection provide important resources to help states and local communities develop and sustain community wildfire protection capacity. Policy-makers should seek opportunities to allocate other federal resources in a way that rewards communities for proactive actions that collectively result in national benefit. For more than ten years, The Nature Conservancy has worked cooperatively with the Forest Service and the Department of the Interior to foster the Fire Learning Network (FLN) that brings communities together and helps them build collaborative, science-based strategies that protect people and ecosystems (Benson and Garmestani 2013). The FLN supports public-private landscape partnerships that engage in collaborative planning and implementation, and it provides a means for sharing the tools and innovations that help them scale up. Locally, the FLN helps federal land managers convene collaborative planning efforts, build trust and understanding among stakeholders, improve community capacity to live with fire, access training that helps fire professionals work with local communities, and address climate change and other emerging threats.

Relatively small federal and state investments in community capacity can have substantial results for lowering wildfire risk, and is the most cost-effective way of reducing harmful impacts to society, while also allowing for enhanced, safe, and controlled use of fire to restore wildlands as appropriate.

The Forest Service and other members of the Fire Adapted Communities Coalition are working to get communities the information and resources they need to successfully live with fire. The website www.fireadapted.org provides access to educational materials and tools in support of community wildfire protection.

RECOMMENDATION 5

Seek policy adjustments that foster innovation and improvement in National Environmental Policy Act (NEPA) implementation, thereby increasing the scale and quality of projects and plans

The administration has established a goal of increasing the pace of restoration and job creation on the National Forests (USDA Forest Service 2012b). The Forest Service acknowledges that the pace and scale of restoration must dramatically increase to get ahead of the growing threats facing America's forest ecosystems, watersheds, and forest-dependent communities. To facilitate this, effective use must be made of all available management tools, and the Forest Service must explore opportunities to increase the efficiency of planning and implementation processes.

There is broad commitment to the principles of public engagement and environmental review embodied in NEPA. The Forest Service is currently testing a variety of innovative NEPA strategies that hold promise for broader application. Adaptive NEPA is an approach by which the official record of decision allows sufficient leeway for some variety of subsequent federal actions, thereby streamlining analysis, allowing for more efficient project implementation, and enabling land managers to more effectively incorporate emerging science. Greater use can be made of the categorical exclusion procedures allowed under NEPA without diminishing the intent of this key environmental law. Full public participation and transparency in federal decision-making, based on science and public discourse, results in better management decisions that in the long run are more effective and efficient. The National Forest System Land Management Planning Rule and draft Directives (Federal Register 2012) emphasize collaborative, science-based adaptive management. Application of this framework will guide a new round of forest planning that is both more meaningful and more efficient, and set the stage for timely implementation of projects that achieve multiple benefits on the ground.

RECOMMENDATION 6

Increase shared commitment to and support for forest restoration by states and local governments

Federal agencies alone cannot prevent the loss of homes, infrastructure, and natural resource values in the wildland-urban interface (WUI). Individuals and communities living in the WUI must meaningfully invest in preparing for and reducing their own risk from fire. Postfire studies show that using fire-resistant building materials and reducing fuels in and around the home ignition zone are the most effective ways to reduce the likelihood that a home will burn (Graham et al. 2012). Similarly, community investments in improved ingress and egress routes, clear evacuation strategies, strategic fuel breaks, and increased firefighting capacity can go a long way toward enabling a community to successfully weather a wildfire event.

Community commitment is also necessary to effectively shift the national approach to wildfire from a costly emphasis on disaster response to a balanced and proactive strategy. Research shows that rising wildfire-suppression costs are directly linked to the growing presence of homes and related infrastructure in the WUI (Stein et al. 2013). A corresponding analysis by Headwaters Economics revealed that 84 percent of the WUI is still undeveloped, so there is great potential for the costs associated with wildfire protection to exponentially increase (Rasker 2013). According to the same study, if just half of the WUI is developed in the future, annual firefighting costs could explode to between $2.3 and $4.3 billion. States and communities need to pay close attention to the ramifications of their planning on the resulting wildfire environment. Federal lands and surrounding communities also need to foster greater partnerships and multilateral cooperation and coordination. There are many opportunities for states and municipalities to directly participate and even help fund forest management activities on nearby federal lands that increase the livability of forest-dependent communities and reduce fire risk. The Eastern Oregon study (Federal Forest Advisory Committee 2012) demonstrates that state investments in federal land management can yield great savings to the state in reduced unemployment costs, reduced social services, and increased tax revenue. Elsewhere, such as in Flagstaff, Arizona, communities are contributing directly to restore forest conditions that reduce fire risk to protect existing watershed and recreation resources (Flagstaff Watershed Protection Project 2012).

RECOMMENDATION 7

Enhance participation of additional sectors of society, such as water and power utilities, recreation and tourism, public health, and industrial users of clean water

There are many opportunities for diverse and sustainable sources of nonfederal funding to provide an effective complement to federal land management resources. There are a number of efforts underway to produce revenue for upstream forest restoration that benefits downstream water users and water companies while enhancing the restoration and maintenance of federal forests.

The Forest Service has been particularly active and innovative in Colorado. Since 2009 the agency has established partnerships with five water utilities

(Denver Water, Aurora Water, Colorado Springs Utilities, Northern Water, and Pueblo Water), several major corporations (such as MillerCoors, Vail Resorts, and Coca-Cola), and several philanthropic entities (Brian Ferebee, USDA Forest Service, personal communication 2013). Such efforts, often spearheaded by the National Forest Foundation, are exciting beginnings for greater shared responsibility that can reduce wildfire risk while enhancing forest health (see National Forest Foundation 2013).

There are additional, important partnerships with forest products industries. Their investments in new biomass and wood products development can play a substantial role in facilitating the removal of overstocked trees, while enhancing the condition of the forest and streams following harvest.

The insurance and reinsurance industries are closely involved in wildland fire issues and are important partners in such efforts as the Fire Adapted Communities Coalition (see website Fireadapted.org). These industries, which have direct contact with citizens, have a desire to see fire risks reduced (Munich Reinsurance America 2013). There may be additional opportunities to bring various compensatory mitigation funds for the support of forest restoration.

Wildfires and even controlled fires can have sizable impacts on public health due to smoke (Knowlton et al. 2011). There is a great need to increase engagement with public health agencies concerning impacts of smoke, and the relative merits of massive, uncontrolled smoke from severe wildfires versus controlled smoke episodes from prescribed burning.

RECOMMENDATION 8

Increase the safe and effective use of wildland fire

The beneficial use of fire as a tool for resource management is another area where greater efficiency and effectiveness could be achieved. By increasing the use of controlled burns and naturally ignited wildland fires to accomplish resource benefit, land managers can achieve ecological and community protection goals on a larger scale and at reduced cost. Congress and the administration should make it clear that the safe and effective use of fire is a priority for land management agencies, and provide the necessary funding, training, and leadership support needed.

Recommendation 9

Increase research on economic, social, and ecological impacts of forest investment

It is essential that the federal government and other sectors invest in monitoring, research, and accountability studies for fuels treatment, wildfire management strategies, and related efforts. This requires relatively small investments when compared to the costs of fire suppression and fire damage, but is essential if scientists are to learn what works and what does not. Furthermore, new technologies, including remote sensing, LIDAR, and focused social science studies can offer creative new perspectives to increase efficiency of action.

The Anthropocene, this new epoch in which *Homo sapiens* has become the predominant force in the global biosphere, is about more than just a changing climate. The climate has always been in a state of flux, but species and communities in most instances have found ways to adapt and survive, through migration, mutation, or other coping mechanisms. One thing that is different this time is the pace of the change. Climate shifts that in past epochs have taken place over millennia are now happening in decades. Natural adaptation strategies are of limited success in today's circumstances, heightening the risk of unprecedented ecological disruptions.

Forest managers with responsibilities for sustaining multiple ecosystem services without interruption or significant decline will be especially challenged. Helping private forest landowners understand how climate change is likely to affect their management objectives is the first step to assisting them in taking actions that—because two-thirds of the nation's forests are in private ownership—can have a major impact on how well US forests adapt to, and also mitigate, climate change.

The second and perhaps greater challenge is to develop the political will to make considerable new public investments in sustaining forests. Efforts should be focused on developing new and alternative budgeting methods for wildfire suppression and mitigation, in addition to making a clear and compelling economic case illustrating costs avoided by investing in proactive forest restoration treatments.

Similarly, public programs aimed at stemming the loss of private forest land to development or conversion are perpetually underfunded relative to the level of interest among private forest landowners who otherwise would utilize these programs for conservation easements and other forms of land protection.

For conservationists, this is a defining era. Meeting human needs for food, shelter, energy, and especially water will continue to alter landscapes at an expanding scale, with direct, indirect, and induced effects that are far too complex for humans to predict or for other species to anticipate. Most important, through decisive actions taken now, there is an opportunity—and responsibility—to change the future, and avoid the projected switch in US forests from providing an important carbon sink to becoming themselves a major net source of carbon emissions. Stemming the loss of private forests to development, restoring public forests to relieve climate-induced environmental stresses, reduce fire risks, and protect essential public values and ecosystem services—these all have substantial environmental, economic, and societal benefits in addition to reducing carbon emissions. These are goals that are already well understood and supported by a broad consensus of Americans. The barriers to achieving these goals are eminently surmountable, making forest conservation in the Anthropocene a remarkable challenge and a transformative prospect for this generation and the next.

Works Cited

Acemoglu, D., and J. A. Robinson. 2012. *Why Nations Fail: The Origins of Power, Prosperity, and Poverty*. New York: Brown Business, Random House.

Adams, H. D., M. Guardiola-Claramonte, G. A. Barron-Gafford, J. C. Villegas, D. D. Breshears, C. B. Zou, P. A. Troch, and T. E. Huxman. 2009. "Temperature Sensitivity of Drought-Induced Tree Mortality Portends Increased Regional Die-Off Under Global-Change-Type Drought." *Proceedings of the National Academy of Sciences of the United States of America* 106 (17): 7063–6. http://dx.doi.org/10.1073/pnas.0901438106.

Adams, S. N., and K.A.M. Engelhardt. 2009. "Diversity Declines in Microstegium Vimineum (Japanese Stiltgrass) Patches." *Biological Conservation* 142 (5): 1003–10. http://dx.doi.org/10.1016/j.biocon.2009.01.009.

Adeney, J. M., N. L. Christensen, and S. L. Pimm. 2009. "Reserves Protect against Deforestation Fires in the Amazon." *PLoS One* 4: e5014.

Ager, A. A., N. M. Vaillant, and M. A. Finney. 2010. "A Comparison of Landscape Fuel Treatment Strategies to Mitigate Wildland Fire Risk in the Urban Interface and Preserve Old Forest Structure." *Forest Ecology and Management* 259 (8): 1556–70. http://dx.doi.org/10.1016/j.foreco.2010.01.032.

Aikio, S., R. P. Duncan, and P. E. Hulme. 2012. "The Vulnerability of Habitats to Plant Invasion: Disentangling the Roles of Propagule Pressure, Time and

Sampling Effort." *Global Ecology and Biogeography* 21 (8): 778–86. http://dx.doi
.org/10.1111/j.1466-8238.2011.00711.x.

Aitken, S. N., S. Yeaman, J. A. Holliday, T. Wang, and S. Curtis-McLane. 2008.
"Adaptation, Migration or Extirpation: Climate Change Outcomes for Tree Populations." *Evolutionary Applications* 1 (1): 95–111. http://dx.doi.org/10.1111/j.1752
-4571.2007.00013.x.

Albert, R. C. 1987. *Damming the Delaware: The Rise and Fall of Tocks Island Dam.* 2nd
ed. University Park, PA: Pennsylvania State University Press.

Allen, C. D. 2007. "Interactions across Spatial Scales among Forest Dieback, Fire,
and Erosion in Northern New Mexico Landscapes." *Ecosystems (New York, N.Y.)*
10 (5): 797–808. http://dx.doi.org/10.1007/s10021-007-9057-4.

Allen, C. D., R. S. Anderson, R. B. Jass, J. L. Toney, and C. H. Baisan. 2008. "Paired
Charcoal and Tree-ring Records of High-frequency Holocene Fire from Two
New Mexico Bog Sites." *International Journal of Wildland Fire* 17 (1): 115–30.
http://dx.doi.org/10.1071/WF07165.

Allen, C. D., and D. D. Breshears. 1998. "Drought-Induced Shift of a Forest-
Woodland Ecotone: Rapid Landscape Response to Climate Variation." *Proceedings of the National Academy of Sciences of the United States of America* 95 (25):
14839–42. http://dx.doi.org/10.1073/pnas.95.25.14839.

Allen, C. D., A. K. Macalady, H. Chenchouni, D. Bachelet, N. McDowell, M.
Vennetier, T. Kitzberger, A. Rigling, D. D. Breshears, E. H. Hogg, et al. 2010. "A
Global Overview of Drought and Heat-induced Tree Mortality Reveals Emerging Climate Change Risks for Forests." *Forest Ecology and Management* 259 (4):
660–84. http://dx.doi.org/10.1016/j.foreco.2009.09.001.

Allen, C. D., M. Savage, D. A. Falk, K. F. Suckling, T. W. Swetnam, T. Schulke, P. B.
Stacey, P. Morgan, M. Hoffman, and J. T. Klingel. 2002. "Ecological Restoration
of Southwestern Ponderosa Pine Ecosystems: A Broad Perspective." *Ecological
Applications* 12 (5): 1418–33. http://dx.doi.org/10.1890/1051-0761(2002)012[1418
:EROSPP]2.0.CO;2.

Anderson, L. J. 2005. "California's Reaction to Caulerpa Taxifolia: A Model for Invasive Species Rapid Response." *Biological Invasions* 7 (6): 1003–16. http://dx.doi.org
/10.1007/s10530-004-3123-z.

Anderson, M. G., M. Clark, and A. Olivero Sheldon. 2014. "Estimating Climate
Resilience for Conservation across Geophysical Settings." *Conservation Biology* 28
(4): 959–739. http://dx.doi.org/10.1111/cobi.12272.

Anderson, M. G., and C. E. Ferree. 2010. "Conserving the Stage: Climate Change
and the Geophysical Underpinnings of Species Diversity." *PLoS One* 5 (7): e11554.
http://dx.doi.org/10.1371/journal.pone.0011554.

Anderson, P. D., and D. J. Chmura. 2009. "Silvicultural Approaches to Maintain
Forest Health and Productivity under Current and Future Climates." *Western
Forester* 54 (1): 6–8.

Anderson, R. S., C. D. Allen, J. L. Toney, R. B. Jass, and A. N. Bair. 2008. "Holocene Vegetation and Fire Regimes in Subalpine and Mixed Conifer Forests, Southern Rocky Mountains, USA." *International Journal of Wildland Fire* 17 (1): 96–114. http://dx.doi.org/10.1071/WF07028.

Anderson, M. G., M. Clark, and A. Olivero Sheldon. 2014. "Estimating Climate Resilience for Conservation across Geophysical Settings." *Conservation Biology* 28 (4): 959–70. http://onlinelibrary.wiley.com/doi/10.1111/cobi.12272/full.

Anderson-Texeira, K., A. Miller, J. Mohan, T. W. Hudiburg, B. D. Duval, and E. H. DeLucia. 2013. "Altered Dynamics of Forest Recovery under a Changing Climate." *Global Change Biology* 19 (7): 1–21.

Andreadis, T. G., and R. M. Weseloh. 1990. "Discovery of Entomophaga Maimaiga in North American Gypsy Moth, Lymantria dispar." *Proceedings of the National Academy of Sciences of the United States of America* 87 (7): 2461–5. http://dx.doi.org/10.1073/pnas.87.7.2461.

Andréassian, V. 2004. "Water and Forests: From Historical Controversy to Scientific Debate." *Journal of Hydrology (Amsterdam)* 291 (1-2): 1–27. http://dx.doi.org/10.1016/j.jhydrol.2003.12.015.

Aplet, G., and J. Gallo. 2012. "Applying Climate Adaptation Concepts to the Landscape Scale: Examples from the Sierra and Stanislaus National Forests." The Wilderness Society. http://wilderness.org/sites/default/files/Sierra_and_Stanislaus_climate_adaptation.pdf.

Archer, D. 2005. "The Fate of Fossil Fuel CO_2 in Geologic Time." *Journal of Geophysical Research* 110 (C9): C09S05. http://dx.doi.org/10.1029/2004JC002625.

Archer, D., and V. Brovkin. 2008. "The Millennial Atmospheric Lifetime of Anthropogenic CO_2." *Climatic Change* 90 (3): 283–97. http://dx.doi.org/10.1007/s10584-008-9413-1.

Archer, D., and A. Ganopolski. 2005. "A Movable Trigger: Fossil Fuel CO2 and the Onset of the Next Glaciation." *Geochemistry Geophysics Geosystems* 6 (5): n/a. http://dx.doi.org/10.1029/2004GC000891.

Arndt, D. S., M. O. Baringer, M. R. Johnson, L. V. Alexander, H. J. Diamond, R. L. Fogt, J. M. Levy, J. Richter-Menge, P. W. Thorne, L. A. Vincent, and A. B. Watkins. 2010. "State of the Climate in 2009." *Bulletin of the American Meteorological Society* 91 (7).

Arno, S. F., and C. E. Fiedler. 2005. *Mimicking Nature's Fire: Restoring Fire-prone Forests in the West*. Washington, DC: Island Press.

Association of Metropolitan Water Architects. 2009. "Confronting Climate Change: An Early Analysis of Water and Wastewater Adaptation Costs." http://www.amwa.net/galleries/climate-change/ConfrontingClimateChangeOct09.pdf.

Astrup, R., K. D. Coates, and E. Hall. 2008. "Recruitment Limitation in Forests: Lessons from an Unprecedented Mountain Pine Beetle Epidemic." *Forest Ecology and Management* 256 (10): 1743–50. http://dx.doi.org/10.1016/j.foreco.2008.07.025.

Aubin, I., C. M. Garbe, S. Colombo, C. R. Drever, D. W. McKenney, C. Messier, J. Pedlar, M. A. Saner, L. Venier, A. M. Wellstead, et al. 2011. "Why We Disagree about Assisted Migration: Ethical Implications of a Key Debate Regarding the Future of Canada's Forests." *Forestry Chronicle* 87 (06): 755–65. http://dx.doi.org /10.5558/tfc2011-092.

Augustin, L., C. Barbante, P.R.F. Barnes, J.-M. Barnola, M. Bigler, E. Castellano, O. Cattani, J. Chappellaz, D. Dahl-Jensen, B. Delmonte, et al., and the EPICA Community Members. 2004. "Eight Glacial Cycles from an Antarctic Ice Core." *Nature* 429 (6992): 623–8. http://dx.doi.org/10.1038/nature02599.

Aukema, J. E., B. Leung, K. Kovacs, C. Chivers, K.O. Britton, J. Englin, S. J. Frankel, R. G. Haight, T. P. Holmes, A. M. Liebhold, et al. 2011. "Economic Impacts of Non-native Forest Insects in the Continental United States." *PLoS One* 6 (9): e24587. http://dx.doi.org/10.1371/journal.pone.0024587.

Aukema, J. E., D. G. McCullough, B. Von Holle, A. M. Liebhold, K. Britton, and S. J. Frankel. 2010. "Historical Accumulation of Nonindigenous Forest Pests in the Continental United States." *Bioscience* 60 (11): 886–97. http://dx.doi.org/10.1525 /bio.2010.60.11.5.

Averyt, K., J. Fisher, A. Huber-Lee, A. Lewis, J. Macknick, N. Madden, J. Rogers, and S. Tellinghuisen. 2011. "Freshwater Use by U.S. Power Plants: Electricity's Thirst for a Precious Resource." A Report of the Energy and Water in a Warming World Initiative, Union of Concerned Scientists. http://www.ucsusa.org/clean_ energy/our-energy-choices/energy-and-water-use/freshwater-use-by-us-power -plants.html.

Backlund, P., A. Janetos, D. Schimel, et al. 2008. "The Effects of Climate Change on Agriculture, Land Resources, Water Resources, and Biodiversity in the United States." http://www.usda.gov/oce/climate_change/SAP4_3/CCSPFinalRepo rt.pdf.

Bagne, K. E., M. M. Friggens, and D. M. Finch. 2011. "A System for Assessing Vulnerability of Species (SAVS) to Climate Change." Gen. Tech. Rep. RMRS-GTR-257. Fort Collins, CO: USDA, Forest Service, Rocky Mountain Research Station.

Bailey, R. G. 1983. "Delineation of Ecosystem Regions." *Environmental Management* 7 (4): 365–73. http://dx.doi.org/10.1007/BF01866919.

Bailey, R. G. 1995. *Description of the Ecoregions of the United States*. 2nd ed., rev. Misc. Publ. No. 1391 (rev.). Washington, DC: USDA, Forest Service.

Bailey, R. G. 1996. *Ecosystem Geography*. New York: Springer-Verlag. http://dx.doi org/10.1007/978-1-4612-2358-0.

Baker, W. L. 2009. *Fire Ecology in Rocky Mountain Landscapes*. Washington, DC: Island Press.

Baldwin, J. L. 1973. *Climates of the United States*. Washington, DC: US Department of Commerce, National Oceanic and Atmospheric Administration, Environmental Data Service.

Balling, R. C., Jr., G. A. Meyer, and S. G. Wells. 1992. "Relation of Surface Climate and Burned Area in Yellowstone National Park." *Agricultural and Forest Meteorology* 60 (3-4): 285–93. http://dx.doi.org/10.1016/0168-1923(92)90043-4.

Ban, N. C., M. Mills, J. Tam, C. C. Hicks, S. Klain, N. Stoeckl, M. C. Bottrill, J. Levine, R. L Pressey, T. Satterfield, et al. 2013. "A Social–ecological Approach to Conservation Planning: Embedding Social Considerations." *Frontiers in Ecology and the Environment* 11 (4): 194–202. http://dx.doi.org/10.1890/110205.

Baral, A., and G. Guha. 2004. "Trees for Carbon Sequestration or Fossil Fuel Substitution: The Issue of Cost vs. Carbon Benefit." *Biomass and Bioenergy* 27 (1): 41–55. http://dx.doi.org/10.1016/j.biombioe.2003.11.004.

Barber, C. P., M. A. Cochrane, C. Souza, Jr., and A. Veríssimo. 2012. "Dynamic Performance Assessment of Protected Areas." *Biological Conservation* 149 (1): 6–14. http://dx.doi.org/10.1016/j.biocon.2011.08.024.

Barlow, C. 2011. "Paleoecology and the Assisted Migration Debate: Why a Deep-time Perspective Is Vital." Accessed March 19, 2016. http://www.torreya guardians.org/assisted_migration_paleoecology.html.

Barnett, T. P., D. W. Pierce, H. G. Hidalgo, C. Bonfils, B. D. Santer, T. Das, G. Bala, A. W. Wood, T. Nozawa, A. A. Mirin, et al. 2008. "Human-Induced Changes in the Hydrology of the Western United States." *Science* 319 (5866): 1080–3. http://dx.doi.org/10.1126/science.1152538.

Barton, A. M. 2002. "Intense Wildfire in Southeastern Arizona: Transformation of a Madrean Oak–Pine Forest to Oak Woodland." *Forest Ecology and Management* 165 (1-3): 205–12. http://dx.doi.org/10.1016/S0378-1127(01)00618-1.

Bates, B. C., P. Hope, B. Ryan, I. Smith, and S. Charles. 2008. "Key Findings from the Indian Ocean Climate Initiative and Their Impact on Policy Development in Australia." *Climatic Change* 89 (3-4): 339–54. http://dx.doi.org/10.1007/s10584-007 9390 9.

Beale, C. M., N. E. Baker, M. J. Brewer, and J. J. Lennon. 2013. "Protected Area Networks and Savannah Bird Biodiversity in the Face of Climate Change and Land Degradation." *Ecology Letters* 16 (8): 1061–8. http://dx.doi.org/10.1111/ele.12139.

Beardmore, T., and R. Winder. 2011. "Review of Science-based Assessments of Species Vulnerability: Contributions to Decision-making for Assisted Migration." *Forestry Chronicle* 87 (6): 745–54. http://dx.doi.org/10.5558/tfc2011-091.

Beaulieu, J. 2009. "Optisource: A Tool for Optimizing Seed Transfer." Branching out, No. 55. Quebec City, QC: Natural Resources Canada, Canadian Forest Service.

Beaulieu, J., and A. Rainville. 2005. "Adaptation to Climate Change: Genetic Variation Is Both a Short- and a Long-term Solution." *Forestry Chronicle* 81 (5): 704–9. http://dx.doi.org/10.5558/tfc81704-5.

Beecher, S., T. Thaler, G. Griffith, A. Perry, T. Crossett, and R. Rasker, eds. 2013. "Adapting to a Changing Climate: Risks and Opportunities for the Upper Delaware River Region." Model Forest Policy Program in Association with

Common Waters Partnership. Sagle, ID: Pinchot Institute for Conservation the Cumberland River Compact and Headwaters Economics.

Beeson, C. E., and P. F. Doyle. 1995. "Comparison of Bank Erosion at Vegetated and Non-vegetated Channel Bends." *Journal of the American Water Resources Association* 31 (6): 983–90. http://dx.doi.org/10.1111/j.1752-1688.1995.tb03414.x.

Beier, C., S. A. Signell, A. Luttman, and A. T. DeGaetano. 2012. "High Resolution Climate Change Mapping with Gridded Historical Climate Products." *Landscape Ecology* 27 (3): 327–42. http://dx.doi.org/10.1007/s10980-011-9698-8.

Beier, P., and B. Brost. 2010. "Use of Land Facets to Plan for Climate Change: Conserving the Arenas, Not the Actors." *Conservation Biology* 24 (3): 701–10. http://dx .doi.org/10.1111/j.1523-1739.2009.01422.x.

Benedict, M., and E. McMahon. 2006. *Green Infrastructure: Linking Landscapes and Communities.* Washington, DC: Conservation Fund.

Bengston, D. N., R. L. Olson, and L. A. DeVaney. 2012. "The Future of Wildland Fire Management in a World of Rapid Change and Great Uncertainty." In *Proceedings of 3rd Human Dimensions of Wildland Fire, April 17–19, 2012, Seattle, Washington.* Missoula, MT: International Association of Wildland Fire.

Bennett, G., N. Carroll, and K. Hamilton. 2013. *Charting New Waters: State of Watershed Payments 2012.* Washington, DC: Forest Trends; http://www.forest-trends .org/documents/files/doc_3308.pdf.

Benson, Melinda Harm, and Ahjond S. Garmestani. 2013. "A Framework for Resilience-Based Governance of Social-Ecological Systems." *Ecology and Society* 18 (1). http://ssrn.com/abstract=2213523.

Bentz, B. J., J. Régnière, C. J. Fettig, E. M. Hansen, J. L. Hayes, J. A. Hicke, R. G. Kelsey, J. F. Negrón, and S. J. Seybold. 2010. "Climate Change and Bark Beetles of the Western United States and Canada: Direct and Indirect Effects." *Bioscience* 60 (8): 602–13. http://dx.doi.org/10.1525/bio.2010.60.8.6.

Berger, A., and M. F. Loutre. 2002. "An Exceptionally Long Interglacial Ahead?" *Science* 297 (5585): 1287–8. http://dx.doi.org/10.1126/science.1076120.

Berkes, F., J. Colding, and C. Folke. 2003. *Navigating Social-ecological Systems: Building Resilience for Complexity and Change.* Cambridge: Cambridge University Press.

Berry, N. J., O. L. Phillips, S. L. Lewis, J. K. Hill, D. P. Edwards, N. B. Tawatao, N. Ahmad, D. Magintan, C. V. Khen, M. Maryati, et al. 2010. "The High Value of Logged Tropical Forests: Lessons from Northern Borneo." *Biodiversity and Conservation* 19 (4): 985–97. http://dx.doi.org/10.1007/s10531-010-9779-z.

Bessie, W. C., and E. A. Johnson. 1995. "The Relative Importance of Fuels and Weather on Fire Behavior in Subalpine Forests." *Ecology* 76 (3): 747–62. http://dx .doi.org/10.2307/1939341.

Betancourt, J. L. 1990. "Late Quaternary Biogeography of the Colorado Plateau." In *Packrat Middens: The Last 40,000 Years of Biotic Change,* ed. J. L. Betancourt, T. R. V. Devender, and P. S. Martin, 259–92. Tucson, AZ: University of Arizona Press.

Bhola, N., J. O. Ogutu, H. P. Piepho, M. Y. Said, R. S. Reid, N. T. Hobbs, and H. Olff. 2012. "Comparative Changes in Density and Demography of Large Herbivores in the Masai Mara Reserve and Its Surrounding Human- dominated Pastoral Ranches in Kenya." *Biodiversity and Conservation* 21 (6): 1509–30. http://dx.doi .org/10.1007/s10531-012-0261-y.

Biastoch, A., C. W. Böning, F. U. Schwarzkopf, and J. R. E. Lutjeharms. 2009. "Increase in Agulhas Leakage Due to Poleward Shift of Southern Hemisphere Westerlies." *Nature* 462 (7272): 495–8. http://dx.doi.org/10.1038/nature08519.

Bierbaum, R. J., B. Smith, A. Lee, Maria Blair, L. Carter, F. S. Chapin, P. Fleming, S. Ruffo, M. Stults, S. McNeeley, et al. 2013. "A Comprehensive Review of Climate Adaptation in the United States: More Than Before, But Less Than Needed." *Mitigation and Adaptation Strategies for Global Change* 18 (3): 361–406. http://dx.doi.org/10.1007/s11027-012-9423-1.

Bigio, E., T. W. Swetnam, and C. I I. Baisan. 2010. "A Comparison and Integration of Tree-ring and Alluvial Records of Fire History at the Missionary Ridge Fire, Durango, Colorado, USA." *Holocene* 20 (7): 1047–61. http://dx.doi.org/10.1177 /0959683610369502.

Bigsby, K., P. Tobin, and E. Sills. 2011. "Anthropogenic Drivers of Gypsy Moth Spread." *Biological Invasions* 13 (9): 2077–90. http://dx.doi.org/10.1007/s10530-011-0027-6.

Blanchon, P., A. Eisenhauer, J. Fietzke, and V. Liebetrau. 2009. "Rapid Sea-level Rise and Reef Back-stepping at the Close of the Last Interglacial High-stand." *Nature* 458 (7240): 881–4. http://dx.doi.org/10.1038/nature07933.

Blunden, J., and D. A. Arndt. 2013. "State of the Climate in 2012." *Bulletin of the American Meteorological Society* 94 (8): S1–S238. http://dx.doi.org/10.1175/2013 BAMSStateoftheClimate.1.

Boer, M. M., R. J. Sadler, R. S. Wittkuhn, L. McCaw, and P. Grierson. 2009. "Long-Term Impacts of Prescribed Burning on Regional Extent and Incidence of Wild-fires—Evidence from Fifty Years of Active Fire Management in SW Australian Forests." *Forest Ecology and Management* 259 (1): 132–42. http://dx.doi.org/10.1016 /j.foreco.2009.10.005.

Bolte, A., C. Ammer, M. Lof, P. Madsen, G.-J. Nabuurs, P. Schall, P. Spathelf, and J. Rock. 2009. "Adaptive Forest Management in Central Europe—Climate Change Impacts, Strategies and Integrative Concepts." *Scandinavian Journal of Forest Research* 24 (6): 473–82. http://dx.doi.org/10.1080/02827580903418224.

Bonnington, C., D. Weaver, and E. Fanning. 2007. "Livestock and Large Wild Mammals in the Kilombero Valley, in Southern Tanzania." *African Journal of Ecology* 45 (4): 658–63. http://dx.doi.org/10.1111/j.1365-2028.2007.00793.x.

Bormann, F. H., and G. E. Likens. 1979. *Pattern and Process in a Forested Ecosystem.* New York: Springer-Verlag. http://dx.doi.org/10.1007/978-1-4612-6232-9.

Bosch, J., L. M. Carrascal, L. Duran, S. Walker, and M. C. Fisher. 2007. "Climate Change and Outbreaks of Amphibian Chytridiomycosis in a Montane Area of

Central Spain: Is There a Link?" *Proceedings. Biological Sciences* 274 (1607): 253–60. http://dx.doi.org/10.1098/rspb.2006.3713.

Botkin, D. B. 1979. "A Grandfather Clock Down the Staircase: Stability and Disturbances in Natural Ecosystems." In *Forests: Fresh Perspectives from Ecosystem Analysis, Proceedings of the 40th Annual Biology Colloquium*, 1–10. Corvallis: Oregon State University Press.

Botkin, D. B. 1991. *Discordant Harmonies: A New Ecology for the Twenty-first Century.* New York: Oxford University Press.

Botkin, D. B., H. Saxe, M. B. Araújo, R. Betts, R. H. W. Bradshaw, T. Cedhagen, P. Chesson, T. P. Dawson, J. R. Etterson, D. P Faith, et al. 2007. "Forecasting the Effects of Global Warming on Biodiversity." *Bioscience* 57 (3): 227–36. http://dx.doi.org/10.1641/B570306.

Bouché, P., I. Douglas-Hamilton, G. Wittemyer, A. J. Nianogo, J.-L. Doucet, P. Lejeune, and C. Vermeulen. 2011. "Will Elephants Soon Disappear from West African Savannahs?" *PLoS ON 6* (6): e20619. http://dx.doi.org/10.1371/journal.pone.0020619.

Bourque, C., and J. Pomeroy. 2001. "Effects of Forest Harvesting on Summer Stream Temperatures in New Brunswick, Canada: An Inter-catchment, Multiple-year Comparison." *Hydrology and Earth System Sciences* 5 (4): 599–614. http://dx.doi.org/10.5194/hess-5-599-2001.

Bowen, G. J., W. C. Clyde, P. L. Koch, S. Ting, J. Alroy, T. Tsubamoto. Y. Wang, and Y. Wang. 2002. "Mammalian Dispersal at the Paleocene/Eocene Boundary." *Science* 295 (5562): 2062–5. http://dx.doi.org/10.1126/science.1068700.

Bowman, D. M. J. S., J. K. Balch, P. Artaxo, W. J. Bond, J. M. Carlson, M. A. Cochrane, C. M. D'Antonio, R. S. DeFries, J. C. Doyle, S. P. Harrison, et al. 2009. "Fire in the Earth System." *Science* 324 (5926): 481–84. http://dx.doi.org/10.1126/science.1163886.

Boyce, S. G. 1995. *Landscape Forestry.* New York: John Wiley & Sons.

Boyle, S. A., and F. B. Samson. 1985. "Effects of Nonconsumptive Recreation on Wildlife: A Review." *Wildlife Society Bulletin* 13:110–6.

Bradley, B. A., D. M. Blumenthal, R. Early, E. D. Grosholz, J. J. Lawler, L. P. Miller, C.J.B. Sorte, C. M. D'Antonio, J. M. Diez, J. S. Dukes, et al. 2011. "Global Change, Global Trade, and the Next Wave of Plant Invasions." *Frontiers in Ecology and the Environment* 10 (1): 20–8. http://dx.doi.org/10.1890/110145.

Bradley, B. A., and D. S. Wilcove. 2009. "When Invasive Plants Disappear: Transformative Restoration Possibilities in the Western United States Resulting from Climate Change." *Restoration Ecology* 17 (5): 715–21. http://dx.doi.org/10.1111/j.1526-100X.2009.00586.x.

Bradley, B. A., D. S. Wilcove, and M. Oppenheimer. 2010. "Climate Change Increases Risk of Plant Invasion in the Eastern United States." *Biological Invasions* 12 (6): 1855–72. http://dx.doi.org/10.1007/s10530-009-9597-y.

Brantley, S., C. Ford, and J. Vose. 2013. "Future Species Composition Will Affect Forest Water Use After Loss of Eastern Hemlock from Southern Appalachian

Forests." *Ecological Applications* 23 (4): 777–90. http://dx.doi.org/10.1890/12
-0616.1.

Brasier, C. M. 2001. "Rapid Evolution of Introduced Plant Pathogens via Interspecific Hybridization." *Bioscience* 51 (2): 123–33. http://dx.doi.org/10.1641/0006
-3568(2001)051[0123:REOIPP]2.0.CO;2.

Brauman, K. A., G. C. Daily, T. K. Duarte, and H. A. Mooney. 2007. "The Nature
and Value of Ecosystem Services: An Overview Highlighting Hydrologic Services." *Annual Review of Environment and Resources* 32 (1): 67–98. http://dx.doi
.org/10.1146/annurev.energy.32.031306.102758.

Breshears, D. D., N. S. Cobb, P. M. Rich, K. P. Price, C. D. Allen, R. G. Balice, W. H.
Romme, J. H. Kastens, M. L. Floyd, J. Belnap, et al. 2005. "Regional Vegetation Die-off in Response to Global-Change-Type Drought." *Proceedings of the
National Academy of Sciences of the United States of America* 102 (42): 15144–8.
http://dx.doi.org/10.1073/pnas.0505734102.

Breshears, D. D., O. B. Myers, C. W. Meyer, F. J. Barnes, C. B. Zou, C. D. Allen,
N. G. McDowell, and W. T. Pockman. 2008. "Tree Die-off in Response to Global
Change-type Drought: Mortality Insights from a Decade of Plant Water Potential Measurements." *Frontiers in Ecology and the Environment* 7 (4): 185–9. http://dx
.doi.org/10.1890/080016.

Briggs, C. M. 2007. "Multicycle Adaptive Simulation of Boiling Water Reactor Core
Simulators." MS thesis, North Carolina State University.

Briggs, D. J., and P. Smithson. 1986. *Fundamentals of Physical Geography*. Totowa, NJ:
Rowman & Littlefield.

Brigham-Grette, J., M. Melles, P. Minyuk, A. Andreev, P. Tarasov, R. DeConto,
S. Koenig, N. Nowaczyk, V. Wennrich, P. Rosen, et al. 2013. "Pliocene Warmth,
Polar Amplification, and Stepped Pleistocene Cooling Recorded in NE Arctic
Russia." *Science* 340 (6139): 1421–7. http://dx.doi.org/10.1126/science.1233137.

Brooks, K. N., D. Current, and D. Wyse. 2003. "Restoring Hydrologic Function
of Altered Landscapes: An Integrated Watershed Management Approach." In
Watershed Management: Water Resources for the Future, 101–14. Food and Agriculture Organization (FAO) of the United Nations.

Brown, P. M., E. K. Heyerdahl, S. G. Kitchen, and M. H. Weber. 2008. "Climate
Effects on Historical Fires (1630–1900) in Utah." *International Journal of Wildland
Fire* 17 (1): 28–39. http://dx.doi.org/10.1071/WF07023.

Brown, P. M., and R. Wu. 2005. "Climate and Disturbance Forcing of Episodic Tree
Recruitment in a Southwestern Ponderosa Pine Landscape." *Ecology* 86 (11):
3030–8. http://dx.doi.org/10.1890/05-0034.

Brown, T. J., B. Hall, and A. Westerling. 2004. "The Impact of Twenty-first Century
Climate Change on Wildland Fire Danger in the Western United States: An
Applications Perspective." *Climatic Change* 62 (1–3): 356–88.

Brubaker, L. 1986. "Responses of Tree Populations to Climatic Change." *Vegetatio* 67
(2): 119–30. http://dx.doi.org/10.1007/BF00037362.

Bruner, A. G., R. E. Gullison, R. E. Rice, and G. A. B. da Fonseca. 2001. "Effectiveness of Parks in Protecting Tropical Biodiversity." *Science* 291 (5501): 125–8. http://dx.doi.org/10.1126/science.291.5501.125.

Brusca, R. C., J. F. Wiens, W. M. Meyer, J. Eble, K. Franklin, J. T. Overpeck, and W. Moore. 2013. "Dramatic Response to Climate Change in the Southwest: Robert Whittaker's 1963 Arizona Mountain Plant Transect Revisited." *Ecology and Evolution* n/a. http://dx.doi.org/10.1002/ece3.720.

Butler, W. H. 2013. "Collaboration at Arm's Length: Navigating Agency Engagement in Landscape-scale Ecological Restoration Collaboratives." *Journal of Forestry* 111 (6): 395–403.

Byrne, M. K., and P. Olwell. 2008. "Seeds of Success: The National Native Seed Collection Program in the United States." *Public Gardens* 23:3–4.

Camacho, A. 2013. "The Law of Ethics of Assisted Migration." Transcribed video presentation from the conference "Assisted Migration: A Primer for Reforestation and Restoration Decision Makers," 21 February 2013, Portland, OR. On file at: USDA, Forest Service, Rocky Mountain Research Station. Moscow, ID: Forestry Sciences Laboratory. Accessed March 19, 2016. http://www.rngr.net/resources/assisted-migration/presentations/law-and-ethics-of-assisted-migration.

Campbell, R. K. 1986. "Mapped Genetic Variation of Douglas-fir to Guide Seed Transfer in Southwest Oregon." *Silvae Genetica* 35:2–3.

Cannon, S. H., J. E. Gartner, R. C. Wilson, J. C. Bowers, and J. L. Laber. 2008. "Storm Rainfall Conditions for Floods and Debris Flows from Recently Burned Areas in Southwestern Colorado and Southern California." *Geomorphology* 96 (3–4): 250–69. http://dx.doi.org/10.1016/j.geomorph.2007.03.019.

Cannon, S. H., and S. L. Reneau. 2000. "Conditions for Generation of Fire-related Debris Flows, Capulin Canyon, New Mexico." *Earth Surface Processes and Landforms* 25 (10): 1103–21. http://dx.doi.org/10.1002/1096-9837(200009)25:10<1103::AID-ESP120>3.0.CO;2-H.

Cantrill, D. J., and I. Poole. 2012. *The Vegetation of Antarctica Through Geological Time.* Cambridge: Cambridge University Press. http://dx.doi.org/10.1017/CBO9781139024990.

Caro, T., J. Darwin, T. Forrester, C. Ledoux-Bloom, and C. Wells. 2011. "Conservation in the Anthropocene." *Conservation Biology* 26 (1): 185–8. http://dx.doi.org/10.1111/j.1523-1739.2011.01752.x.

Caro, T., A. Dobson, A. J. Marshall, and C. A. Peres. 2014. "Compromise Solutions between Conservation and Road Building in the Tropics." *Current Biology* 24 (16): R722–5. http://dx.doi.org/10.1016/j.cub.2014.07.007.

Carpe Diem West. 2011. "Watershed Investment Programs in the American West: An Updated Look; Linking Upstream Watershed Health and Downstream Security." http://www.carpediemwest.org/sites/carpediemwest.org/files/WIP%20Report%20Design%20FINAL%2011.15.11.pdf.

Carpenter, S. R., and W. A. Brock. 2006. "Rising Variance: A Leading Indicator of Ecological Transition." *Ecology Letters* 9 (3): 311–8. http://dx.doi.org/10.1111/j.1461-0248.2005.00877.x.

Carroll, C., R. F. Noss, P. C. Paquet, and N. H. Schumaker. 2004. "Extinction Debt of Protected Areas in Developing Landscapes." *Conservation Biology* 18 (4): 1110–20. http://dx.doi.org/10.1111/j.1523-1739.2004.00083.x.

Cayan, D. R., T. Das, D. W. Pierce, T. P. Barnett, M. Tyree, and A. Gershunov. 2010. "Future Dryness in the Southwest U.S. and the Hydrology of the Early 21st Century Drought." *Proceedings of the National Academy of Sciences of the United States of America* 107 (50): 21271–6. http://dx.doi.org/10.1073/pnas.0912391107.

Council for Environmental Quality and U.S. Department of the Interior (CEQ/USDOI). 2009. "Conference Report to the FY 2010 Interior and Related Agencies Appropriations Bill."

CFLR Steering Committee. 2012. "People Restoring America's Forests: 2012 Report on the Collaborative Forest Landscape Restoration Program." http://www.fs.fed.us/restoration/documents/cflrp/CoalitionReports/CFLRP2012AnnualReport20130108.pdf.

Chape, S., J. Harrison, M. Spalding, and I. Lysenko. 2005. "Measuring the Extent and Effectiveness of Protected Areas as an Indicator for Meeting Global Biodiversity Targets." *Philosophical Transactions of the Royal Society of London. Series B, Biological Sciences* 360 (1454): 443–55. http://dx.doi.org/10.1098/rstb.2004.1592.

Chapin, F. S., III, J. P. Kofinas, and C. Folke, eds. 2009. *Principles of Ecosystem Stewardship: Resilience-Based Natural Resource Management in a Changing World.* New York: Springer.

Chapin, F. S., III, B. H. Walker, R. J. Hobbs, D. U. Hooper, J. H. Lawton, O. E. Sala, and D. Tilman. 1997. "Biotic Control over the Functioning of Ecosystems." *Science* 277 (5325): 500–4. http://dx.doi.org/10.1126/science.277.5325.500.

Chapin, F. S., III, E. S. Zavaleta, V. T. Eviner, R. L. Naylor, P. M. Vitousek, H. L. Reynolds, D. U. Hooper, S. Lavorel, O. E. Sala, S. E. Hobbie, et al. 2000. "Consequences of Changing Biodiversity." *Nature* 405 (6783): 234–42. http://dx.doi.org/10.1038/35012241.

Charnley, S., A. P. Fischer, and E. T. Jones. 2008. "Traditional and Local Ecological Knowledge about Forest Biodiversity in the Pacific Northwest." *Gen.Tech. Rep. PNW-GTR-751.* Portland: US Department of Agriculture and Forest Service, Pacific Northwest Research Station.

Cheetham, R., and C. Billett. 2003. "SmartConservation™: Automating a Conservation Value Assessment Model." http://www.smartconservationsoftware.org/.

Chen, F. H., M. R. Qiang, Z. D. Feng, H. B. Wang, and J. Bloemendal. 2003. "Stable East Asian Monsoon Climate During the Last Interglacial (Eemian) Indicated by Paleosol S1 in the Western Part of the Chinese Loess Plateau." *Global and Planetary Change* 36 (3): 171–9. http://dx.doi.org/10.1016/S0921-8181(02)00183-2.

Cissel, J., F. Swanson, and P. Weisberg. 1999. "Landscape Management Using Historical Fire Regimes: Blue River, Oregon." *Ecological Applications* 9 (4): 1217–31. http://dx.doi.org/10.1890/1051-0761(1999)009[1217:LMUHFR]2.0.CO;2.

Clark, P. U., and P. Huybers. 2009. "Interglacial and Future Sea Level." *Nature* 462 (7275): 856–7. http://dx.doi.org/10.1038/462856a.

Cleveland, C. 2012. "Ecoregions of the United States (Bailey)." http://www.eoearth.org/view/article/152244.

Climate Central. 2012. "The Age of Western Wildfires." http://www.climatecentral.org/wgts/wildfires/Wildfires2012.pdf.

Cobb, R. C., D. A. Orwig, and S. Currie. 2006. "Decomposition of Green Foliage in Eastern Hemlock Forests of Southern New England Impacted by Hemlock Woolly Adelgid Infestations." *Canadian Journal of Forest Research* 36 (5): 1331–41. http://dx.doi.org/10.1139/x06-012.

Colborn, T., D. Dumanoski, and J. P. Myers. 1996. *Our Stolen Future: Are We Threatening Our Fertility, Intelligence, and Survival? A Scientific Detective Story.* New York: Dutton.

Cole, K. L., K. Ironside, J. Eischeid, G. Garfin, P. B. Duffy, and C. Toney. 2011. "Past and Ongoing Shifts in Joshua Tree Distribution Support Future Modeled Range Contraction." *Ecological Applications* 21 (1): 137–49. http://dx.doi.org/10.1890/09-1800.1.

Collins, B. M., J. D. Miller, A. E. Thode, M. Kelly, J. W. van Wagtendonk, and S. L. Stephens. 2009. "Interactions among Wildland Fires in a Long Established Sierra Nevada Natural Fire Area." *Ecosystems (New York, N.Y.)* 12 (1): 114–28. http://dx.doi.org/10.1007/s10021-008-9211-7.

Collins, B. M., P. Omi, and P. Chapman. 2006. "Regional Relationships between Climate and Wildfire-burned Area in the Interior West, USA." *Canadian Journal of Forest Research* 36 (3): 699–709. http://dx.doi.org/10.1139/x05-264.

Collins, B. M., and G. B. Roller. 2013. "Early Forest Dynamics in Stand-replacing Fire Patches in the Northern Sierra Nevada, California, USA." *Landscape Ecology* 28 (9): 1801–13. http://dx.doi.org/10.1007/s10980-013-9923-8.

Collins, B. M., and S. L. Stephens. 2010. "Stand-replacing Patches within a Mixed Severity Fire Regime: Quantitative Characterization Using Recent Fires in a Long-established Natural Fire Area." *Landscape Ecology* 25 (6): 927–39. http://dx.doi.org/10.1007/s10980-010-9470-5.

Collins, B. M., S. L. Stephens, J. J. Moghaddas, and J. Battles. 2010. "Challenges and Approaches in Planning Fuel Treatments across Fire-Excluded Forested Landscapes." *Journal of Forestry* 108 (1): 24–31.

Coltman, D. W., P. O'Donoghue, J. T. Jorgenson, J. T. Hogg, C. Strobeck, and M. Festa-Bianchet. 2003. "Undesirable Evolutionary Consequences of Trophy Hunting." *Nature* 426 (6967): 655–8. http://dx.doi.org/10.1038/nature02177.

Colunga-Garcia, M., R. A. Magarey, R. A. Haack, Stuart H. Gage, and Jiaquo Qi. 2010. "Enhancing Early Detection of Exotic Pests in Agricultural and Forest

Ecosystems Using an Urban-gradient Framework." *Ecological Applications* 20 (2): 303–10. http://dx.doi.org/10.1890/09-0193.1.

Colwell, R., S. Avery, J. Berger, G. E. Davis, H. Hamilton, T. Lovejoy, S. Malcom, A. McMullen, M. Novacek, R. J. Roberts, et al. 2012. *Revisiting Leopold: Resource Stewardship in the National Parks. A Report of the National Park System Advisory Board, Science Committee.* Washington, DC: National Park Foundation.

Compton, B. W., K. McGarigal, S. A. Cushman, and L. G. Gamble. 2007. "A Resistant-Kernel Model of Connectivity for Amphibians that Breed in Vernal Pools." *Conservation Biology* 21 (3): 788–99. http://dx.doi.org/10.1111/j.1523-1739 .2007.00674.x.

Conn, C., S. Claggett, B. Drake, et al. 2010. "Forests and Terrestrial Ecosystems." In *Comprehensive Strategy for Reducing Maryland's Vulnerability to Climate Change, Phase II: Building Societal, Economic, and Ecological Resilience, Report of the Maryland Commission on Climate Change, Adaptation and Response and Scientific and Technical Working Groups*, ed. K. Boicor and Z. P. Johnson. Cambridge, MD: University of Maryland Center for Environmental Science. Annapolis, MD: Maryland Department of Natural Resources.

Cooper, C. F. 1960. "Changes in Vegetation, Structure, and Growth of Southwestern Pine Forests Since White Settlement." *Ecological Monographs* 30 (2): 129–64. http://dx.doi.org/10.2307/1948549.

Cornett, M., and M. White. 2013. "Forest Restoration in a Changing World: Complexity and Adaptation Examples from the Great Lakes Region of North America." In *Managing Forests as Complex Adaptive Systems: Building Resilience to the Challenge of Global Change*, ed. C. Messier, K. Puettmann, and K. D. Coates, 113–32. New York: Routledge Press.

Covington, W. W., and M. M. Moore. 1994. "Southwestern Ponderosa Pine Forest Structure: Changes Since Euro-American Settlement." *Journal of Forestry* 92:39–47.

Craigie, I. D., J. E. M. Baillie, A. Balmford, C. Carbone, Ben Collen, Rhys E. Green, and Jon M. Hutton. 2010. "Large Mammal Population Declines in Africa's Protected Areas." *Biological Conservation* 143 (9): 2221–8. http://dx.doi.org/10.1016/j .biocon.2010.06.007.

Crimmins, M. A. 2006. "Synoptic Climatology of Extreme Fire-weather Conditions across the Southwest United States." *International Journal of Climatology* 26 (8): 1001–16. http://dx.doi.org/10.1002/joc.1300.

Crimmins, M. A., and A. C. Comrie. 2004. "Interactions between Antecedent Climate and Wildfire Variability across Southeastern Arizona." *International Journal of Wildland Fire* 13 (4): 455–66. http://dx.doi.org/10.1071/WF03064.

Cross, M. S., P. D. McCarthy, G. Garfin, D. Gori, and C.A.F. Enquist. 2013. "Accelerating Adaptation of Natural Resource Management to Address Climate Change." *Conservation Biology* 27 (1): 4–13. http://dx.doi.org/10.1111/j.1523-1739.2012.01954.x.

Crowe, K. A., and W. H. Parker. 2008. "Using Portfolio Theory to Guide Reforestation and Restoration under Climate Change Scenarios." *Climatic Change* 89 (3–4):
355–70. http://dx.doi.org/10.1007/s10584-007-9373-x.

Crutsinger, G. M., M. D. Collins, J. A. Fordyce, Z. Compert, C. C. Nice, and N. J.
Sanders. 2006. "Plant Genotypic Diversity Predicts Community Structure and
Governs and Ecosystem Process." *Science* 313 (5789): 966–8. http://dx.doi.org
/10.1126/science.1128326.

Crutzen, P., and E. F. Stoermer. 2000. "The Anthropocene." *Global Change Newsletter*
41:12–3.

Currano, E. D., P. Wilf, S. L. Wing, C. C. Labandeira, E. C. Lovelock, and D. L.
Royer. 2008. "Sharply Increased Insect Herbivory During the Paleoecene-
Eocene Thermal Maximum." *Proceedings of the National Academy of Sciences of
the United States of America* 105 (6): 1960–64. http://dx.doi.org/10.1073/pnas
.0708646105.

Cushing, T. L., and T. J. Straka. 2011. "Extension Forestry and Family Forest Owners: A Data Source." *Journal of Extension* 49 (2): 2–6.

Cusick, D. 2012. "Rapid Climate Changes Turn North Woods into Moose Grave-
yard." *Climatewire*. Accessed May 18, 2012. http://www.scientificamerican.com
/article/rapid-climate-changes-turn-north-woods-into-moose-graveyard/.

Dahm, C., L. Sherson, D. Van Horn, et al. 2013. "Continuous Measurement of
Carbon and Nutrient Dynamics in Streams in Forested Catchments." Paper pre-
sented at the Symposium for European Freshwater Sciences, July 1–5, Münster,
Germany.

Dai, A. 2011. "Drought under Global Warming: A Review." *WIREs Climate Change* 2
(1): 45–65. http://dx.doi.org/10.1002/wcc.81.

Dale, V. H., L. A. Joyce, S. McNulty, R. P. Neilson, M. P. Ayres, M. D. Flannigan, P. J.
Hanson, L. C. Irland, A. E. Lugo, C. J. Peterson, et al. 2001. "Climate Change
and Forest Disturbances." *Bioscience* 51 (9): 723–34. http://dx.doi.org/10.1641
/0006-3568(2001)051[0723:CCAFD]2.0.CO;2.

Davis, L. S., K. N. Johnson, P. Bettinger, and T. E. Howard. 2001. *Forest Management
to Sustain Ecological, Economic, and Social Values*. 4th ed. Long Grove, IL: Wave-
land Press.

Davis, M. B. 1981. "Quaternary History and the Stability of Forest Communities."
In *Forest Succession: Concepts and Application*, ed. D. C. West, H. H. Shugart, and
D. B. Botkin, 132–53. New York: Springer-Verlag. http://dx.doi.org/10.1007/978
-1-4612-5950-3_10.

de la Cretaz, A. L., and P. K. Barten. 2007. *Land Use Effects on Streamflow and Water
Quality in the Northeastern United States*. Boca Raton, FL: CRC Press.

DeFries, R., A. Hansen, A. C. Newton, and M. C. Hansen. 2005. "Increasing Isolation of Protected Areas in Tropical Forests over the Past Twenty Years." *Ecological Applications* 15 (1): 19–26. http://dx.doi.org/10.1890/03-5258.

DeGraaf, R. M., M. Yamasaki, W. B. Leak, and J. W. Lanier. 1992. "New England Wildlife: Management Forested Habitats." General Technical Report NE-144, USDA Forest Service, Northeastern Forest Experiment Station, Radnor, PA.

De Lattre-Gasquet, M. 2006. "The Use of Foresight in Agricultural Research." In *Science and Technology Policy for Development, Dialogues at the Interface*, ed. L. Box and R. Engelhard, 191–214. London: Anthem Press.

Delaware River Basin Conservation Areas and Recommended Strategies. 2011. *Final Report for the National Fish and Wildlife Foundation.* Submitted by The Nature Conservancy, Partnership for the Delaware Estuary, and the Natural Lands Trust.

Delaware River Basin Commission (DRBC). 1982. Interstate Water Management Recommendations of the Parties to the US Supreme Court Decree of 1954 to the Delaware River Basin Commission Pursuant to Commission Resolution 78-20. http://www.nj.gov/drbc/library/documents/regs/GoodFaithRec.pdf.

Delgado, J. A., P. M. Groffman, M. A. Nearing, T. Goddard, D. Reicosky, R. Lal, N. R. Kitchen, C. W. Rice, D. Towery, and P. Salon. 2011. "Conservation Practices to Mitigate and Adapt to Climate Change." *Journal of Soil and Water Conservation* 66: 118A–129A.

DeMenocal, P., J. Ortiz, T. Guilderson, J. Adkins, M. Sarnthein, L. Baker, and M. Yarusinsky. 2000. "Abrupt Onset and Termination of the African Humid Period: Rapid Climate Responses to Gradual Insolation Forcing." *Quaternary Science Reviews* 19 (1-5): 347–61. http://dx.doi.org/10.1016/S0277-3791(99)00081-5.

Denver Water Authority. 2013. "From Forests to Faucets: U.S. Forest Service and Denver Water Watershed Management Partnership." http://www.denverwater.org/supplyplanning/watersupply/partnershipuSFS/.

Derr, T., E. Margolis, M. Savage, et al. 2009. "Santa Fe Municipal Watershed Plan: 2009–2029." New Mexico: Santa Fe.

Dessai, S., M. Hulme, R. Lempert, and R. Jr. Pielke. 2009. "Climate Prediction: A Limit to Adaptation." In *Adapting to Climate Change: Thresholds, Values, Governance*, ed. W. N. Adger, I. Lorenzoni, and K. L. O'Brien, 64–78. Cambridge, UK: Cambridge University Press. http://dx.doi.org/10.1017/CBO9780511596667.006.

Dettinger, M. D. 2006. "A Component-Resampling Approach for Estimating Probability Distributions from Small Forecast Ensembles." *Climatic Change* 76 (1-2): 149–68. http://dx.doi.org/10.1007/s10584-005-9001-6.

de Vernal, A., and C. Hillaire-Marcel. 2008. "Natural Variability of Greenland Climate, Vegetation, and Ice Volume During the Past Million Years." *Science* 320 (5883): 1622–5. http://dx.doi.org/10.1126/science.1153929.

Devine, W., C. Aubry, J. Miller, K. Potter, and A. Bower. 2012. *Climate Change and Forest Trees in the Pacific Northwest: Guide to Vulnerability Assessment Methodology.* Olympia, WA: USDA, Forest Service, Pacific Northwest Region.

Dewhirst, S., and F. Lutscher. 2009. "Dispersal in Heterogeneous Habitats: Thresholds, Spatial Scales, and Approximate Rates of Spread." *Ecology* 90 (5): 1338–45. http://dx.doi.org/10.1890/08-0115.1.

Dickens, G. R. 2011. "Down the Rabbit Hole: Toward Appropriate Discussion of Methane Release from Gas Hydrate Systems During the Paleocene-Eocene Thermal Maximum and Other Past Hyperthermal Events." *Climate of the Past* 7 (3): 831–46. http://dx.doi.org/10.5194/cp-7-831-2011.

Dietze, M., and P. Moorcroft. 2011. "Tree Mortality in the Eastern and Central United States: Patterns and Drivers." *Global Change Biology* 17 (11): 3312–26. http://dx.doi.org/10.1111/j.1365-2486.2011.02477.x.

Diez, J. M., C. M. D'Antonio, J. S. Dukes, Edwin D Grosholz, Julian D Olden, Cascade JB Sorte, Dana M Blumenthal, Bethany A Bradley, Regan Early, Inés Ibáñez, et al. 2012. "Will Extreme Climatic Events Facilitate Biological Invasions?" *Frontiers in Ecology and the Environment* 10 (5): 249–57. http://dx.doi.org/10.1890/110137.

Dilling, L., and M. C. Lemos. 2011. "Creating Usable Science: Opportunities and Constraints for Climate Knowledge Use and Their Implications for Science Policy." *Global Environmental Change* 21 (2): 680–9. http://dx.doi.org/10.1016/j.gloen vcha.2010.11.006.

Dingle, D. H. 1996. *Migration: The Biology of Life on the Move*. New York: Oxford University Press.

Dodds, K. J., P. de Groot, and D. A. Orwig. 2010. "The Impact of Sirex Noctilio in Pinus Resinosa and Pinus Sylvestris Stands in New York and Ontario." *Canadian Journal of Forest Research* 40 (2): 212–23. http://dx.doi.org/10.1139/X09-181.

Dodds, W. K., W. W. Bouska, J. L. Eitzmann, T. J. Pilger, K. L. Pitts, A. J. Riley, J. T. Schloesser, and D. J. Thornbrugh. 2009. "Eutrophication of U.S. Freshwaters: Analysis of Potential Economic Damages." *Environmental Science & Technology* 43 (1): 12–9. http://dx.doi.org/10.1021/es801217q.

Dohrenwend, R. E. 1977. *Raindrop Erosion in the Forest*. Research Note No. 24. Ford Forestry Center.

Donnegan, J. A., T. T. Veblen, and J. S. Sibold. 2001. "Climatic and Human Influences on Fire History in Pike National Forest, Central Colorado." *Canadian Journal of Forest Research* 31 (9): 1526–39. http://dx.doi.org/10.1139/x01-093.

Dorcas, M. E., J. D. Willson, R. N. Reed, R. W. Snow, M. R. Rochford, M. A. Miller, W. E. Meshaka, P. T. Andreadis, F. J. Mazzotti, C. M. Romagosa, et al. 2012. "Severe Mammal Declines Coincide with Proliferation of Invasive Burmese Pythons in Everglades National Park." *Proceedings of the National Academy of Sciences of the United States of America* 109 (7): 2418–22. http://dx.doi.org/10.1073/pnas.1115226109.

Dowdle, B. 1984. "Why Have We Retained the Federal Lands? An Alternative Hypothesis." In *Rethinking the Federal Lands*, ed. S. Brubaker, 61–73. Washington, DC: Resources for the Future, Inc.

Drummond, M. A., and T. R. Loveland. 2010. "Land-use Pressure and a Transition to Forest-Cover Loss in the Eastern United States." *Bioscience* 60 (4): 286–98. http://dx.doi.org/10.1525/bio.2010.60.4.7.

Dudley, N., and S. Stolton. 2003. "Running Pure: The Importance of Forest Protection to Drinking Water: A Research Report for the World Bank/WWF Alliance for Forest Conservation and Sustainable Use." http://d2ouvy59p0dg6k.cloud front.net/downloads/runningpurereport.pdf.

Eby, M., K. Zickfeld, A. Montenegro, D. Archer, K. J. Meissner, and A. J. Weaver. 2008. "Lifetime of Anthropogenic Climate Change: Millennial Time Scales of Potential CO_2 and Surface Temperature Perturbations." *Journal of Climate* 22 (10): 2501–11. http://dx.doi.org/10.1175/2008JCLI2554.1.

Ecological Restoration Institute. 2013. "The Efficacy of Hazardous Fuels Treatments: A Rapid Assessment of the Economic and Ecological Consequences of Alternative Hazardous Fuels Treatments." Northern Arizona University. http://library.eri.nau.edu/gsdl/collect/erilibra/index/assoc/D2013004.dir/doc.pdf.

Egan, T. 2009. *The Big Burn: Teddy Roosevelt and the Fire that Saved America*. New York: Houghton Mifflin Harcourt.

Ellison, A. M., M. S. Bank, D. B. Clinton, E. A. Colburn, K. Elliott, C. R. Ford, D. R. Foster, B. D. Kloeppel, J. D. Knoepp, G. M. Lovett, et al. 2005. "Loss of Foundation Species: Consequences for the Structure and Dynamics of Forested Ecosystems." *Frontiers in Ecology and the Environment* 3 (9): 479–86. http://dx.doi.org/10.1890/1540-9295(2005)003[0479:LOFSCF]2.0.CO;2.

Ellum, D. 2009. "Proactive Coevolution: Staying Ahead of Invasive Species in the Face of Climate Change and Uncertainty." *Forest Wisdom* (13): 1–3.

Elsner, M., L. Cuo, N. Voisin, J. S. Deems, A. F. Hamlet, J. A. Vano, K.E.B. Mickelson, S.-Y. Lee, and D. P. Lettenmaier. 2010. "Implications of 21st Century Climate Change for the Hydrology of Washington State." *Climatic Change* 102 (1-2): 225–60. http://dx.doi.org/10.1007/s10584-010-9855-0.

Erickson, V., C. Aubry, P. Berrang, R. Blush, A. Bower, B. Crane, T. DeSpain, J. Hamlin, M. Horning, R. Johnson, et al. 2012. *Genetic Resource Management and Climate Change: Genetic Options for Adapting National Forests to Climate Change*. Washington, DC: USDA Forest Service; http://www.fs.usda.gov/Internet/FSE _DOCUMENTS/stelprdb5368468.pdf.

Eschtruth, A. K., N. L. Cleavitt, J. J. Battles, Richard A Evans, and Timothy J Fahey. 2006. "Vegetation Dynamics in Declining Eastern Hemlock Stands: 9 Years of Forest Response to Hemlock Woolly Adelgid Infestation." *Canadian Journal of Forest Research* 36 (6): 1435–50. http://dx.doi.org/10.1139/x06-050.

Evans, A. M. 2008. "Growth and Infestation by Hemlock Woolly Adelgid of Two Exotic Hemlock Species in a New England Forest." *Journal of Sustainable Forestry* 26 (3): 223–40. http://dx.doi.org/10.1080/10549810701879735.

Evans, A. M., and A. J. Finkral. 2010. "A New Look at Spread Rates of Exotic Diseases in North American Forests." *Forest Science* 56 (5): 453–9.

Evans, A. M., and T. G. Gregoire. 2007. "A Geographically Variable Model of Hemlock Woolly Adelgid Spread." *Biological Invasions* 9 (4): 369–82. http://dx.doi.org/10.1007/s10530-006-9039-z.

Executive Office of the President. 2003. "President Bush Signs Healthy Forests Restoration Act into Law." Washington, DC: White House archives. Accessed March 19, 2016. http://georgewbush-whitehouse.archives.gov/infocus/healthyforests/.

"Executive Order 13112 of February 3, 1999: Establishing the National Invasive Species Council." 1999. GPO: Federal Register vol. 64, no. 25, Feb 8.

"Executive Order 13653 of Nov. 1, 2013: Preparing the United States for the Impacts of Climate Change." 2013. Washington, DC: The White House, Office of the Press Secretary. Accessed Jan 29, 2014. http://www.whitehouse.gov/the-press-office/2013/11/01/executive-order-preparing-united-states-impacts-climate-change.

Fa, J. E., C. A. Peres, and J. Meeuwig. 2002. "Bushmeat Exploitation in Tropical Forests: An Intercontinental Comparison." *Conservation Biology* 16 (1): 232–7. http://dx.doi.org/10.1046/j.1523-1739.2002.00275.x.

Falk, D. A., E. K. Heyerdahl, P. M. Brown, C. Farris, P. Z Fulé, D. McKenzie, T. W. Swetnam, A. H. Taylor, and M. L. Van Horne. 2011. "Multi-scale Controls of Historical Forest-fire Regimes: New Insights from Fire-scar Networks." *Frontiers in Ecology and the Environment* 9 (8): 446–54. http://dx.doi.org/10.1890/100052.

Fankhauser, S., J. B. Smith, and R. S. Tol. 1999. "Weathering Climate Change: Some Simple Rules to Guide Adaptation Decisions." *Ecological Economics* 30 (1): 67–78. http://dx.doi.org/10.1016/S0921-8009(98)00117-7.

Federal Forest Advisory Committee. 2012. "National Forest Health Restoration: An Economic Assessment of Forest Restoration on Oregon's Eastside National Forests." Prepared for Governor John Kitzhaber and Oregon's Legislative Leaders. http://www.oregon.gov/odf/BOARD/docs/2013_January/BOFATTCH_20130109_08_03.pdf.

Federal Register. 2012. "Final Rule and Record of Decision." 36 CFR Part 219. USDA, Forest Service, National Forest System Land Management Planning." Federal Register 77 (68) 9 April.

Feulner, G. 2009. "Climate Modeling of Mass-extinction Events: A Review." *International Journal of Astrobiology* 8 (03): 207–12. http://dx.doi.org/10.1017/S1473550409990061.

Finley, B. 2013. "Feds Project Climate Change Will Double Wildfire Risk in Forests." *Denver Post*, April 3, 2013. http://www.denverpost.com/ci_22943189/feds-project-climate-change-will-double wildfire-risk#ixzz2mvvPfrniU.

Finney, M. A., C. W. McHugh, and I. C. Grenfell. 2005. "Stand- and Landscape-level Effects of Prescribed Burning on Two Arizona Wildfires." *Canadian Journal of Forest Research* 35 (7): 1714–22. http://dx.doi.org/10.1139/x05-090.

Flagstaff Watershed Protection Project. 2012. "Project Background." http://www.flagstaffwatershedprotection.org/about/background/.

Flannery, T. 2001. *The Eternal Frontier: An Ecological History of North America and Its Peoples*. New York: Atlantic Monthly Press.

Flombaum, P., and O. E. Sala. 2008. "Higher Effect of Plant Species Diversity on Productivity in Natural than Artificial Ecosystems." *Proceedings of the National Academy of Sciences of the United States of America* 105 (16): 6087–90. http://dx.doi.org/10.1073/pnas.0704801105.

Forseth, I. N., and A. F. Innis. 2004. "Kudzu (Pueraria Montana): History, Physiology, and Ecology Combine to Make a Major Ecosystem Threat." *Critical Reviews in Plant Sciences* 23 (5): 401–13. http://dx.doi.org/10.1080/07352680490505150.

Foxcroft, L. C., V. Jarošík, P. Pyšek, D. M. Richardson, and M. Rouget. 2011. "Protected-Area Boundaries as Filters of Plant Invasions." *Conservation Biology* 25:400–5.

Francis, J. A., and S. J. Vavrus. 2012. "Evidence Linking Arctic Amplification to Extreme Weather in Mid-latitudes." *Geophysical Research Letters* 39 (6): L06801. http://dx.doi.org/10.1029/2012GL051000.

Franklin, J. F., R. J. Mitchell, and B. J. Palik. 2007. "Natural Disturbance and Stand Development Principles for Ecological Forestry." Gen.Tech. Rep. NRS-19. Newton Square, PA: USDA, Forest Service, Northern Research Station.

Frechette, J. D., and G. A. Meyer. 2009. "Holocene Fire-Related Alluvial-Fan Deposition and Climate in Ponderosa Pine and Mixed-Conifer Forests, Sacramento Mountains, New Mexico, USA." *The Holocene* 19 (4): 639–51.

Freed, L. A., R. L. Cann, M. L. Goff, W. A. Kuntz, and G. R. Bodner. 2005. "Increase in Avian Malaria at Upper Elevation in Hawai'i." *Condor* 107 (4): 753–64. http://dx.doi.org/10.1650/7820.1.

Fulé, P. Z. 2008. "Does It Make Sense to Restore Wildland Fire in Changing Climate?" *Restoration Ecology* 16 (4): 526–31. http://dx.doi.org/10.1111/j.1526-100X.2008.00489.x.

Fulé, P. Z., W. W. Covington, and M. M. Moore. 1997. "Determining Reference Conditions for Restoration of Southwestern Ponderosa Pine Forests." *Ecological Applications* 7 (3): 895–908. http://dx.doi.org/10.1890/1051-0761(1997)007[0895:DRCFEM]2.0.CO;2.

Fulé, P. Z., J. E. Crouse, T. A. Heinlein, M. M. Moore, W. W. Covington, and G. Verkamp. 2003. "Mixed-Severity Fire Regime in a High-Elevation Forest: Grand Canyon, Arizona." *Landscape Ecology* 18 (5): 465–86. http://dx.doi.org/10.1023/A:1026012118011.

Fulé, P.Z., J. E. Crouse, J. P. Roccaforte, and E. L. Kalies. 2012. "Do Thinning and/or Burning Treatments in Western USA Ponderosa or Jeffrey Pine-Dominated Forests Help Restore Natural Fire Behavior?" *Forest Ecology and Management* 269: 68–81.

Fuller, R. A., E. McDonald-Madden, K. A. Wilson, J. Carwardine, H. S. Grantham, J. E. M. Watson, C. J. Klein, D. C. Green, and H. P. Possingham. 2010. "Replacing

Underperforming Protected Areas Achieves Better Conservation Outcomes." *Nature* 466: 365–67.

Funtowicz, S. O., and J. R. Ravetz. 1993. "Science for a Post-Normal Age." *Futures* 25 (7): 739–55. http://dx.doi.org/10.1016/0016-3287(93)90022-L.

Gabbert, B. 2010. "Cerro Grande Fire, 10 Years Ago Today." *Wildfire Today blog*: http://wildfiretoday.com/2010/05/10/cerro-grande-fire-10-years-ago-today/.

Gaines, W. L., D. W. Peterson, C. A. Thomas, and R. J. Harrod. 2012. "Adaptations to Climate Change: Colville and Okanogan-Wenatchee National Forests." Gen. Tech. Rep. PNW-GTR-862. Portland, OR: USDA, Forest Service, Pacific Northwest Research Station.

Galatowitsch, S., L. Frelich, and Laura Phillips-Mao. 2009. "Regional Climate Change Adaptation Strategies for Biodiversity Conservation in a Midcontinental Region of North America." *Biological Conservation* 142 (10): 2012–22. http://dx.doi.org/10.1016/j.biocon.2009.03.030.

Gandhi, K., and D. Herms. 2010a. "Direct and Indirect Effects of Alien Insect Herbivores on Ecological Processes and Interactions in Forests of Eastern North America." *Biological Invasions* 12 (2): 389–405. http://dx.doi.org/10.1007/s10530-009-9627-9.

Gandhi, K., and D. Herms. 2010b. "North American Arthropods at Risk Due to Widespread Fraxinus Mortality Caused by the Alien Emerald Ash Borer." *Biological Invasions* 12 (6): 1839–46. http://dx.doi.org/10.1007/s10530-009-9594-1.

Government Accountability Office (GAO). 2004. "Wildfire Suppression Funding Transfers Cause Project Cancellations and Delays, Strained Relationships, and Management Disruptions." GAO-04-612. Washington, DC: US Government Accountability Office.

Government Accountability Office (GAO). 2007. "Climate Change: Agencies Should Develop Guidance for Addressing the Effects on Federal Land and Water Resources." GAO-07-863. Washington, DC: US Government Accountability Office.

Government Accountability Office (GAO). 2009. "Climate Change Adaptation: Strategic Federal Planning Could Help Government Officials Make More Informed Decisions." GAO-10-113. Washington, DC: US Government Accountability Office.

Gartner, M. H., T. T. Veblen, R. L. Sherriff, and T. L. Schoennagel. 2012. "Proximity to Grasslands Influences Fire Frequency and Sensitivity to Climate Variability in Ponderosa Pine Forests of the Colorado Front Range." *International Journal of Wildland Fire* 21 (5): 562–71. http://dx.doi.org/10.1071/WF10103.

Gartner, T., J. Mulligan, R. Schmidt, and J. Gunn. 2013. *Natural Infrastructure: Investing in Forested Landscapes for Source Water Protection in the United States.* Washington, DC: World Resources Institute.

Garzione, C. N. 2008. "Surface Uplift of Tibet and Cenozoic Global Cooling." *Geology* 36 (12): 1003–4. http://dx.doi.org/10.1130/focus122008.1.

Gavier-Pizarro, G. I., V. C. Radeloff, S. I. Stewart, Cynthia D. Huebner, and Nicholas S. Keuler. 2010. "Housing Is Positively Associated With Invasive Exotic Plant Species Richness in New England, USA." *Ecological Applications* 20 (7): 1913–25. http://dx.doi.org/10.1890/09-2168.1.

Geldmann, J., M. Barnes, L. Coad, I. D. Craigie, M. Hockings, and N. D. Burgess. 2013. "Effectiveness of Terrestrial Protected Areas in Reducing Habitat Loss and Population Declines." *Biological Conservation* 161:230–8. http://dx.doi.org/10.1016/j.biocon.2013.02.018.

Gereta, E., E. Mwangomo, and E. Wolanski. 2009. "Ecohydrology as a Tool For the Survival of the Threatened Serengeti Ecosystem." *Ecohydrology & Hydrobiology* 9 (1): 115–24. http://dx.doi.org/10.2478/v10104-009-0035-7.

Gerlach, T. 2011. "Volcanic Versus Anthropogenic Carbon Dioxide." *EOS* 92 (24): 201–2. http://dx.doi.org/10.1029/2011EO240001.

Gershunov, A., B. Rajagopalan, J. Overpeck, K. Guirguis, D. Cayan, M. Hughes, M. Dettinger, C. Castro, R. E. Schwartz, M. Anderson, et al. 2013. "Future Climate: Projected Extremes." In *Assessment of Climate Change in the Southwest United States: A Report Prepared for the National Climate Assessment, Southwest Climate Alliance*, ed. G. Garfin, A. Jardine, R. Merideth, M. Black, and S. LeRoy, 126–47. Washington, DC: Island Press. http://dx.doi.org/10.5822/978-1-61091-484-0_7.

Geyer, W. R., J. T. Morris, F. G. Prahl, and D. A. Jay. 2000. "Interaction between Physical Processes and Ecosystem Structure: A Comparative Approach." In *Estuarine Science: A Synthetic Approach to Research and Practice*, ed. J. E. Hobbie, 177–206. Washington, DC: Island Press.

Gibbs, J. N., and D. Wainhouse. 1986. "Spread of Forest Pests and Pathogens in the Northern Hemisphere." *Forestry* 59 (2): 141–53. http://dx.doi.org/10.1093/forestry/59.2.141.

Gill, R.M.A., and V. Beardall. 2001. "The Impact of Deer on Woodlands: The Effects of Browsing and Seed Dispersal on Vegetation Structure and Composition." *Forestry* 74 (3): 209–18. http://dx.doi.org/10.1093/forestry/74.3.209.

Gillett, N. P., A. J. Weaver, F. W. Zwiers, and M. D. Fannigan. 2004. "Detecting the Effect of Climate Change on Canadian Forest Fires." *Geophysical Research Letters* 31 (18): L18211. http://dx.doi.org/10.1029/2004GL020876.

Girvetz, E. H., C. Zganjar, G. T. Raber, E. P. Maurer, P. Kareiva, and J. J. Lawler. 2009. "Applied Climate-Change Analysis: The Climate Wizard Tool." *PLoS ONE* 4 (12): e8320. http://dx.doi.org/10.1371/journal.pone.0008320.

Glick, P., B. Stein, and N. Edelson. 2011. *Scanning the Conservation Horizon: A Guide to Climate Change Vulnerability Assessment*. Washington, DC: National Wildlife Federation.

Goforth, B. R., and R. A. Minnich. 2008. "Densification, Stand-Replacement Wildfire, and Extirpation of Mixed Conifer Forest in Cuyamaca Rancho State Park, Southern California." *Forest Ecology and Management* 256 (1-2): 36–45. http://dx.doi.org/10.1016/j.foreco.2008.03.032.

Goldman-Benner, R. L., S. Benitez, A. Calvache, A. Ramos, and F. Veiga. 2013. "Water Funds: A New Ecosystem Service and Biodiversity Conservation Strategy." In *Encyclopedia of Biodiversity*, 2nd ed., vol. 7. ed. S. A. Levin, 352–66. Waltham, MA: Academic Press. http://dx.doi.org/10.1016/B978-0-12-384719-5.00330-0.

Gómez-Aparicio, L., and C. D. Canham. 2008. "Neighbourhood Analyses of the Allelopathic Effects of the Invasive Tree Ailanthus Altissima in Temperate Forests." *Journal of Ecology* 96 (3): 447–58. http://dx.doi.org/10.1111/j.1365-2745.2007.01352.x.

Gong, G., L. Wang, L. Condon, A. Shearman, and U. Lall. 2010. "A Simple Framework for Incorporating Seasonal Streamflow Forecasts into Existing Water Resource Management Practices." *Journal of the American Water Resources Association* 46 (3): 574–85.

Gotham, D. J., J. R. Angel, and S. C. Pryor. 2012. "Vulnerability of the Electricity and Water Sectors to Climate Change in the Midwest." In *Climate Change in the Midwest: Impacts, Risks, Vulnerability and Adaptation*, ed. S. C. Pryor, 158–77. Indianapolis: Indiana University Press.

Graham, R., M. Finney, C. McHugh, J. Cohen, D. Calkin, R. Stratton, L. Bradshaw, and N. Nikolov. 2012. "Fourmile Canyon Fire Findings." Gen. Tech. Rep. RMRS-GTR-289. Fort Collins, CO: USDA, Forest Service, Rocky Mountain Research Station; http://www.fs.fed.us/rm/pubs/rmrs_gtr289.pdf.

Granoszewski, W., D. Demske, M. Nita, G. Heumann, and A. A. Andreev. 2004. "Vegetation and Climate Variability During the Last Interglacial Evidenced in the Pollen Record from Lake Baikal." *Global and Planetary Change* 46 (1-4): 187–98. http://dx.doi.org/10.1016/j.gloplacha.2004.09.017.

Grant, G. E., C. L. Tague, and C. D. Allen. 2013. "Watering the Forest for the Trees: An Emerging Priority for Managing Water in Forest Landscapes." *Frontiers in Ecology and the Environment* 11 (6): 314–21. http://dx.doi.org/10.1890/120209.

Gray, L. K., T. Gylander, M. S. Mbogga, Pei-yu Chen, and Andreas Hamann. 2011. "Assisted Migration to Address Climate Change: Recommendations for Aspen Reforestation in Western Canada." *Ecological Applications* 21 (5): 1591–603. http://dx.doi.org/10.1890/10-1054.1.

Gray, L. K., and A. Hamann. 2013. "Tracking Suitable Habitat for Tree Populations under Climate Change in Western North America." *Climatic Change* 117 (1-2): 289–303. http://dx.doi.org/10.1007/s10584-012-0548-8.

Grégoire, J. M., and D. Simonetti. 2010. "Interannual Changes of Fire Activity in the Protected Areas of the SUN Network and Other Parks and Reserves of the West and Central Africa Region Derived from MODIS Observations." *Remote Sensing* 2 (2): 446–63. http://dx.doi.org/10.3390/rs2020446.

Grissino-Mayer, H., W. Romme, M. Floyd, and D. Hanna. 2004. "Long-term Climatic and Human Influences on Fire Regimes of the San Juan National Forest, Southwestern Colorado, USA." *Ecology* 85:1708–24. http://dx.doi.org/10.1890/02-0425.

Grissino-Mayer, H. D., and T. W. Swetnam. 2000. "Century-Scale Climate Forcing of Fire Regimes in the American Southwest." *Holocene* 10 (2): 213–20. http://dx .doi.org/10.1191/095968300668451235.

Groom, J., L. Dent, and L. Madsen. 2011. "Stream Temperature Change Detection for State and Private Forests in the Oregon Coast Range." *Water Resources Research* 47 (1): 1–12. http://dx.doi.org/10.1029/2009WR009061.

Gunderson, L. H. 2000. "Ecological Resilience—In Theory and Application." *Annual Review of Ecology and Systematics* 31 (1): 425–39. http://dx.doi.org/10.1146 /annurev.ecolsys.31.1.425.

Gunn, J. S., J. M. Hagan, and A. A. Whitman. 2009. *Forestry Adaptation and Mitigation in a Changing Climate—A Forest Resource Manager's Guide For the Northeastern United States*. Brunswick, ME: Natural Capital Initiative at Manomet.

Gurevitch, J., and D. K. Padilla. 2004. "Are Invasive Species a Major Cause of Extinctions?" *Trends in Ecology & Evolution* 19 (9): 470–4. http://dx.doi.org/10.1016/j .tree.2004.07.005.

Gutzler, D. S., and T. O. Robbins. 2011. "Climate Variability and Projected Change in the Western United States: Regional Downscaling and Drought Statistics." *Climate Dynamics* 37 (5-6): 835–49. http://dx.doi.org/10.1007/s00382-010-0838-7.

Haag, T., A. S. Santos, D. A. Sana, R. G. Morato, L. Cullen, Jr., P. G. Crawshaw, Jr., C. De Angelo, M. S. Di Bitetti, F. M. Salzano, and E. Eizirik. 2010. "The Effect of Habitat Fragmentation on the Genetic Structure of a Top Predator: Loss of Diversity and High Differentiation among Remnant Populations of Atlantic Forest Jaguars (Panthera onca)." *Molecular Ecology* 19 (22): 4906–21. http://dx.doi .org/10.1111/j.1365-294X.2010.04856.x.

Hall, M.H.P., and D. B. Fagre. 2003. "Modeled Climate-induced Glacier Change in Glacier National Park, 1850–2100." *Bioscience* 53 (2): 131–40. http://dx.doi.org /10.1641/0006-3568(2003)053[0131:MCIGCI]2.0.CO;2.

Halofsky, J. E., D. L. Peterson, K. O'Halloran, and C. Hoffman, eds. 2011. "Adapting to Climate Change at Olympic National Forest and Olympic National Park." General Technical Report PNW-GTR-844. Portland, OR: USDA Forest Service.

Hannah, L. ed. 2012. *Saving a Million Species: Extinction Risk from Climate Change*. Washington, DC: Island Press.

Hannah, L., G. Midgley, T. Lovejoy, W. J. Bond, M. Bush, J. C. Lovett, D. Scott, and F. I. Woodward. 2002. "Conservation of Biodiversity in a Changing Climate." *Conservation Biology* 16 (1): 264–68. http://dx.doi.org/10.1046/j.1523-1739.2002 .00465.x.

Hansen, A. J., and R. DeFries. 2007. "Ecological Mechanisms Linking Protected Areas to Surrounding Lands." *Ecological Applications* 17 (4): 974–88. http://dx.doi .org/10.1890/05-1098.

Hanson, C., L. Yonavjak, C. Clarke, S. Minnemeyer, L. Boisrobert, A. Leach, and K. Schleeweis. 2010. *Southern Forests for the Future*. Washington, DC: World Resources Institute.

Harrison, S. 1994. "Metapopulations and Conservation." In *Large Scale Ecology and Conservation Biology*, ed. P. Edwards, R. May, and N. Webb, 118–28. Oxford: Blackwell Scientific Publications.

Hartsough, B. R., S. Abrams, R. J. Barbour, E. S. Drews, J. D. McIver, J. J. Moghaddas, D. W. Schwilk, and S. L. Stephens. 2008. "The Economics of Alternative Fuel Reduction Treatments in Western United States Dry Forests: Financial and Policy Implications from the National Fire and Fire Surrogate Study." *Forest Economics and Policy* 10 (6): 344–54. http://dx.doi.org/10.1016/j.forpol.2008.02.001.

Hayden, K. J., A. Nettel, R. S. Dodd, and M. Garbelotto. 2011. "Will All the Trees Fall? Variable Resistance to an Introduced Forest Disease in a Highly Susceptible Host." *Forest Ecology and Management* 261 (11): 1781–91. http://dx.doi.org/10.1016/j.foreco.2011.01.042.

Hayes, J. L., and I. Ragenovich. 2001. "Non-native Invasive Forest Insects of Eastern Oregon and Washington." *Northwest Science* 75 (Spec. Issue): 77–84.

Hebblewhite, M. 2011. "Effects of Energy Development on Ungulates." In *Energy Development and Wildlife Conservation in Western North America*, ed. D. E. Naugle, 71–94. Washington, DC: Island Press. http://dx.doi.org/10.5822/978-1-61091-022-4_5.

Hebda, R. J. 2008. "Climate Change, Forests, and the Forest Nursery Industry." In *National Proceedings: Forest and Conservation Nursery Associations—2007*, ed. R. K. Dumroese and L. E. Riley, 81–82. RMRS-P-57. Fort Collins, CO: USDA, Forest Service Rocky Mountain Research Station.

Hedde, C. 2012. *U.S. Natural Catastrophe Update*. Princeton, NJ: Munich Reinsurance America, Inc.

Heffernan, O. 2010. "Earth Science: The Climate Machine." *Nature* 463 (7284): 1014–6. http://dx.doi.org/10.1038/4631014a.

Heilig, G. K. 2003. "Multifunctionality of Landscapes and Ecosystem Services with Respect to Rural Development." In *Sustainable Development of Multifunctional Landscapes*, ed. K. Helming and H. Wiggering, 39–51. Berlin: Springer-Verlag. http://dx.doi.org/10.1007/978-3-662-05240-2_3.

Helford, R. M. 2000. "Constructing Nature as Constructing Science: Expertise, Activist Science, and Public Conflict In the Chicago Wilderness." In *Restoring Nature: Perspectives from the Social Sciences and Humanities*, ed. P. H. Gobster and R. B. Hull, 119–42. Washington, DC: Island Press.

Heller, N. E., and E. S. Zavaleta. 2009. "Biodiversity Management in the Face of Climate Change: A Review of 22 Years of Recommendations." *Biological Conservation* 142 (1): 14–32. http://dx.doi.org/10.1016/j.biocon.2008.10.006.

Hessburg, P. F., K. M. Reynolds, R. B. Salter, J. Dickinson, W. Gaines, and R. Harrod. 2013. "Landscape Evaluation for Restoration Planning on the Okanogan-Wenatchee National Forest, USA." *Sustainability* 5 (3): 805–40. http://dx.doi.org/10.3390/su5030805.

Hewitt, N., N. Klenk, A. L. Smith, D. R. Bazely, N. Yan, S. Wood, J. I. MacLellan, C. Lipsig-Mumme, and I. Henriques. 2011. "Taking Stock of the Assisted Migration Debate." *Biological Conservation* 144 (11): 2560–72. http://dx.doi.org/10.1016/j.biocon.2011.04.031.

Hicke, J. A., A. J. H. Meddens, C. D. Allen, and C. A. Kolden. 2013. "Carbon Stocks of Trees Killed by Bark Beetles and Wildfire in the Western United States." *Environmental Research Letters* 8 (3): 035032. http://dx.doi.org/10.1088/1748-9326/8/3/035032.

Hidalgo, H. G., T. Das, M. D. Dettinger, D. R. Cayan, D. W. Pierce, T. P. Barnett, G. Bala, A. Mirin, A. W. Wood, C. Bonfils, et al. 2009. "Detection and Attribution of Streamflow Timing Changes to Climate Change in the Western United States." *Journal of Climate* 22 (13): 3838–55. http://dx.doi.org/10.1175/2009JCLI2470.1.

Higuera, P. E., C. Whitlock, and J. A. Gage. 2010. "Linking Tree-Ring and Sediment-Charcoal Records to Reconstruct Fire Occurrence and Area Burned in Subalpine Forests of Yellowstone National Park, USA." *Holocene* 21 (2): 327–41. http://dx.doi.org/10.1177/0959683610374882.

Hobbs, R. J., S. Arico, J. Aronson, J. S. Baron, P. Bridgewater, V. A. Cramer, P. R. Epstein, J. J. Ewel, C. A. Klink, A. E. Lugo, et al. 2006. "Novel Ecosystems: Theoretical and Management Aspects of the New Ecological World Order." *Global Ecology and Biogeography* 15 (1): 1–7. http://dx.doi.org/10.1111/j.1466-822X.2006.00212.x.

Hoegh-Guldberg, O., L. Hughes, S. McIntyre, D. B. Lindenmayer, C. Parmesan, H. P. Possingham, and C. D. Thomas. 2008. "Assisted Colonization and Rapid Climate Change." *Science* 321 (5887): 345–6. http://dx.doi.org/10.1126/science.1157897.

Holden, Z. A., P. Morgan, M. A. Crimmins, R. K. Steinhorst, and A.M.S. Smith. 2007. "Fire Season Precipitation Variability Influences Fire Extent and Severity in a Large Southwestern Wilderness Area, United States." *Geophysical Research Letters* 34 (16): n/a. http://dx.doi.org/10.1029/2007GL030804.

Hole, D. G., S. G. Willis, D. J. Pain, L. D. Fishpool, S.H.M. Butchart, Y. C. Collingham, C. Rahbek, and B. Huntley. 2009. "Projected Impacts of Climate Change on a Continent-wide Protected Area Network." *Ecology Letters* 12 (5): 420–31. http://dx.doi.org/10.1111/j.1461-0248.2009.01297.x.

Holling, C. S. 1973. "Resilience and Stability of Ecological Systems." *Annual Review of Ecology and Systematics* 4 (1): 1–23. http://dx.doi.org/10.1146/annurev.es.04.110173.000245.

Holmes, T. P., J. E. Aukema, B. Von Holle, A. Liebhold, and E. Sills. 2009. "Economic Impacts of Invasive Species in Forests." *Annals of the New York Academy of Sciences* 1162 (1): 18–38. http://dx.doi.org/10.1111/j.1749-6632.2009.04446.x.

Hooper, D. U., F. S. Chapin, III, J. J. Ewel, A. Hector, P. Inchausti, S. Lavorel, J. H. Lawton, D. M. Lodge, M. Loreau, S. Naeem, et al. 2005. "Effects of Biodiversity on Ecosystem Functioning: A Consensus of Current Knowledge." *Ecological Monographs* 75 (1): 3–35. http://dx.doi.org/10.1890/04-0922.

Hopkins, A. D. 1920. "The Bioclimatic Law." *Monthly Weather Review* 48 (6): 355. http://dx.doi.org/10.1175/1520-0493(1920)48<355a:TBL>2.0.CO;2.

Hornbeck, J. W., M. B. Adams, E. S. Corbett, E. S. Verry, and J. A. Lynch. 1995. "A Summary of Water Yield Experiments on Hardwood Forested Watersheds in Northeastern United States." In *Proceedings, 10th Central Hardwood Forest Conference*, ed. K. W. Gottschalk, and S. L. C. Fosbroke, 282–95. Gen. Tech. Rep. NE-197. Radnor, PA: US Department of Agriculture, Forest Service, Northeastern Forest Experiment Station.

Horvitz, C. C., J. B. Pascarella, S. McMann, A. Freedman, and R. H. Hofstetter. 1998. "Functional Roles of Invasive Non-indigenous Plants in Hurricane-affected Subtropical Hardwood Forests." *Ecological Applications* 8 (4): 947–74. http://dx.doi.org/10.1890/1051-0761(1998)008[0947:FROINI]2.0.CO;2.

Houghton, J. T., Y. Ding, D. J. Griggs, M. Nouger, P. J. van der Linden, X. Dai, K. Maskell, and C. A. Johnson. 2001. *Climate Change 2001: The Scientific Basis: Contribution of Working Group I to the Third Assessment Report of the Intergovernmental Panel on Climate Change*. London: Cambridge University Press.

Howe, G. T., J. B. St. Clair, and R. Beloin. 2009. "Seedlot Selection Tool." Accessed March 19, 2016. http://sst.forestry.oregonstate.edu/.

Huebner, C. D., and P. C. Tobin. 2006. "Invasibility of Mature and 15-year-old Deciduous Forests by Exotic Plants." *Plant Ecology* 186 (1): 57–68. http://dx.doi.org/10.1007/s11258-006-9112-9.

Hulme, M., R. Pielke, and S. Dessai. 2009. "Keeping Prediction in Perspective." *Nature Climate Change* 3 (0911): 126–7. http://dx.doi.org/10.1038/climate.2009.110.

Hunter, M. L., Jr., G. L. Jacobson, Jr., and T. Webb, III. 1988. "Paleoecology and the Coarse-Filter Approach to Maintaining Biological Diversity." *Conservation Biology* 2 (4): 375–85. http://dx.doi.org/10.1111/j.1523-1739.1988.tb00202.x.

Huntington, T. G. 2003. "Climate Warming Could Reduce Runoff Significantly in New England, USA." *Agricultural and Forest Meteorology* 117 (3–4): 193–201. http://dx.doi.org/10.1016/S0168-1923(03)00063-7.

Huntington, T. G., A. D. Richardson, K. J. McGuire, and K. Hayhoe. 2009. "Climate and Hydrological Changes in the Northeastern United States: Recent Trends and Implications for Forested and Aquatic Ecosystems." *Canadian Journal of Forest Research* 39 (2): 199–212. http://dx.doi.org/10.1139/X08-116.

Hurd, B. H., M. Callaway, J. Smith, and P. Kirshen. 2004. "Climatic Change and US Water Resources: From Modeled Watershed Impacts to National Estimates." *Journal of the American Water Resources Association* 40 (1): 129–48. http://dx.doi.org/10.1111/j.1752-1688.2004.tb01015.x.

Huston, M. A. 1994. *Biological Diversity: The Coexistence of Species*. New York: Cambridge University Press.

Hutto, R. L., and R. T. Belote. 2013. "Distinguishing Four Types of Monitoring Based on the Questions They Address." *Forest Ecology and Management* 289:183–9. http://dx.doi.org/10.1016/j.foreco.2012.10.005.

Iniguez, J. M., T. W. Swetnam, and C. H. Baisan. 2009. "Spatially and Temporally Variable Fire Regime on Rincon Peak, Arizona, USA." *Fire Ecology* 5 (1): 3–21. http://dx.doi.org/10.4996/fireecology.0501003.

Intergovernmental Panel on Climate Change. 2001. *Climate Change 2001: Impacts, Adaptation, and Vulnerability*. Cambridge: Cambridge University Press.

Intergovernmental Panel on Climate Change. 2007a. *Climate Change 2007: Impacts, Adaptation, and Vulnerability*. Cambridge: Cambridge University Press.

Intergovernmental Panel on Climate Change. 2007b. *Climate Change 2007: The Physical Science Basis*. Geneva: IPCC Secretariat.

Intergovernmental Panel on Climate Change. 2013a. *Climate Change 2013: The Physical Science Basis*. 2216.

Intergovernmental Panel on Climate Change. 2013b. "Working Group I (WGI): Summary for Policymakers." In *Climate Change 2013: The Physical Science Basis*, 1–29. Contribution of Working Group I to the Fifth Assessment Report of the Intergovernmental Panel on Climate Change (I). Cambridge: Cambridge University Press.

International Association of Fire Chiefs. 2013. "WUI Fact Sheet." http://www.iaw fonline.org/pdf/WUI_Fact_Sheet_08012013.pdf.

Iverson, L. R., S. N. Matthews, A. N. Prasad, M. P. Peters, and G. Yohe. 2012. "Development of Risk Matrices for Evaluating Climatic Change Responses of Forested Habitats." *Climatic Change* 114 (2): 231–43. http://dx.doi.org/10.1007/s10584-012-0412-x.

Iverson, L. R., A. M. Prasad, B. J. Hale, and E. K. Sutherland. 1999. "An Atlas of Current and Potential Future Distributions of Common Trees of the Eastern United States." General Technical Report NE-265. Radnor, PA: Northeastern Research Station, USDA Forest Service. http://treesearch.fs.fed.us/pubs/7662.

Jackson, S. T. 2006. "Vegetation, Environment, and Time: The Origination and Termination of Ecosystems." *Journal of Vegetation Science* 17 (5): 549–57. http://dx.doi.org/10.1111/j.1654-1103.2006.tb02478.x.

Jackson, S. T. 2012. "Conservation and Resource Management in a Changing World: Extending Historical Range of Variation beyond the Baseline." Chap. 7 in *Historical Environmental Variation in Conservation and Natural Resource Management*, ed. J. A. Wiens, G. D. Hayward, H. D. Safford, and C. M. Giffen, 92–110. Hoboken, NJ: John Wiley & Sons. http://dx.doi.org/10.1002/9781118329726.ch7.

Jackson, S. T., J. L. Betancourt, R. K. Booth, and S. T. Gray. 2009. "Ecology and the Ratchet of Events: Climate Variability, Niche Dimensions, and Species Distributions." *Proceedings of the National Academy of Sciences of the United States of America* 106 (Supplement_2): 19685–92. http://dx.doi.org/10.1073/pnas.0901644106.

Jactel, H., E. Brockerhoff, and P. Duelli. 2005. "A Test of the Biodiversity–Stability Theory: Meta-analysis of Tree Species Diversity Effects on Insect Pest Infestations, and Re-examination of Responsible Factors." In *Forest Diversity and Function. Temperate and Boreal Systems*, ed. M. Scherer-Lorenzen, C. Körner, and E. D. Schulze, 235–62. Berlin, Germany: Springer. http://dx.doi.org/10.1007/3-540 -26599-6_12.

Jarnevich, C., and T. Stohlgren. 2009. "Near Term Climate Projections for Invasive Species Distributions." *Biological Invasions* 11 (6): 1373–9. http://dx.doi.org /10.1007/s10530-008-9345-8.

Jim, C. Y., and S. S. Xu. 2003. "Getting Out of the Woods: Quandaries of Protected Area Management in China." *Mountain Research and Development* 23 (3): 222–6. http://dx.doi.org/10.1659/0276-4741(2003)023[0222:GOOTW]2.0.CO;2.

Johnson, G. R., F. C. Sorensen, J. B. St. Clair, and R. C. Cronn. 2004. "Pacific Northwest Forest Tree Seed Zones: A Template for Native Plants?" *Native Plants Journal* 5 (2): 131–40. http://dx.doi.org/10.2979/NPJ.2004.5.2.131.

Johnson, R. C., S. Boyce, L. Brandt, V. Erickson, L. Iverson, G. Kujawa, L. Stritch, and B. Tkacz. 2013. "Policy and Strategy Considerations for Assisted Migration on USDA Forest Service Lands." In *Proceedings of the 60th Western International Forest Disease Work Conference*, comp. J. Browning and P. Palacios, 35–41. Lake Tahoe, CA.

Johnson, S., and S. Wondzell. 2005. "Keeping It Cool: Unraveling the Influences on Stream Temperature." *Science Findings* 73:2–5.

Jones, T. A., and T. A. Monaco. 2009. "A Role for Assisted Evolution in Designing Native Plant Materials for Domesticated Landscapes." *Frontiers in Ecology and the Environment* 7 (10): 541–7. http://dx.doi.org/10.1890/080028.

Joppa, L. N., and A. Pfaff. 2010. "Global Protected Area Impacts." *Proceedings. Biological Sciences* 278 (1712): 1633–8. http://dx.doi.org/10.1098/rspb.2010.1713.

Joyce, L. A., G. M. Blate, J. S. Littell, S. G. McNulty, C. I. Millar, S. C. Moser, R. P. Neilson, K. O'Halloran, and D. L. Peterson. 2008. "National Forests. Preliminary Review of Adaptation Options for Climate-sensitive Ecosystems and Resources." In *A Report by the U.S. Climate Change Science Program and the Subcommittee on Global Change Research*, ed. S. H. Julius and J. M. West, 3-1 to 3-127. Washington, DC: US Environmental Protection Agency.

Joyce, L. A., G. M. Blate, S. G. McNulty, C. I. Millar, S. Moser, R. P. Neilson, and D. L. Peterson. 2009. "Managing for Multiple Resources under Climate Change." *Environmental Management* 44 (6): 1022–32. http://dx.doi.org/10.1007/s00267-009 -9324-6.

Joyce, L. A., and C. I. Millar. 2014. "Improving the Role of Vulnerability Assessments in Decision Support for Effective Climate Adaptation." In *Forest Conservation and Management in the Anthropocene: Conference Proceedings*, ed. V. A. Sample and R. P. Bixler, 245–71. Proceedings. RMRS-P-71. Fort Collins, CO: US Department of Agriculture, Forest Service, Rocky Mountain Research Station.

Kareiva, P., R. Lalasz, and M. Marvier. 2011. "Conservation in the Anthropocene." *Breakthrough Journal* 2 (Fall). http://breakthroughjournal.org/content/authors/peter-kareiva-robert-lalasz-an-1/conservation-in-the-anthropoce.shtml.

Kareiva, P., S. Watts, R. McDonald, and T. Boucher. 2007. "Domesticated Nature: Shaping Landscapes and Ecosystems for Human Welfare." *Science* 316 (5833): 1866–9. http://dx.doi.org/10.1126/science.1140170.

Kareiva, P., and M. Marvier. 2012. "What is Conservation Science?" *Bioscience* 62 (11): 962–9. http://dx.doi.org/10.1525/bio.2012.62.11.5.

Kareiva, P. 2014. "New Conservation: Setting the Record Straight and Finding Common Ground." *Conservation Biology* 28 (3):634–6. http://dx.doi.org/10.1111/cobi.12295.

Karl, T. R., J. M. Melillo, and T. C. Peterson, eds. 2009. *Global Climate Change Impacts in the United States. United States Global Change Research Program (USGCRP).* New York: Cambridge University Press.

Kauffman, G. J. 2011. *Socioeconomic Value of the Delaware River Basin in Delaware, New Jersey, New York, and Pennsylvania: The Delaware River Basin, an Economic Engine for over 400 Years.* Newark, DE: University of Delaware.

Kaufman, D. S., T. A. Ager, N. J. Anderson, P. M. Anderson, J. T. Andrews, P. J. Bartlein, L. B. Brubaker, L. L. Coats, L. C. Cwynar, M. L. Duvall, et al. 2004. "Holocene Thermal Maximum in the Western Arctic (0–180°W)." *Quaternary Science Reviews* 23 (5-6): 529–60. http://dx.doi.org/10.1016/j.quascirev.2003.09.007.

Keeley, J. E., G. H. Aplet, N. L. Christensen, S. G. Conard, E. A. Johnson, P. N. Omi, D. L. Peterson, and T. W. Swetnam. 2009. "Ecological Foundations for Fire Management in North American Forest and Shrubland Ecosystems." Gen. Tech. Rep. PNW-GTR-779. Portland, OR: USDA, Forest Service, Pacific Northwest Research Station.

Kelly, A. B., C. J. Small, and G. D. Dreyer. 2009. "Vegetation Classification and Invasive Species Distribution in Natural Areas of Southern New England." *Journal of the Torrey Botanical Society* 136 (4): 500–519. http://dx.doi.org/10.3159/09-RA-007.1.

Kenis, M., M. A. Auger-Rozenberg, A. Roques, A. Roques, L. Timms, C. Péré, M. J. W. Cook, J. Settele, S. Augustin, and C. Lopez-Vaamonde. 2009. "Ecological Effects of Invasive Alien Insects." In *Ecological Impacts of Non-native Invertebrates and Fungi on Terrestrial Ecosystems*, ed. D. Langor and J. Sweeney, 21–45. The Netherlands: Springer. http://dx.doi.org/10.1007/978-1-4020-9680-8_3.

Kitzberger, T., P. M. Brown, E. K. Hyerdahl, T. W. Swetnam, and T. T. Veblen. 2007. "Contingent Pacific–Atlantic Ocean Influence on Multicentury Wildfire Synchrony over Western North America." *Proceedings of the National Academy of Sciences of the United States of America* 104 (2): 543–8. http://dx.doi.org/10.1073/pnas.0606078104.

Kleczewski, N. M., and S. L. Flory. 2010. "Leaf Blight Disease on the Invasive Grass Microstegium vimineum Caused By a Bipolaris sp." *Plant Disease* 94 (7): 807–11. http://dx.doi.org/10.1094/PDIS-94-7-0807.

Knapp, P. A. 1995. "Intermountain West Lightning-caused Fires: Climatic Predictors of Area Burned." *Journal of Range Management* 48 (1): 85–91. http://dx.doi.org /10.2307/4002510.

Knight, D. K. 2013. "The Health and Integrity of the Current Supply Chain: Logging." Paper presented at The State and Future of US Forestry and the Forest Industry. Sponsored by the US Endowment for Forestry and Communities and the US Forest Service. Washington, DC: Resources for the Future.

Knight, M. H. 1995. "Drought-Related Mortality of Wildlife in the Southern Kalahari and the Role of Man." *African Journal of Ecology* 33 (4): 377–94. http://dx.doi .org/10.1111/j.1365-2028.1995.tb01047.x.

Knorn, J., T. Kuemmerle, V. C. Radeloff, W. S. Keeton, V. Gancz, I.-A. Biriş, M. Svoboda, P. Griffiths, A. Hagatis, and P. Hostert. 2013. "Continued Loss of Temperate Old-growth Forests in the Romanian Carpathians Despite an Increasing Protected Area Network." *Environmental Conservation* 40 (2): 182–93. http://dx .doi.org/10.1017/S0376892912000355.

Knowlton, K., M. Rotkin-Ellman, L. Geballe, W. Max, and G. M. Solomon. 2011. "Six Climate Change-Related Events in the United States Accounted for about $14 Billion in Lost Lives and Health Costs." *Health Affairs* 30 (11): 2167–76. http://dx.doi.org/10.1377/hlthaff.2011.0229.

Koch, F., D. Yemshanov, M. Colunga-Garcia, R. D. Magarey, and W. D. Smith. 2011. "Potential Establishment of Alien-invasive Forest Insect Species in the United States: Where and How Many?" *Biological Invasions* 13 (4): 969–85. http://dx.doi .org/10.1007/s10530-010-9883-8.

Koch, F. H., D. Yemshanov, R. D. Magarey, and W. D. Smith. 2012. "Dispersal of Invasive Forest Insects via Recreational Firewood: A Quantitative Analysis." *Journal of Economic Entomology* 105 (2): 438–50. http://dx.doi.org/10.1603/EC11270.

Koch, J. L., D. W. Carey, M. E. Mason, and C. D. Nelson. 2010. "Assessment of Beech Scale Resistance in Full- and Half-Sibling American Beech Families." *Canadian Journal of Forest Research* 40 (2): 265–72. http://dx.doi.org/10.1139 /X09-189.

Kogan, M. 1998. "Integrated Pest Management: Historical Perspectives and Contemporary Developments." *Annual Review of Entomology* 43 (1): 243–70. http://dx.doi.org/10.1146/annurev.ento.43.1.243.

Krawchuk, M. A., and M. A. Moritz. 2011. "Constraints on Global Fire Activity Vary Across a Resource Gradient." *Ecology* 92 (1): 121–32. http://dx.doi .org/10.1890/09-1843.1.

Kroshy, M., J. Tewksbury, N. M. Haddad, and J. Hoekstra. 2010. "Ecological Connectivity for a Changing Climate." *Conservation Biology* 24 (6): 1686–9. http://dx.doi.org/10.1111/j.1523-1739.2010.01585.x.

Kuhman, T., S. Pearson, and M. Turner. 2010. "Effects of Land-Use History and the Contemporary Landscape on Non-native Plant Invasion at Local and Regional

Scales in the Forest-dominated Southern Appalachians." *Landscape Ecology* 25 (9): 1433–45. http://dx.doi.org/10.1007/s10980-010-9500-3.

Kundzewicz, Z. W., S. Budhakooncharoen, A. Bronstert, H. Hoff, D. Lettenmaier, L. Menzel, and R. Schulze. 2002. "Coping with Variability and Change: Floods and Droughts." *Natural Resources Forum* 26 (4): 263–74. http://dx.doi.org/10.1111/1477-8947.00029.

Langlet, O. 1971. "Two Hundred Years Genecology." *Taxon* 20 (5/6): 653–721. http://dx.doi.org/10.2307/1218596.

Larson, A. J., R. T. Belote, C. A. Cansler, S. A. Parks, and M. S. Dietz. 2013a. "Latent Resilience in Ponderosa Pine Forest: Effects of Resumed Frequent Fire." *Ecological Applications* 23 (6): 1243–9. http://dx.doi.org/10.1890/13-0066.1.

Larson, A. J., R. T. Belote, M. Williamson, and G. Aplet. 2013b. "Making Monitoring Count: Project Design for Active Adaptive Management." *Journal of Forestry* 111 (5): 348–56. http://dx.doi.org/10.5849/jof.13-021.

Larson, A. J., and D. Churchill. 2012. "Tree Spatial Patterns in Fire-frequent Forests of Western North America, Including Mechanisms of Pattern Formation and Implications for Designing Fuel Reduction and Restoration Treatments." *Forest Ecology and Management* 267:74–92. http://dx.doi.org/10.1016/j.foreco.2011.11.038.

La Sorte, F. A., T. M. Lee, H. Wilman, and W. Jetz. 2012. "Disparities between Observed and Predicted Impacts of Climate Change on Winter Bird Assemblages." *Journal of Animal Ecology* 81 (4): 914–25. http://rspb.royalsocietypublishing.org/content/276/1670/3167.

Laurance, W. F., M. Goosem, and S. G. Laurance. 2009. "Impacts of Roads and Linear Clearings on Tropical Forests." *Trends Ecol. Evol.* 24: 659–69.

Lavine, A., G. A. Kuyumjian, S. L. Reneau, D. Katzman, and D. V. Malmon. 2005. "A Five-Year Record of Sedimentation in the Los Alamos Reservoir, New Mexico, Following the Cerro Grande Fire." Los Alamos Technical Publication LA-UR-05-7526. http://catalog.lanl.gov.

Lawler, J. J., and J. D. Olden. 2011. "Reframing the Debate over Assisted Colonization." *Frontiers in Ecology and the Environment* 9 (10): 569–74. http://dx.doi.org/10.1890/100106.

Ledig, F. T., and J. H. Kitzmiller. 1992. "Genetic Strategies for Reforestation in the Face of Global Climate Change." *Forest Ecology and Management* 50 (1-2): 153–69. http://dx.doi.org/10.1016/0378-1127(92)90321-Y.

Leendertz, F. H., H. Ellerbrok, C. Boesch, E. Couacy-Hymann, Kerstin Mätz-Rensing, R. Hakenbeck, C. Bergmann, P. Abaza, S. Junglen, Y. Moebius, et al. 2004. "Anthrax Kills Wild Chimpanzees in a Tropical Rainforest." *Nature* 430 (6998): 451–2. http://dx.doi.org/10.1038/nature02722.

Lempert, R. J., and M. T. Collins. 2007. "Managing the Risk of Uncertain Threshold Responses: Comparison of Robust, Optimum, and Precautionary Approaches." *Risk Analysis* 27 (4): 1009–26. http://dx.doi.org/10.1111/j.1539-6924.2007.00940.x.

Lempert, R. J., N. Kalra, S. Peyraud, S. Peyraud, Z. Mao, S. Bach Tan, D. Circa, and A. Lotsch. 2013. "Ensuring Robust Flood Risk Management in Ho Chi Minh City." World Bank Policy Research Working Paper #6465. Washington, DC. http://dx.doi.org/10.1596/1813-9450-6465.

Lenihan, J. M., D. Bachelet, R. P. Neilson, and R. Drapek. 2008. "Response of Vegetation Distribution, Ecosystem Productivity, and Fire to Climate Change Scenarios for California." Climatic Change 87 (S1): 215–30. http://dx.doi.org/10.1007/s10584-007-9362-0.

Leopold, L. 1997. Water, Rivers and Creeks. Sausalito, CA: University Science Books.

Liebhold, A. M., W. L. MacDonald, D. Bergdahl, and V. C. Mastro. 1995. "Invasion by Exotic Forest Pests: A Threat to Forest Ecosystems." Forest Science Monograph 30 (2): 1–38.

Liebhold, A. M., D. G. McCullough, L. M. Blackburn, S. J. Frankel, B. Von Holle, and J. E. Aukema. 2013. "A Highly Aggregated Geographical Distribution of Forest Pest Invasions in the USA." Diversity & Distributions 19 (9): 1208–16. http://dx.doi.org/10.1111/ddi.12112.

Lindenmayer, D., R. J. Hobbs, D. Montague, A. J. Bennett, A. Burgman, M. Cale, P. Calhoun, A. Gramer, V. Cullen, P. Driscoll, et al. 2008. "A Checklist for Ecological Management of Landscapes for Conservation." Ecology Letters 11 (1): 78–91.

Linder, M. 2000. "Developing Adaptive Forest Management Strategies to Cope with Climate Change." Tree Physiology 20 (5–6): 299–307.

Lindgren, D., and C. C. Ying. 2000. "A Model Integrating Seed Source Adaptation and Seed Use." New Forests 20 (1): 87–104. http://dx.doi.org/10.1023/A:100678213824.

Littell, J. S., D. McKenzie, D. L. Peterson, and A. L. Westerling. 2009. "Climate and Wildfire Area Burned in Western U.S. Ecoprovinces, 1916–2003." Ecological Applications 19 (4): 1003–21. http://dx.doi.org/10.1890/07-1183.1.

Littell, J. S., D. L. Peterson, C. I. Millar, and K. O'Halloran. 2012. "U.S. National Forests Adapt to Climate Change through Science-management Partnerships." Climatic Change 110 (1-2): 269–96. http://dx.doi.org/10.1007/s10584-011-0066-0.

Liu, J., M. Linderman, Z. Ouyang, L. An, J. Yang, and H. Zhang. 2001. "Ecological Degradation in Protected Areas: The Case of Wolong Nature Reserve for Giant Pandas." Science 292 (5514): 98–101. http://dx.doi.org/10.1126/science.1058104.

Loarie, S. R., P. B. Duffy, H. Hamilton, Gregory P. Asner, Christopher B. Field, and David D. Ackerly. 2009. "The Velocity of Climate Change." Nature 462 (7276): 1052–5. http://dx.doi.org/10.1038/nature08649.

Locke, H., and G. M. Tabor. 2005. "The Future of Y2Y." In Yellowstone to Yukon: Freedom to Roam, ed. F. Schulz. Seattle, WA: The Mountaineers Books.

Lodge, D. M., S. Williams, H. J. MacIsaac, K. R. Hayes, B. Leung, S. Reichard, R. N. Mack, P. B. Moyle, M. Smith, D. A. Andow, et al. 2006. "Biological Invasions: Recommendations for U.S. Policy and Management." Ecological Applications 16 (6): 2035–54. http://dx.doi.org/10.1890/1051-0761(2006)016[2035:BIRFUP]2.0.CO;2.

Lorenzen, E. D., D. Nogues-Bravo, L. Orlando, J. Weinstock, J. Binladen, K. A. Marske, A. Ugan, M. K. Borregaard, M.T.P. Gilbert, R. Nielsen, et al. 2011. "Species-Specific Responses of Late Quaternary Megafauna to Climate and Humans." *Nature* 479 (7373): 359–64. http://dx.doi.org/10.1038/nature10574.

Lovejoy, T., and L. Hannah, eds. 2005. *Climate Change and Biodiversity*. New Haven: Yale University Press.

Lundgren, M. R., C. J. Small, and G. D. Dreyer. 2004. "Influence of Land Use and Site Characteristics on Invasive Plant Abundance in the Quinebaug Highlands of Southern New England." *Northeastern Naturalist* 11 (3): 313–32. http://dx.doi.org/10.1656/1092-6194(2004)011[0313:IOLUAS]2.0.CO;2.

Luoto, M., and R. K. Heikkinen. 2008. "Disregarding Topographical Heterogeneity Biases Species Turnover Assessments Based on Bioclimatic Models." *Global Change Biology* 14 (3): 483–94. http://dx.doi.org/10.1111/j.1365-2486.2007.01527.x.

Maasch, K., P. A. Mayewski, E. Rohling, C. Stager, K. Karlen, L. D. Meeker, and E. Meyerson. 2005. "Climate of the Past 2000 Years." *Geografiska Annaler* 87 (A): 7–15.

MacDougall, A. A., S. McCann, G. Gellner, and R. Turkington. 2013. "Diversity Loss with Persistent Human Disturbance Increases Vulnerability to Ecosystem Collapse." *Nature* 494 (7435): 86–9. http://dx.doi.org/10.1038/nature11869.

Mack, R. N., D. Simberloff, W. M. Lonsdale, H. Evans, M. Clout, and F. A. Bazzaz. 2000. "Biotic Invasions: Causes, Epidemiology, Global Consequences, and Control." *Ecological Applications* 10 (3): 689–710. http://dx.doi.org/10.1890/1051-0761(2000)010[0689:BICEGC]2.0.CO;2.

Mandryk, A. M., and R. W. Wein. 2006. "Exotic Vascular Plant Invasiveness and Forest Invasibility in Urban Boreal Forest Types." *Biological Invasions* 8 (8): 1651–62. http://dx.doi.org/10.1007/s10530-005-5874-6.

Manea, A., and M. Leishman. 2011. "Competitive Interactions between Native and Invasive Exotic Plant Species Are Altered under Elevated Carbon Dioxide." *Oecologia* 165 (3): 735–44. http://dx.doi.org/10.1007/s00442-010-1765-3.

Margolis, E. Q., and J. Balmat. 2009. "Fire History and Fire–climate Relationships along a Fire Regime Gradient in the Santa Fe Municipal Watershed, NM, USA." *Forest Ecology and Management* 258 (11): 2416–30. http://dx.doi.org/10.1016/j.foreco.2009.08.019.

Margolis, E. Q., T. W. Swetnam, and C. D. Allen. 2007. "A Stand-Replacing Fire History in Upper Montane Forests of the Southern Rocky Mountains." *Canadian Journal of Forest Research* 37 (11): 2227–41.

Margolis, E. Q., T. W. Swetnam, and C. D. Allen. 2011. "Historical Stand-replacing Fire in Upper Montane Forests of the Madrean Sky Islands and Mogollon Plateau, Southwestern USA." *Fire Ecology* 7 (3): 88–107. http://dx.doi.org/10.4996/fireecology.0703088.

Marks, D., J. Kimball, D. Tingey, and Tim Link. 1998. "The Sensitivity of Snowmelt Processes to Climate Conditions and Forest Cover During Rain-on-Snow: A

Case Study of the 1996 Pacific Northwest Flood." *Hydrological Processes* 12 (10-11): 1569–87. http://dx.doi.org/10.1002/(SICI)1099-1085(199808/09)12:10/11<1569 ::AID-HYP682>3.0.CO;2-L.

Marland, G., and S. Marland. 1992. "Should We Store Carbon in Trees?" *Water, Air, and Soil Pollution* 64 (1-2): 181–95. http://dx.doi.org/10.1007/BF00477101.

Marland, G., B. Schlamadinger, and P. Leiby. 1997. "Forest/Biomass-Based Mitigation Strategies: Does the Timing of Carbon Reductions Matter?" *Critical Reviews in Environmental Science and Technology* 27 (sup001): 213–26. http://dx.doi .org/10.1080/10643389709388521.

Marlon, J. R., P. J. Bartlein, D. G. Gavin, C. J. Long, R. S. Anderson, C. E. Briles, K. J. Brown, D. Colombaroli, D. J. Hallett, M. J. Power, et al. 2012. "Long-term Perspective on Wildfires in the Western USA." *Proceedings of the National Academy of Sciences of the United States of America* 109 (9): E535–43. http://dx.doi.org/10.1073 /pnas.1112839109.

Marris, E. 2009. "Planting the Forest for the Future." *Nature* 459 (7249): 906–8. http://dx.doi.org/10.1038/459906a.

Martin, D. 2013. "Wildfire Effects on Water Supplies: Understanding the Impacts on the Wuantity and Timing of Post-Fire Runoff and Stream Flows." Water Research Foundation: Wildfire Readiness and Response Workshop.

Martinson, E. J., and P. N. Omi. 2013. "Fuel Treatments and Fire Severity: A Meta-analysis." Res. Pap. RMRS-RP-103WWW. Fort Collins, CO: USDA, Forest Service, Rocky Mountain Research Station.

Mascaro, J., R. F. Hughes, and S. A. Schnitzer. 2012. "Novel Forests Maintain Ecosystem Processes after the Decline of Native Tree Species." *Ecological Monographs* 82 (2): 221–8. http://dx.doi.org/10.1890/11-1014.1.

Mascia, M. B., and S. Pailler. 2011. "Protected Area Downgrading, Downsizing, and Degazettement (PADDD) and Its Conservation Implications." *Conservation Letters* 4 (1): 9–20. http://dx.doi.org/10.1111/j.1755-263X.2010.00147.x.

Mascia, M. B., S. Pailler, R. Krithivasan, V. Roshchanka, D. Burns, Mc. J. Mlotha, D. R. Murray, and N. Peng. 2014. "Protected Area Downgrading, Downsizing, and Degazettement (PADDD) in Africa, Asia, and Latin America and the Caribbean, 1900–2010." *Biological Conservation* 169:355–61. http://dx.doi.org/10.1016/j .biocon.2013.11.021.

Mattangkilang, T. 2013. "High Coal Prices Spur Illegal Mining in Indonesia, Police Say." *Jakarta Globe*, May 30. http://jakartaglobe.beritasatu.com/news /high-coal-prices-encouraging-illegal-mining-police-say/.

Matusick, G., K. X. Ruthrof, N. C. Brouwers, B. Dell, and G.St.J. Hardy. 2013. "Sudden Forest Canopy Collapse Corresponding with Extreme Drought and Heat in a Mediterranean-Type Eucalypt Forest in Southwestern Australia." *European Journal of Forest Research* 132 (3): 497–510. http://dx.doi.org/10.1007/s10342 -013-0690-5.

Matyas, C. 1994. "Modeling Climate Change Effects with Provenance Test Data." *Tree Physiology* 14 (7-8-9): 797–804. http://dx.doi.org/10.1093/treephys/14.7-8 -9.797.

Mawdsley, J. R., R. O'Malley, and D. S. Ojima. 2009. "A Review of Climate-Change Adaptation Strategies for Wildlife Management and Biodiversity Conservation." *Conservation Biology* 23 (5): 1080–9. http://dx.doi.org/10.1111/j.1523-1739.2009 .01264.x.

Mayewski, P. A., L. D. Meeker, S. Whitlow, M. S. Twickler, M. C. Morrison, R. B. Alley, P. Bloomfield, and K. Taylor. 1993. "The Atmosphere During the Younger Dryas." *Science* 261 (5118): 195–7. http://dx.doi.org/10.1126/science.261.5118.195.

Mayewski, P. A., E. E. Rohling, J. C. Stager, W. Karlén, K. A. Maasch, L. D. Meeker, E. A. Meyerson, F. Gasse, S. van Kreveld, K. Holmgren, et al. 2004. "Holocene Climate Variability." *Quaternary Research* 62 (3): 243–55. http://dx.doi.org /10.1016/j.yqres.2004.07.001.

McCabe, G. J., and M. A. Ayers. 1989. "Hydrologic Effects of Climate Change in the Delaware River Basin." *Journal of the American Water Resources Association* 25 (6): 1231–42. http://dx.doi.org/10.1111/j.1752-1688.1989.tb01335.x.

McCabe, G. J., J. L. Betancourt, S. T. Gray, M. A. Palecki, and H. G. Hidalgo. 2008. "Associations of Multi-Decadal Sea-Surface Temperature Variability with U.S. Drought." *Quaternary International* 188 (1): 31–40. http://dx.doi.org/10.1016/j .quaint.2007.07.001.

McCarthy, L. F. 2004. *Snapshot: State of the National Fire Plan.* Santa Fe, NM: Forest Guild; http://www.forestguild.org/publications/research/2004/national_fire _plan.pdf.

McCarthy, P. D. 2012. "Climate Change Adaptation for People and Nature: A Case Study from the U.S. Southwest." *Advances in Climate Change Research* 3 (1): 22–37. http://dx.doi.org/10.3724/SP.J.1248.2012.00022.

McClure, M. S. 1991. "Density-dependent Feedback and Population Cycles in Adelges tsugae (Homoptera: Adelgidae) on Tsuga canadensis." *Environmental Entomology* 20 (1): 258–64. http://dx.doi.org/10.1093/ee/20.1.258.

McCune, B. H., and J. B. Grace. 2002. *Analysis of Ecological Communities.* Corvallis, OR: MjM Software Design; www.pcord.com.

McDowell, N., D. J. Beerling, D. D. Breshears, R. A. Fisher, K. F. Raffa, and M. Stitt. 2011. "The Interdependence of Mechanisms Underlying Climate-driven Vegetation Mortality." *Trends in Ecology & Evolution* 26 (10): 523–32. http://dx.doi.org /10.1016/j.tree.2011.06.003.

McIver, J., A. Youngblood, and S. L. Stephens. 2009. "The National Fire and Fire Surrogate Study: Ecological Consequences of Fuel Reduction Methods in Seasonally Dry Forests." *Ecological Applications* 19 (2): 283–4. http://dx.doi.org/10 .1890/07-1785.1.

McKay, J. K., C. E. Christian, S. Harrison, and K. J. Rice. 2005. "'How Local Is Local?'—A Review of Practical and Conceptual Issues in Genetics of

Restoration." *Restoration Ecology* 13 (3): 432–40. http://dx.doi.org/10.1111/j.1526 -100X.2005.00058.x.

McKenney, D. W., B. G. Mackey, and D. Joyce. 1999. "Seedwhere: A Computer Tool to Support Seed Transfer and Ecological Restoration Decisions." *Environmental Modelling* 14 (6): 589–95. http://dx.doi.org/10.1016/S1364-8152(98)00095-4.

McKenney, D. W., J. Pedlar, and G. A. O'Neill. 2009. "Climate Change and Forest Seed Zones: Past Trends, Future Prospects and Challenges to Ponder." *Forestry Chronicle* 85 (2): 258–66. http://dx.doi.org/10.5558/tfc85258-2.

McKenzie, D., Z. Gedalof, D. L. Peterson, and P. Mote. 2004. "Climatic Change, Wildfire, and Conservation." *Conservation Biology* 18 (4): 890–902. http://dx.doi .org/10.1111/j.1523-1739.2004.00492.x.

McKinney, M., and S. Johnson. 2013. *"Large Landscape Conservation in the Rocky Mountain West: An Inventory and Status Report."* Missoula. Center for Natural Resources and Environmental Policy, University of Montana.

McKinney, M. J., L. Scarlett, and D. Kemmis. 2010. *Large Landscape Conservation: A Strategic Framework for Policy and Action.* Cambridge, MA: Lincoln Institute of Land Policy.

McLachlan, J. S., J. J. Hellmann, and M. W. Schwartz. 2007. "A Framework for Debate of Assisted Migration in an Era of Climate Change." *Conservation Biology* 21 (2): 297–302. http://dx.doi.org/10.1111/j.1523-1739.2007.00676.x.

McLane, S. C., and S. N. Aitken. 2012. "Whitebark Pine (*Pinus albicaulis*) Assisted Migration Potential: Testing Establishment North of the Species Range." *Ecological Applications* 22 (1): 142–53. http://dx.doi.org/10.1890/11-0329.1.

McRae, B. H., and V. B. Shah. 2009. *Circuitscape User's Guide.* Santa Barbara: University of California; http://www.circuitscape.org.

Means, E., R. Patrick, L. Ospina, and Nicole West. 2005. "Scenario Planning: A Tool to Manage Future Water Utility Uncertainty." *Journal—American Water Works Association* 97 (10): 68.

Meiklejohn, K., R. Ament, and G. Tabor. 2010. *Habitat Corridors and Landscape Connectivity: Clarifying the Terminology.* New York: Center for Large Landscape Conservation.

Melillo, J., T. Richmond, and G. Yohe, eds. 2014. *Climate Change Impacts in the United States: The Third National Climate Assessment.* U.S. Global Change Research Program; 10.7930/J0Z31WJ2.

Merkle, S. A., G. M. Andrade, C. J. Nairn, W. A. Powell, and C. A. Maynard. 2007. "Restoration of Threatened Species: A Noble Cause for Transgenic Trees." *Tree Genetics & Genomes* 3 (2): 111–8. http://dx.doi.org/10.1007/s11295-006-0050-4.

Merino, M., B. Carpinetti, and A. Abba. 2009. "Invasive Mammals in the National Parks System of Argentina." *Natural Areas Journal* 29 (1): 42–9. http://dx.doi .org/10.3375/043.029.0105.

Metz, D., S. Byerly, and G. Lewi. 2011. "Findings from Recent Survey of City of Santa Fe Voters." Santa Fe, New Mexico: Poll conducted by Fairbank, Maslin, Maulin and Metz for The Nature Conservancy, February 28.

Metz, M. R., J. M. Varner, K. M. Frangioso, R. K. Meentemeyer, and D. M. Rizzo. 2013. "Unexpected Redwood Mortality from Synergies between Wildfire and an Emerging Infectious Disease." *Ecology* 94 (10): 2152–9. http://dx.doi.org/10.1890/13-0915.1.

MGB. 2012. *The Case for Tall Wood Buildings: How Mass Timber Offers a Safe, Economical, and Environmentally Friendly Alternative for Tall Building Structures.* Vancouver, BC: MGB Architecture and Design.

Millar, C. I., N. L. Stephenson, and S. L. Stephens. 2007. "Climate Change and Forests of the Future: Managing in the Face of Uncertainty." *Ecological Applications* 17 (8): 2145–51. http://dx.doi.org/10.1890/06-1715.1.

Miller, J. D., B. M. Collins, J. A. Lutz, S. L. Stephens, J. W. van Wagtendonk, and D. A. Yasuda. 2012. "Differences in Wildfires among Ecoregions and Land Management Agencies in the Sierra Nevada region, California, USA." *Ecosphere* 3 (9): art80. http://dx.doi.org/10.1890/ES12-00158.1.

Miller, J. D., and H. D. Safford. 2012. "Trends in Wildfire Severity 1984–2010 in the Sierra Nevada, Modoc Plateau, and Southern Cascades, California, USA." *Fire Ecology* 8 (3): 41–57. http://dx.doi.org/10.4996/fireecology.0803041.

Miller, J. D., H. D. Safford, M. Crimmins, and A. E. Thode. 2009. "Quantitative Evidence for Increasing Forest Fire Severity in the Sierra Nevada and Southern Cascade Mountains, California and Nevada, USA." *Ecosystems (New York, N.Y.)* 12 (1): 16 32. http://dx.doi.org/10.1007/s10021-008-9201-9.

Millspaugh, S. H., C. Whitlock, and P. J. Bartlein. 2004. "Postglacial Fire, Vegetation, and Climate History of the Yellowstone-Lamar and Central Plateau Provinces, Yellowstone National Park." In *After the Fires: The Ecology of Change in Yellowstone National Park*, ed. L. L. Wallace, 10–28. New Haven, CT: Yale University Press. http://dx.doi.org/10.12987/yale/9780300100488.003.0002.

Milly, P., J. Betancourt, M. Falkenmark, R. M. Hirsch, Z. W. Kundzewicz, D. P. Lettenmaier, and R. J. Stouffer. 2008. "Stationarity Is Dead: Whither Water Management?" *Science* 319 (5863): 573–74. http://dx.doi.org/10.1126/science.1151915.

Milly, P. C. D., K. A. Dunne, and A. V. Vecchia. 2005. "Global Pattern of Trends Instream Flow and Water Availability in a Changing Climate." *Nature* 438 (7066): 347–50. http://dx.doi.org/10.1038/nature04312.

Minteer, B. A., and T. R. Miller. 2011. "The New Conservation Debate: Ethical Foundations, Strategic Trade-offs, and Policy Opportunities." *Biological Conservation* 144 (3): 945–7. http://dx.doi.org/10.1016/j.biocon.2010.07.027.

Mitchell, S., M. Harmon, and K. O'Connell. 2009. "Forest Fuel Reduction Alters Fire Severity and Long-term Carbon Storage in Three Pacific Northwest Ecosystems." *Ecological Applications* 19 (3): 643–55. http://dx.doi.org/10.1890/08-0501.1.

Morelli, T. L., and S. C. Carr. 2011. "A Review of the Potential Effects of Climate Change on Quaking Aspen (*Populus tremuloides*) in the Western United States and a New Tool for Surveying Sudden Aspen Decline." Gen. Tech. Rep. PSW-GTR-235. Albany, CA: USDA Forest Service, Pacific Southwest Research Station.

Morelli, T. L., M. C. McGlinchy, and R. P. Neilson. 2011. "A Climate Change Primer for Land Managers: An Example from the Sierra Nevada." In *Res. Pap. PSW-RP-262*. Albany, CA: USDA Forest Service, Pacific Southwest Research Station.

Morelli, T. L., S. Yeh, N. Smith, M. B. Hennessy, and C. I. Millar. 2012. "Climate Project Screening Tool: An Aid for Climate Change Adaptation." In *Res. Pap. PSW-RP-263*. Albany, CA: USDA, Forest Service, Pacific Southwest Research Station.

Morin, R. S., A. M. Liebhold, P. C. Tobin, K. W. Gottschalk, and E. Luzader. 2007. "Spread of Beech Bark Disease in the Eastern United States and Its Relationship to Regional Forest Composition." *Canadian Journal of Forest Research* 37 (4): 726–36. http://dx.doi.org/10.1139/X06-281.

Moritz, M. A., M. A. Parisien, E. Batllori, M. A. Krawchuk, J. Van Dorn, D. J. Ganz, and K. Hayhoe. 2012. "Climate Change and Disruptions to Global Fire Activity." *Ecosphere* 3 (6): 1–22. http://dx.doi.org/10.1890/ES11-00345.1.

Morris, D. F., M. K. Macauley, R. J. Koop, and R. Morgenstern. 2011. *Summary Report. Reforming Institutions and Managing Extremes; U.S. Policy Approaches for Adapting to a Changing Climate*. Washington, DC: Resources for the Future.

Moritz, C., J. L. Patton, C. J. Conroy, J. L. Parra, G. C. White, and S. R. Beissinger. 2008. "Impact of a Century of Climate Change on Small-mammal Communities in Yosemite National Park, USA." *Science* 322 (5899): 261–4. http://dx.doi.org/10.1126/science.1163428.

Moy, C. M., G. O. Seltzer, D. T. Rodbell, and D. M. Anderson. 2002. "Variability of El Niño/Southern Oscillation Activity at Millennial Timescales During the Holocene Epoch." *Nature* 420 (6912): 162–5. http://dx.doi.org/10.1038/nature01194.

Mtahiko, M. G. G., E. Gereta, A. R. Kajuni, E. A. T. Chiombola, G. Z. Ng'umbi, P. Coppolillo, and E. Wolanski. 2006. "Towards an Ecohydrology-Based Restoration of the Usangu Wetlands and the Great Ruaha River, Tanzania." *Wetlands Ecology and Management* 14 (6): 489–503. http://dx.doi.org/10.1007/s11273-006-9002-x.

Mueller, J. M., and J. J. Hellmann. 2008. "An Assessment of Invasion Risk from Assisted Migration." *Conservation Biology* 22 (3): 562–7. http://dx.doi.org/10.1111/j.1523-1739.2008.00952.x.

Muller, E. H., L. Sirkin, and J. L. Craft. 1993. "Stratigraphic Evidence of a Pre-Wisconsinan Interglaciation in the Adirondack Mountains, New York." *Quaternary Research* 40 (2): 163–8. http://dx.doi.org/10.1006/qres.1993.1068.

Munich Reinsurance America. 2013. "Focus on Wildfire Firefighting." http://www
.munichreamerica.com/mram/en/publications-expertise/research-spotlight
/wildfire-firefighters/index.html.

Murdoch, P. S., J. S. Baron, and T. L. Miller. 2000. "Potential Effects of Climate
Change on Surface Water Quality in North America." *Journal of the American
Water Resources Association* 36 (2): 347–66. http://dx.doi.org/10.1111/j.1752-1688
.2000.tb04273.x.

Murdoch, P. S., C. L. Bonitz, K. W. Eakin, A. J. Ranalli, and E. C. Witt. 1991. *Episodic
Acidification and Associated Fish and Aquatic Invertebrate Responses in Four Catskill
Mountain Streams: An Interim Report of the Episodic Response Project*. Albany, NY:
US Geological Survey.

Najjar, R., A. Ross, D. Kreeger, and S. Kilham. 2012. "Climate Change in the Technical Report for the Delaware Estuary & Basin." PDE Report No. 12–01, 225–41.

Nash, C. H., and E. A. Johnson. 1996. "Synoptic Climatology of Lightning-caused
Forest Fires in Subalpine and Boreal Forests." *Canadian Journal of Forest Research*
26 (10): 1859–74. http://dx.doi.org/10.1139/x26-211.

National Academy of Sciences. 2008. *Hydrologic Effects of a Changing Forest Landscape*. Washington, DC: The National Academies Press.

National Climate Assessment Development Advisory Committee [NCADAC]. 2013.
"U.S. Global Change Research Program." National Climate Assessment, Chapter
10.

National Climate Assessment. 2013. "Draft v.11." http://www.globalchange.gov
/ncadac.

National Fish, Wildlife and Plants Climate Adaptation Partnership. 2012. "Climate
Adaptation Strategy." doi: 10.3996/082012-FWSReport-1.

National Forest Foundation. 2013. "Where the Water Begins." *Your National Forests*
(Winter/Spring). https://www.nationalforests.org/our-forests/your-national
-forests-magazine/where-the-water-begins

National Interagency Fire Center. 2001. "Review and Update of the 1995 Federal
Wildland Fire Management Policy." Washington, DC: Various governmental
agency partners. https://www.nifc.gov/PIO_bb/Policy/FederalWildlandFire
ManagementPolicy_2001.pdf.

National Interagency Fire Center (NIFC). 2006. "Wildland Fire Statistics." Boise,
ID: National Interagency Fire Center. Accessed February 15, 2012. www.nifc
.gov/fire_info/fire_stats.htm.

National Interagency Fire Center. 2009. "Guidance for Implementation of Federal
Wildland Fire Management Policy." Washington, DC: Various governmental
agency partners. http://www.nifc.gov/policies/policies_documents/GIFW
FMP.pdf.

National Research Council (NRC). 2011. *Climate Stabilization Targets: Emissions, Concentrations, and Impacts over Decades to Millennia*. Washington, DC: The National
Academies Press.

Natural Resources Canada. 2016. "Assisted Migration." Accessed March 19, 2016. http://www.nrcan.gc.ca/forests/climate-change/adaptation/13121.

Nature Conservancy. 1982. "Natural Heritage Program Operations Manual." Unpublished. Arlington, VA.

Nature Conservancy. 2012. "Water Funds: Conserving Green Infrastructure. A Guide for Design, Creation and Operation." Accessed August 25, 2013. http://www.nature.org/media/freshwater/latin-america-water-funds.pdf.

Nature Conservancy. 2014. "Rio Grande Water Fund." www.nature.org/riogrande.

NatureServe. 2015. "Climate Change Vulnurability Index." Accessed March 19, 2016. http://www.natureserve.org/conservation-tools/climate-change-vulnerabili ty-index .

Naugle, D. E., K. E. Doherty, B. L. Walker, H. E. Copeland, M. J. Holloran, and J. D. Tack. 2011. "Sage-Grouse and Cumulative Impacts of Energy Development." In *Energy Development and Wildlife Conservation in Western North America*, ed. D. E. Naugle, 55–70. Washington, DC: Island Press. http://dx.doi.org/10.5822/978 -1-61091-022-4_4.

NLCD (National Landcover Database). 2001. U.S. Department of the Interior, U.S. Geological Survey. http://www.mrlc.gov/.

Neff, R., H. Chang, C. G. Knight, R. G. Najjar, B. Yarnal, and H. A. Walker. 2000. "Impact of Climate Variation and Change on Mid-Atlantic Region Hydrology and Water Resources." *Climate Research* 14 (3): 207–18. http://dx.doi.org/10.3354 /cr014207.

New Jersey Department of Environmental Protection. 2010. "Statewide Forest Resource Assessment and Resource Strategies." http://www.state.nj.us/dep /parksandforests/forest/docs/NJFSassessment.pdf.

New York City Panel on Climate Change. 2009. *Climate Risk Information*. New York: MORE.

New York State Department of Environmental Conservation. 2010. "Forest Resource Assessment & Strategy 2010–2015: Keeping New York's Forests as Forests." http://www.dec.ny.gov/docs/lands_forests_pdf/fras070110.pdf.

Newmark, W. D. 2008. "Isolation of African Protected Areas." *Frontiers in Ecology and the Environment* 6 (6): 321–8. http://dx.doi.org/10.1890/070003.

Nielsen-Pincus, M., and C. Moseley. 2010. *The Employment and Economic Impacts of Forest and Watershed Restoration in Oregon*. University of Oregon: Institute for Sustainable Environment.

Nørgaard-Pedersen, N., M. Mikkelsen, and Y. Kristoffersen. 2009. "The Last Inter-glacial Warm Period Record of the Arctic Ocean: Proxy-data Support a Major Reduction of Sea Ice." *IOP Conference Series: Earth and Environmental Science* 6: http://dx.doi.org/10.1088/1755-1307/6/7/072002.

North, M., P. Stine, K. O'Hara, W. Zielinski, and S. Stephens. 2009. "An Ecosystem Management Strategy for Sierran Mixed-conifer Forests." 2nd printing, with addendum. Gen. Tech. Rep. PSW-GTR-220. Albany, CA: USDA, Forest Service, Pacific Southwest Research Station.

North, M. P., B. M. Collins, and S. L. Stephens. 2012. "Using Fire to Increase the Scale, Benefits and Future Maintenance of Fuels Treatments." *Journal of Forestry* 110 (7): 392–401. http://dx.doi.org/10.5849/jof.12-021.

Noss, R. F. 1983. "A Regional Landscape Approach to Maintain Diversity." *Bioscience* 33 (11): 700–6. http://dx.doi.org/10.2307/1309350.

Noss, R. F., and A. Cooperrider. 1994. *Saving Nature's Legacy: Protecting and Restoring Biodiversity*. Washington, DC: Island Press.

Notaro, M., A. Mauss, and J. W. Williams. 2012. "Projected Vegetation Changes for the American Southwest: Combined Dynamic Modeling and Bioclimatic-envelope Approach." *Ecological Applications* 22 (4): 1365–88. http://dx.doi.org/10.1890/11-1269.1.

O'Neill, G. A., N. K. Ukrainetz, M. Carlson, B. Jaquish, C. Cartwright, J. King, M. Stoehr, C-Y Xie, A. Yanchuk, J. Krakowski, et al. 2008. "Assisted Migration to Address Climate Change in British Columbia: Recommendations for Interim Seed Transfer Standards." Tech. Rept. 048. Victoria, BC: BC Ministry of Forest and Range, Research Branch.

O'Brien, S. R., P. A. Mayewski, L. D. Meeker, D. A. Meese, M. S. Twickler, and S. I. Whitlow. 1995. "Complexity of Holocene Climate as Reconstructed from a Greenland Ice Core." *Science* 270 (5244): 1962–4. http://dx.doi.org/10.1126/science.270.5244.1962.

O'Connor, C. D., D. A. Falk, A. M. Lynch, and T. W. Swetnam. 2014. "Fire Severity, Size, and Climate Associations Diverge from Historical Precedent along an Ecological Gradient in the Pinaleño Mountains, Arizona, USA." *Forest Ecology and Management* 329:264–78. http://dx.doi.org/10.1016/j.foreco.2014.06.032.

Oliver, C. D. 1992. "A Landscape Approach: Achieving and Maintaining Biodiversity and Economic Productivity." *Journal of Forestry* 90:20–5.

Oliver, C. D. 2014. "Functional Restoration of Social-forestry Systems Across Spatial and Temporal Scales." *Journal of Sustainable Forestry* 33 (sup1 Supplement 1): S123–48. http://dx.doi.org/10.1080/10549811.2014.884003.

Oliver, C. D., A. B. Adams, and R. J. Zasoski. 1985. *Disturbance Patterns and Forest Development in a Recently Deglaciated Valley in the Northwestern Cascade Mountains of Washington. U.S.A.*

Oliver, C. D., K. Covey, D. Larsen, J. B. McCarter, A. Niccolai, and J. Wilson. 2012. "Landscape Management." In *Forest Landscape Restoration: Integrating Natural and Social Sciences*, ed. J. Stanturf, D. Lamb, and P. Madsen, 39–65. New York: Springer Publishing. http://dx.doi.org/10.1007/978-94-007-5326-6_3.

Oliver, C. D., and B. C. Larson. 1996. *Forest Stand Dynamics, Update edition*. New York: John Wiley.

Oliver, C. D., N. T. Nassar, B. R. Lippke, and J. B. McCarter. 2014. "Carbon, Fossil Fuel, and Biodiversity Mitigation with Wood and Forests." *Journal of Sustainable Forestry* 33 (3): 248–75. http://dx.doi.org/10.1080/10549811.2013.839386.

Oliver, C. D., and K. L. O'Hara. 2004. "Effects of Restoration at the Stand Level." In *Restoration of Boreal and Temperate Forests*, ed. J. A. Stanturf and P. Marsden, 31–59. New York: CRC Press. http://dx.doi.org/10.1201/9780203497784.ch3.

Omernik, J. M. 1987. "Ecoregions of the Conterminous United States." *Annals of the Association of American Geographers* 77 (1): 118–25. http://dx.doi.org/10.1111/j.1467-8306.1987.tb00149.x.

Opdam, P. 1991. "Metapopulation Theory and Habitat Fragmentation: A Review of Holarctic Breeding Bird Studies." *Landscape Ecology* 5 (2): 93–106. http://dx.doi.org/10.1007/BF00124663.

Osti, M., L. Coad, J. B. Fisher, B. Bomhard, and J. M. Hutton. 2011. "Oil and Gas Development in the World Heritage and Wider Protected Area Network in Sub-Saharan Africa." *Biodiversity and Conservation* 20 (9): 1863–77. http://dx.doi.org/10.1007/s10531-011-0056-6.

Ostrom, E. 1993. "Design Principles in Long-Enduring Irrigation Institutions." *Water Resources Research* 29 (7): 1907–12. http://dx.doi.org/10.1029/92WR02991.

Oswalt, C. M., S. N. Oswalt, and W. K. Clatterbuck. 2007. "Effects of Microstegium Vimineum (Trin.) A. Camus on Native Woody Species Density and Diversity in a Productive Mixed-Hardwood Forest in Tennessee." *Forest Ecology and Management* 242 (2–3): 727–32. http://dx.doi.org/10.1016/j.foreco.2007.02.008.

Pace, M. L., and J. J. Cole. 2002. "Synchronous Variation of Dissolved Organic Carbon and Color in Lakes." *Limnology and Oceanography* 47 (2): 333–42. http://dx.doi.org/10.4319/lo.2002.47.2.0333.

Packer, C., H. Brink, B. M. Kissui, H. Maliti, H. Kushnir, and T. Caro. 2011. "Effects of Trophy Hunting on Lion and Leopard Populations in Tanzania." *Conservation Biology* 25 (1): 142–53. http://dx.doi.org/10.1111/j.1523-1739.2010.01576.x.

Paerl, H. W., and J. Huisman. 2008. "Blooms Like It Hot." *Science* 320 (5872): 57–8. http://dx.doi.org/10.1126/science.1155398.

Pain, D. J., A. Sánchez, and A. A. Meharg. 1998. "The Doñana Ecological Disaster: Contamination of a World Heritage Estuarine Marsh Ecosystem with Acidified Pyrite Mine Waste." *Science of the Total Environment* 222 (1-2): 45–54. http://dx.doi.org/10.1016/S0048-9697(98)00290-3.

Pan, Z., D. Andrade, M. Segal, J. Wimberley, N. McKinney, and E. Takle. 2010. "Uncertainty in Future Soil Carbon Trends at a Central US Site under an Ensemble of GCM Scenario Climates." *Ecological Modelling* 221: 876–81 http://dx.doi.org/10.1016/j.ecolmodel.2009.11.013.

Park, A., and C. Talbot. 2012. "Assisted Migration: Uncertainty, Risk and Opportunity." *Forestry Chronicle* 88 (4): 412–9. http://dx.doi.org/10.5558/tfc2012-077.

Parmesan, C. 2006. "Ecological and Evolutionary Responses to Recent Climate Change." *Annual Review of Ecology and Systematics* 37 (1): 637–69. http://dx.doi.org/10.1146/annurev.ecolsys.37.091305.110100.

Partnership for the Delaware Estuary (PDE). 2013. Technical Report for the Delaware Estuary and Basin. PDE Report No. 12-01. www.delawareestuary.org/science_programs_state_of_the_estuary.asp.

Pattison, R., C. D'Antonio, T. Dudley, K. K. Allander, and B. Rice. 2011. "Early Impacts of Biological Control on Canopy Cover and Water Use of the Invasive Saltcedar Tree (Tamarix spp.) in Western Nevada, USA." *Oecologia* 165 (3): 605–16. http://dx.doi.org/10.1007/s00442-010-1859-y.

Pauchard, A., P. B. Alaback, and E. G. Edlund. 2003. "Plant Invasions in Protected Areas at Multiple Scales: Linaria vulgaris (Scrophulariaceae) in the West Yellowstone Area." *Western North American Naturalist* 63:416–28.

Pearson, P. N., and M. R. Palmer. 2000. "Atmospheric Carbon Dioxide Concentrations over the Past 60 Million Years." *Nature* 406 (6797): 695–9. http://dx.doi.org/10.1038/35021000.

Pechony, O., and D. T. Schindell. 2010. "Driving Forces of Global Wildfires over the Past Millennium and the Forthcoming Century." *Proceedings of the National Academy of Sciences of the United States of America* 107 (45): 19167–70. http://dx.doi.org/10.1073/pnas.1003669107.

Pederson, G. T., J. L. Betancourt, and G. J. McCabe. 2013. "Regional Patterns and Proximal Causes of the Recent Snowpack Decline in the Rocky Mountains, US." *Geophysical Research Letters* 40 (9): 1811–6. http://dx.doi.org/10.1002/grl.50424.

Pedlar, J., D. W. McKenney, I. Aubin, T. Beardmore, J. Beaulieu, L. Iverson, G. A. O'Neill, R. S. Winder, and C. Ste-Marie. 2012. "Placing Forestry in the Assisted Migration Debate." *Bioscience* 62 (9): 835–42. http://dx.doi.org/10.1525/bio.2012.62.9.10.

Pedlar, J., D. W. McKenney, J. Beaulieu, S. Colombo, J. McLachlan, and G. O'Neill. 2011. "The Implementation of Assisted Migration in Canadian Forests." *Forestry Chronicle* 87 (06): 766–77. http://dx.doi.org/10.5558/tfc2011-093.

Pennsylvania Department of Conservation and Natural Resources, Bureau of Forestry. 2010. "Pennsylvania Statewide Forest Resource Assessment." http://www.dcnr.state.pa.us/cs/groups/public/documents/document/dcnr_007864.pdf.

Perlack, R., L. Wright, A. Turholow, R. L. Graham, B. J. Stokes, and D. C. Erbach. 2005. *Biomass as Feedstock for Bioenergy and Bioproducts Industry: The Technical Feasibility of a Billion-ton Annual Supply.* Oak Ridge, TN: US Department of Energy, Oak Ridge National Laboratory.

Peters, R. L., and J.D.S. Darling. 1985. "The Greenhouse-Effect and Nature Reserves." *Bioscience* 35 (11): 707–17. http://dx.doi.org/10.2307/1310052.

Peterson, D., C. Millar, L. Joyce, M. Furniss, J. Halofsky, R. Neilson, and T. Morelli. 2011. "Responding to Climate Change in National Forests: A Guidebook for

Developing Adaptation Options." General Technical Report PNW-GTR-855. Portland, Oregon: USDA Forest Service.

Peterson, D. L., J. M. Vose, and T. Patel-Weynand. 2014. *Climate Change and United States Forests. Advances in Global Change Research.* vol. 57. Dordrecht, The Netherlands: Springer. http://dx.doi.org/10.1007/978-94-007-7515-2.

Peterson, T. C., R. R. Heim, Jr., R. Hirsch, D. P. Kaiser, H. Brooks, N. S. Diffenbaugh, R. M. Dole, J. P. Giovannettone, K. Guirguis, T. R. Karl, et al. 2013. "Monitoring and Understanding Changes in Heat Waves, Cold Waves, Floods, and Droughts in the United States: State of Knowledge." *Bulletin of the American Meteorological Society* 94 (6): 821–34. http://dx.doi.org/10.1175/BAMS-D-12 -00066.1.

Petrides, G. A., and O. Petrides. 1998. *A Field Guide to Western Trees: Western United States and Canada.* 1st ed. Boston: Houghton Mifflin Company.

Pettorelli, N., A. L. M. Chauvenet, J. P. Duffy, W. A. Cornforth, A. Meillere, and J.E.M. Baillie. 2012. "Tracking the Effect of Climate Change on Ecosystem Functioning Using Protected Areas: Africa as a Case Study." *Ecological Indicators* 20:269–76. http://dx.doi.org/10.1016/j.ecolind.2012.02.014.

Phillips, B. L. C., L. Kelehear, L. Pizzatto, G. P. Brown, D. Barton, and R. Shine. 2010. "Parasites and Pathogens Lag behind Their Host During Periods of Host Range Advance." *Ecology* 91 (3): 872–81. http://dx.doi.org/10.1890/09-0530.1.

Phillips, S. J., R. P. Anderson, and R. E. Schapire. 2006. "Maximum Entropy Modeling of Species Geographic distributions." *Ecological Modelling* 190 (3): 231–59.

Piana, R., and S. Marsden. 2014. "Impacts of Cattle Grazing on Forest Structure and Raptor Distribution within a Neotropical Protected Area." *Biodiversity and Conservation* 23 (3): 559–72. http://dx.doi.org/10.1007/s10531-013-0616-z.

Pierce, J. L., G. A. Meyer, and A. J. Jull. 2004. "Fire-induced Erosion and Millennial-scale Climate Change in Northern Ponderosa Pine Forests." *Nature* 432 (7013): 87–90. http://dx.doi.org/10.1038/nature03058.

Pimentel, D., R. Zuniga, and D. Morrison. 2005. "Update on the Environmental and Economic Costs Associated with Alien-Invasive Species in the United States." *Ecological Economics* 52 (3): 273–88. http://dx.doi.org/10.1016/j.ecolecon.2004 .10.002.

Pinchot Institute. 2012. "The Role of Communities in Stewardship Contracting: Programmatic Monitoring Report to the US Forest Service for Fiscal Year 2011." Washington, DC: Pinchot Institute for Conservation. http://www.pinchot.org /gp/Stewardship_Contracting.

Pires, M. 2004. "Watershed Protection for a World City: The Case of New York." *Land Use Policy* 21 (2): 161–75. http://dx.doi.org/10.1016/j.landusepol.2003.08.001.

Pitelka, L. F., and the Plant Migration Workshop Group. 1997. "Plant Migration and Climate Change." *American Scientist* 85:464–73.

Poland, T. M., and D. G. McCullough. 2006. "Emerald Ash Borer: Invasion of the Urban Forest and the Threat to North Americas Ash Resource." *Journal of Forestry* 104 (3): 118–24.

Post, E., C. Pedersen, C. C. Wilmers, and M. C. Forchhammer. 2008. "Warming, Plant Phenology and the Spatial Dimension of Trophic Mismatch for Large Herbivores." *Proceedings. Biological Sciences* 275 (1646): 2005–13. http://dx.doi.org/10.1098/rspb.2008.0463.

Potter, K. M., and W. W. Hargrove. 2012. "Determining Suitable Locations for Seed Transfer under Climate Change: A Global Quantitative Method." *New Forests* 43 (5-6): 581–99. http://dx.doi.org/10.1007/s11056-012-9322-z.

Prasad, A. M., L. R. Iverson, S. Mathews, and M. Peters. 2007. *A Climate Change Atlas for 134 Forest Tree Species of the Eastern United States.* Delaware, OH: USDA Forest Service Northern Research Station; www.nrs.fs.fed.us/atlas/tree, Accessed Sept . 13, 2013.

Prasad, A., L. Iverson, M. Peters, J. M. Bossenbroek, S. N. Matthews, T. D. Sydnor, and M. W. Schwartz. 2010. "Modeling the Invasive Emerald Ash Borer Risk of Spread Using a Spatially Explicit Cellular Model." *Landscape Ecology* 25 (3): 353–69. http://dx.doi.org/10.1007/s10980-009-9434-9.

Prato, T., and D. Fagre. 2007. *Sustaining Rocky Mountain Landscapes: Science, Policy, and Management for the Crown of the Continent Ecosystem.* Washington, DC: Resources for the Future.

Preisler, H.K.A.L. Westerling, K. M. Gebert, F. Munoz-Arriola, and T. P. Holmes. 2011. "Spatially Explicit Forecasts of Large Wildland Fire Probability and Suppression Costs for California." *International Journal of Wildland Fire* 20 (4): 508–17. http://dx.doi.org/10.1071/WF09087.

Proctor, M. F., B. N. McLellan, C. Strobeck, and R.M.R. Barclay. 2005. "Genetic Analysis Reveals Demographic Fragmentation of Grizzly Bears Yielding Vulnerably Small Populations." *Proceedings. Biological Sciences* 272 (1579): 2409–16. http://dx.doi.org/10.1098/rspb.2005.3246.

Puettmann, K. J. 2011. "Silvicultural Challenges and Options in the Context of Global Change: 'Simple' Fixes and Opportunities for New Management Approaches." *Journal of Forestry* 109 (6): 321–31.

Raffa, K., B. H. Aukema, B. J. Bentz, A. L. Carroll, J. A. Hicke, M. G. Turner, and W. H. Romme. 2008. "Cross-scale Drivers of Natural Disturbances Prone to Anthropogenic Amplification: The Dynamics of Bark Beetle Eruptions." *Bioscience* 58 (6): 501–17. http://dx.doi.org/10.1641/B580607.

Randin, C. F., R. Engler, S. Normand, M. Zappa, N. E. Zimmermann, P. B. Pearman, P. Vittoz, W. Thuiller, and A. Guisan. 2008. "Climate Change and Plant Distribution: Local Models Predict High-Elevation Persistence." *Global Change Biology* 15 (6): 1557–69. http://dx.doi.org/10.1111/j.1365-2486.2008.01766.x.

Rasker, R. 2013. "Wildfire Costs, New Development, and Rising Temperatures. Headwaters Economics." http://headwaterseconomics.org/wildfire/fire -research-summary.

Ray, A. J., J. J. Barsugli, K. Wolter, and J. Eischeid. 2010. *Rapid-Response Climate Assessment to Support the FWS Status Review of the American Pika*. Boulder, CO: NOAA Earth Systems Research Laboratory; http://www.esrl.noaa.gov/psd /news/2010/pdf/pika_report_final.pdf.

Raymond, C. L., D. L. Peterson, and R. M. Rochefort. 2013. "The North Cascadia Adaptation Partnership: A Science-management Collaboration for Responding to Climate Change." *Sustainability* 5 (1): 136–59. http://dx.doi.org/10.3390 /su5010136.

Raymond, C. L., D. L. Peterson, and R. M. Rochefort. 2014. "Climate Change Vulnerability and Adaptation in the North Cascades Region, Washington." Gen. Tech. Rep. PNW-GTR-892. Portland, OR: US Department of Agriculture, Forest Service, Pacific Northwest Research Station.

Raymond, P. A., and J. E. Saiers. 2010. "Event Controlled DOC Export from Forested Watersheds." *Biogeochemistry* 100 (1-3): 197–209. http://dx.doi.org/10.1007 /s10533-010-9416-7.

Redondo-Brenes, A. 2007. "Implementation of Conservation Approaches in Human-dominated Landscapes: The Path of the Tapir Biological Corridor Case Study, Costa Rica." *Tropical Resources* 26:7–14.

Reeder, T., and N. Ranger. 2010. "How Do You Adapt in an Uncertain World." Lessons from the Thames Estuary 2100 project. Washington, DC: World Resources Report. http://www.worldresourcesreport.org.

Regional Plan Association. 2012. "Landscapes: Improving Conservation Practice in the Northeast Megaregion." http://www.rpa.org/library/pdf/RPA-Northeast -Landscapes.pdf.

Régnière, J., V. Nealis, and K. Porter. 2009. "Climate Suitability and Management of the Gypsy Moth Invasion into Canada." *Biological Invasions* 11 (1): 135–48. http://dx.doi.org/10.1007/s10530-008-9325-z.

Regniere, J., and R. Saint-Amant. 2008. "BioSIM 9—User's Manual." Information Report LAU-X-134. Natural Resources Canada, Canadian Forest Service, Laurentian Forestry Centre.

Rehfeldt, G. E. 1983. "Ecological Adaptations in Douglas-fir (*Pseudotsuga menziesii* var. *glauca*) Populations. III. Central Idaho." *Canadian Journal of Forest Research* 13 (4): 626–32. http://dx.doi.org/10.1139/x83-090.

Rehfeldt, G. E., and B. C. Jaquish. 2010. "Ecological Impacts and Management Strategies for Western Larch in the Face of Climate Change." *Mitigation and Adaptation Strategies for Global Change* 15 (3): 283–306. http://dx.doi.org/10.1007 /s11027-010-9217-2.

Reichard, S. H., and C. W. Hamilton. 1997. "Predicting Invasions of Woody Plants Introduced into North America." *Conservation Biology* 11 (1): 193–203. http://dx .doi.org/10.1046/j.1523-1739.1997.95473.x.

Reichard, S. H., and P. White. 2001. "Horticulture as a Pathway of Invasive Plant Introductions in the United States." *Bioscience* 51 (2): 103–13. http://dx.doi.org/10 .1641/0006-3568(2001)051[0103:HAAPOI]2.0.CO;2.

Renkin, R. A., and D. G. Despain. 1992. "Fuel Moisture, Forest Type, and Light-ning-caused Fire in Yellowstone National Park." *Canadian Journal of Forest Research* 22 (1): 37–45. http://dx.doi.org/10.1139/x92-005.

Rhoades, C., D. Loftis, J. Lewis, and S. Clark. 2009. "The Influence of Silvicultural Treatments and Site Conditions on American Chestnut (Castanea dentata) Seed-ling Establishment in Eastern Kentucky, USA." *Forest Ecology and Management* 258 (7): 1211–8. http://dx.doi.org/10.1016/j.foreco.2009.06.014.

Ricciardi, A., and D. Simberloff. 2009. "Assisted Colonization is Not a Viable Con-servation Strategy." *Trends in Ecology & Evolution* 24 (5): 248–53. http://dx.doi .org/10.1016/j.tree.2008.12.006.

Rice, J., A. Tredennick, and L. A. Joyce. 2012. "Climate Change on the Shoshone National Forest, Wyoming: A Synthesis of Past Climate, Climate Projections, and Ecosystem Implications." Gen. Tech. Rep. RMRS-GTR-264. Fort Collins, CO: USDA, Forest Service, Rocky Mountain Research Station.

Roberts, S. L., J. W. van Wagtendonk, A. K. Miles, and D. A. Kelt. 2011. "Effects of Fire on Spotted Owl Site Occupancy in a Late-successional Forest." *Biological Conservation* 144 (1): 610–9. http://dx.doi.org/10.1016/j.biocon.2010.11.002.

Robinson, G. R., Jr. 1997. "Portraying Chemical Properties of Bedrock for Water Quality and Ecosystem Analysis: An Approach for New England." Open-file Report 97-154. Reston, VA: US Geological Survey.

Rodríguez, A., and M. Delibes. 2003. "Population Fragmentation and Extinction in the Iberian Lynx." *Biological Conservation* 109 (3): 321–31. http://dx.doi.org/10 .1016/S0006-3207(02)00158-1.

Rodríguez, N., D. Armenteras, and J. Retana. 2013. "Effectiveness of Protected Areas in the Colombian Andes: Deforestation, Fire and Land-use Changes." *Regional Environmental Change* 13 (2): 423–35. http://dx.doi.org/10.1007/s10113 -012-0356-8.

Rogala, J. K., M. Hebblewhite, J. Whittington, C. A. White, J. Coleshill, and M. Musiani. 2011. "Human Activity Differentially Redistributes Large Mammals in the Canadian Rockies National Parks." *Ecology and Society* 16 (3): 16. http://dx .doi.org/10.5751/ES-04251-160316.

Rollins, M. G., T. W. Swetnam, and P. Morgan. 2001. "Evaluating a Century of Fire Patterns in Two Rocky Mountain Wilderness Areas Using Digital Fire Atlases." *Canadian Journal of Forest Research* 31 (12): 2107–23. http://dx.doi.org/10.1139 /x01-141.

Romme, W. H., C. D. Allen, J. D. Bailey, W. L. Baker, B. T. Bestelmeyer, P. M. Brown, K. S. Eisenhart, M. L. Floyd, D. W. Huffman, B. F. Jacobs, and R. R. Miller. 2009. "Historical and Modern Disturbance Regimes, Stand Structures, and Landscape Dynamics in Pinon–Juniper Vegetation of the Western United States." *Rangeland Ecology & Management* 62 (3): 203–22.

Romme, W. H., and D. G. Despain. 1989. "Historical Perspective on the Yellowstone Fires of 1988." *Bioscience* 39 (10): 695–9. http://dx.doi.org/10.2307/1311000.

Roos, C. I., and T. W. Swetnam. 2011. "A 1416-Year Reconstruction of Annual, Multi-decadal, and Centennial Variability in Area Burned for Ponderosa Pine Forests of the Southern Colorado Plateau Region, Southwest USA." *The Holocene*: 0959683611423694.

Ross, R. M., R. M. Bennett, C. D. Snyder, J. A. Young, D. R. Smith, and D. P. Lema-rie. 2003. "Influence of Eastern Hemlock (Tsuga canadensis L.) on Fish Community Structure and Function in Headwater Streams of the Delaware River Basin." *Ecology Freshwater Fish* 12 (1): 60–5. http://dx.doi.org/10.1034/j.1600-0633.2003.00006.x.

Rustad, L., J. Campbell, J. S. Dukes, T. Huntington, K. F. Lambert, J. Mohan, and N. Rodenhouse. 2012. "Changing Climate, Changing Forests: The Impacts of Climate Change on Forests of the Northeastern United States and Eastern Canada." Gen. Tech. Rep. NRS-99. Newton Square, PA: USDA, Forest Service, Northern Research Station.

Safford, H. D., D. A. Schmidt, and C. H. Carlson. 2009. "Effects of Fuel Treatments on Fire Severity in an Area of Wildland-urban Interface, Angora Fire, Lake Tahoe Basin, California." *Forest Ecology and Management* 258 (5): 773–87. http://dx.doi.org/10.1016/j.foreco.2009.05.024.

Salzer, M. W., and K. F. Kipfmueller. 2005. "Reconstructed Temperature and Precipitation on a Millennial Timescale from Tree-Rings in the Southern Colorado Plateau, USA." *Climatic Change* 70: 465–87.

Sample, V. A. 1990. *The Impact of the Federal Budget Process on National Forest Planning.* Westport, CT: Greenwood Press.

Sample, V., and T. Tipple. 2001. "Improving Performance and Accountability at the Forest Service: Overcoming the Politics of the Budgetary Process and Improving Budget Execution." *Journal of Public Budgeting, Accounting, and Financial Management* 13 (2).

Sasek, T., and B. Strain. 1990. "Implications of Atmospheric CO_2 Enrichment and Climatic Change for the Geographical Distribution of Two Introduced Vines in the U.S.A." *Climatic Change* 16 (1): 31–51. http://dx.doi.org/10.1007/BF00137345.

Sathre, R., and J. O'Connor. 2008. "A Synthesis of Research on Wood Products and Greenhouse Gas Impacts." Technical Report TR-19. Vancouver, BC: FPInnovations, Forintek Division.

Savage, M., P. M. Brown, and J. Feddema. 1996. "The Role of Climate in a Pine Forest Regeneration Pulse in the Southwestern United States." *Ecoscience* 3:310–8.

Savage, M., and J. N. Mast. 2005. "How Resilient Are Southwestern Ponderosa Pine Forests after Crown Fires?" *Canadian Journal of Forest Research* 35 (4): 967–77. http://dx.doi.org/10.1139/x05-028.

Savage, M., J. N. Mast, and J. J. Feddema. 2013. "Double Whammy: High-Severity Fire and Drought in Ponderosa Pine Forests of the Southwest." *Canadian Journal of Forest Research* 43 (6): 570–83. http://dx.doi.org/10.1139/cjfr-2012-0404.

Schiermeier, Q. 2010. "The Real Holes in Climate Science." *Nature* 463 (7279): 284–7. http://dx.doi.org/10.1038/463284a.

Schlamadinger, B., and G. Marland. 1996. "The Role of Forest Bioenergy Strategies in the Global Carbon Cycle." *Biomass and Bioenergy* 10: 275–300.

Schmidt, K. A., and C. J. Whelan. 1999. "Effects of Exotic Lonicera and Rhamnus on Songbird Nest Predation." *Conservation Biology* 13 (6): 1502–6. http://dx.doi.org/10.1046/j.1523-1739.1999.99050.x.

Schmidtling, R. C. 2001. "Southern Pine Seed Sources." GTR-SRS-44. Asheville, NC: USDA, Forest Service, Southern Research Station.

Schmittner, A., A. Oschlies, H. D. Matthews, and E. D. Galbraith. 2008. "Future Changes in Climate, Ocean Circulation, Ecosystems, and Biogeochemical Cycling Simulated for a Business-as-usual CO_2 Emission Scenario until Year 4000 AD." *Global Biogeochemical Cycles* 22 (1): GB1013. http://dx.doi.org/10.1029/2007 GB002953.

Schneider, R. R., B. Price, P. J. Müller, D. Kroon, and I. Alexander. 1997. "Monsoon Related Variations in Zaire (Congo) Sediment Load and Influence of Fluvial Silicate Supply on Marine Productivity in the East Equatorial Atlantic During the Last 200,000 Years." *Paleoceanography* 12 (3): 463–81. http://dx.doi.org/10.1029 /96PA03640.

Schoennagel, T., R. L. Sherriff, and T. T. Veblen. 2011. "Fire History and Tree Recruitment in the Colorado Front Range Upper Montane Zone: Implications for Forest Restoration." *Ecological Applications* 21:2210–22. http://dx.doi.org/10 .1890/10-1222.1.

Schoennagel, T., T. T. Veblen, and W. H. Romme. 2004. "The Interaction of Fire, Fuels and Climate across Rocky Mountain Forests." *Bioscience* 54 (7): 661–76. http://dx.doi.org/10.1641/0006-3568(2004)054[0661:TIOFFA]2.0.CO;2.

Schultz, C. A., T. Jedd, and R. D. Beam. 2012. "The Collaborative Forest Landscape Restoration Program: A History and Overview of the First Projects." *Journal of Forestry* 110 (7): 381–91. http://dx.doi.org/10.5849/jof.11-082.

Schulz, B., and A. Gray. 2013. "The New Flora of Northeastern USA: Quantifying Introduced Plant Species Occupancy in Forest Ecosystems." *Environmental Monitoring and Assessment* 185 (5): 3931–57. http://dx.doi.org/10.1007/s10661-012-2841-4.

Schwalm, C. R., C. A. Williams, K. Schaefer, D. Baldocchi, T. A. Black, A. H. Goldstein, B. E. Law, W. C. Oechel, K. T. Paw U, and R. L. Scott. 2012. "Reduction in Carbon Uptake During Turn of the Century Drought in Western North America." *Nature Geoscience* 5 (8): 551–6. http://dx.doi.org/10.1038/ngeo1529.

Schwandt, J. W., I. B. Lockman, J. T. Kliejunas, and J. A. Muir. 2010. "Current Health Issues and Management Strategies for White Pines in the Western United States and Canada." *Forest Pathology* 40 (3–4): 226–50. http://dx.doi.org/10.1111/j.1439-0329.2010.00656.x.

Schwartz, M. W. 1994. "Conflicting Goals for Conserving Biodiversity: Issues of Scale and Value." *Natural Areas Journal* 14 (3): 213–6.

Schwartz, M. W., J. J. Hellmann, J. M. McLachlan, D. F. Sax, J. O. Borevitz, J. Brennan, A. E. Camacho, G. Ceballos, J. R. Clark, H. Doremus, et al. 2012. "Managed Relocation: Integrating the Scientific, Regulatory, and Ethical Challenges." *Bioscience* 62 (8): 732–43. http://dx.doi.org/10.1525/bio.2012.62.8.6.

Schwilk, D. W., J. E. Keeley, E. E. Knapp, J. McIver, J. D. Bailey, C. J. Fettig, C. E. Fiedler, R. J. Harrod, J. J. Moghaddas, K. W. Outcalt, et al. 2009. "The National Fire and Fire Surrogate Study: Effects of Fuel Reduction Methods on Forest Vegetation Structure and Fuels." *Ecological Applications* 19 (2): 285–304. http://dx.doi.org/10.1890/07-1747.1.

Scott, D. 2013. "The Health and Integrity of the Current Supply Chain: Labor." Paper presented at The State and Future of US Forestry and the Forest Industry. Sponsored by the US Endowment for Forestry and Communities and the US Forest Service. Washington, DC: Resources for the Future.

Scott, D. W. 2012. *"Pine Butterfly."* Forest Insect and Disease Leaflet 66. USDA Forest Service.

Seager, R., M. Ting, I. Held, Y. Kushnir, J. Lu, G. Vecchi, H.-P. Huang, N. Harnik, A. Leetmaa, N.-C. Lau, et al. 2007. "Model Projections of an Imminent Transition to a More Arid Climate in Southwestern North America." *Science* 316 (5828): 1181–4. http://dx.doi.org/10.1126/science.1139601.

Seager, R., and G. A. Vecchi. 2010. "Greenhouse Warming and the Twenty-first-century Hydroclimate of Southwestern North America." *Proceedings of the National Academy of Sciences of the United States of America* 107 (50): 21277–82. http://dx.doi.org/10.1073/pnas.0910856107.

Seidel, D. J., and W. J. Randel. 2007. "Recent Widening of the Tropical Belt: Evidence from Tropopause Observations." *Journal of Geophysical Research* 112 (D20): D20113. http://dx.doi.org/10.1029/2007JD008861.

Sendak, P. E., R. C. Abt, and R. J. Turner. 2003. "Timber Supply Projections for Northern New England and New York: Integrating a Market Perspective." *Northern Journal of Applied Forestry* 20 (4): 175–85.

Sexton, T. 2013. "Wildfire in the United States: Recent History, Context, and Trends." Wildfire Symposium 2013. Denver, CO: Water Research Foundation.

Seymour, R. S., and M. L. Hunter, Jr. 1999. "Principles of Ecological Forestry." In *Maintaining Biodiversity in Forest Ecosystems*, ed. M. L. Hunter, Jr., 22–62. Cambridge: Cambridge University Press. http://dx.doi.org/10.1017/CBO9780511613029.004.

Shafroth, P., J. Cleverly, T. Dudley, J. P. Taylor, C. Van Riper, E. P. Weeks, and James N. Stuart. 2005. "Control of Tamarix in the Western United States: Implications for Water Salvage, Wildlife Use, and Riparian Restoration." *Environmental Management* 35 (3): 231–46. http://dx.doi.org/10.1007/s00267-004-0099-5.

Sheffield, J., G. Goteti, F. H. Wen, and E. F. Wood. 2004. "A Simulated Soil Moisture Based Drought Analysis for the United States." *Journal of Geophysical Research* 109 (D24): D24108. http://dx.doi.org/10.1029/2004JD005182.

Sherriff, R. S., and T. T. Veblen. 2008. "Variability in Fire-climate Relationships in Ponderosa Pine Forests in the Colorado Front Range." *International Journal of Wildland Fire* 17 (1): 50–9. http://dx.doi.org/10.1071/WF07029.

Sibold, J. S., and T. T. Veblen. 2006. "Relationships of Subalpine Forest Fires in the Colorado Front Range to Interannual and Multi-decadal Scale Climatic Variation." *Journal of Biogeography* 33 (5): 833–42. http://dx.doi.org/10.1111/j.1365-2699 .2006.01456.x.

Silvas-Bellanca, K. 2011. "Ecological Burning in the Sierra Nevada: Actions to Achieve Restoration." White paper of The Sierra Forest Legacy. Accessed June 29, 2013. www.sierraforestlegacy.org.

Sisk, T. D., M. Savage, D. Falk, C. D. Allen, E. Muldavin, and P. McCarthy. 2005. "A Landscape Perspective for Forest Restoration." *Journal of Forestry* 103:319–20.

Sloto, R. A., and D. E. Buxton. 2005. "Water Budgets for Selected Watersheds in the Delaware River Basin, Eastern Pennsylvania and Western New Jersey." US Geological Survey Scientific Investigations Report 2005–5113.

Smith, H., G. Sheridan, P. Lane, P. Nyman, and S. Haydon. 2011. "Wildfire Effects on Water Quality in Forested Catchments: A Review with Implications for Water Supply." *Journal of Hydrology (Amsterdam)* 396 (1-2): 170–92. http://dx.doi .org/10.1016/j.jhydrol.2010.10.043.

Smith, S. B., S. A. DeSando, and T. Pagano. 2013. "The Value of Native and Invasive Fruit-bearing Shrubs for Migrating Songbirds." *Northeastern Naturalist* 20 (1): 171–84. http://dx.doi.org/10.1656/045.020.0114.

Smith, S. D., T. E. Huxman, S. F. Zitzer, T. N. Charlet, D. C. Housman, J. S. Coleman, L. K. Fenstermaker, J. R. Seemann, and R. S. Nowak. 2000. "Elevated CO_2 Increases Productivity and Invasive Species Success in an Arid Ecosystem." *Nature* 408 (6808): 79–82. http://dx.doi.org/10.1038/35040544.

Smith, T. K., D. Rose, and P. D. Gingerich. 2006. "Rapid Asia-Europe-North America Geographic Dispersal of Earliest Eocene Primate *Teilhardia* during the Palocene-Eocene Thermal Maximum." *Proceedings of the National Academy of Sciences of the United States of America* 103:11223–7. http://dx.doi.org/10.1073/pnas .0511296103.

Smith, W. B., P. D. Miles, C. H. Perry, and S. A. Pugh. 2009. "Forest Resources of the United States, 2007." Gen. Tech. Rep. WO–78. Washington, DC: USDA, Forest Service.

Sneeuwjagt, R. J., T. S. Kline, and S. L. Stephens. 2013. "Opportunities for Improved Fire Use and Management in California: Lessons from Western Australia." *Fire Ecology* 9 (2): 14–25. http://dx.doi.org/10.4996/fireecology.0902014.

Snitzer, J., D. Boucher, and K. Kyde. 2005. "Response of Exotic Invasive Plant Species to Forest Damage Caused by Hurricane Isabel." CRC Publication 05–160. Edgewater, MD: Chesapeake Research Consortium.

Snyder, B. 2008. "How to Reach a Compromise on Drilling in AWNR." *Energy Policy* 36 (3): 937–9. http://dx.doi.org/10.1016/j.enpol.2007.10.016.

Snyderman, D., and C. D. Allen. 1997. "Fire on the Mountain: Analysis of Historical Fires for Bandelier National Monument, Santa Fe National Forest, and Surrounding Areas—1909 through 1996." Unpublished report on file at USGS Jemez Mountains Field Station.

Soja, A. J., N. M. Tchebakova, N. H. F. French, M. D. Flannigan, H. H. Shugart, B. J. Stocks, A. I. Sukhinin, E. I. Parfenova, F. Stuart Chapin, III, and P. W. Stackhouse, Jr. 2007. "Climate-induced Boreal Forest Change: Predictions Versus Current Observations." *Global and Planetary Change* 56 (3-4): 274–96. http://dx.doi.org/10.1016/j.gloplacha.2006.07.028.

Soon, W.W.H. 2005. "Variable Solar Irradiance as a Plausible Agent for Multidecadal Variations in the Arctic-wide Surface Air Temperature Record of the Past 130 Years." *Harvard-Smithsonian Center for Astrophysics Geophysical Research Letters* 32 (16): L16712. http://dx.doi.org/10.1029/2005GL023429.

Soulé, M. E., and J. Terborgh. 1999. "Conserving Nature at Regional and Continental Scales—A Scientific Program for North America." *Bioscience* 49 (10): 809–17. http://dx.doi.org/10.2307/1313572.

Spies, T. A., M. A. Hemstrom, A. Youngblood, and S. Hummel. 2006. "Conserving Old Growth Forest Diversity in Disturbance-Prone Landscapes." *Conservation Biology* 20 (2): 351–62. http://dx.doi.org/10.1111/j.1523-1739.2006.00389.x.

Spooner, D., M. Xenopoulos, C. Schneider, and D. A. Woolnough. 2011. "Coextirpation of Host-Affiliate Relationships in Rivers: The Role of Climate Change, Water Withdrawal and Host Specificity." *Global Change Biology* 17 (4): 1720–32. http://dx.doi.org/10.1111/j.1365-2486.2010.02372.x.

Spracklen, D. V., L. J. Mickley, J. A. Logan, R. C. Hudman, R. Yevich, M. D. Flannigan, and A. L. Westerling. 2009. "Impacts of Climate Change from 2000 to 2050 on Wildfire Activity and Carbonaceous Aerosol Concentrations in the Western United States." *Journal of Geophysical Research* 114 (D20): D20301. http://dx.doi.org/10.1029/2008JD010966.

Stager, J. C. 2011. *Deep Future: The Next 100,000 Years of Life on Earth*. New York: St. Martin's Press.

Stager, J. C., B. Cumming, and L. D. Meeker. 2003. "A 10,000-Year High-resolution Diatom Record from Pilkington Bay, Lake Victoria, East Africa." *Quaternary Research* 59 (2): 172–81. http://dx.doi.org/10.1016/S0033-5894(03)00008-5.

Stager, J. C., P. A. Mayewski, and L. D. Meeker. 2002. "Cooling Cycles, Heinrich Event 1, and the Desiccation of Lake Victoria." *Palaeogeography, Palaeoclimatology, Palaeoecology* 183 (1-2): 169–78. http://dx.doi.org/10.1016/S0031-0182(01)00468-0.st

Stager, J. C., P. A. Mayewski, J. White, B. M. Chase, F. H. Neumann, M. E. Meadows, C. D. King, and D. A. Dixon. 2012. "Precipitation Variability in the Winter Rainfall Zone of South Africa During the Last 1400 Years Linked to the Austral Westerlies." *Climate of the Past* 8 (3): 877–87. http://dx.doi.org/10.5194/cp-8-877-2012.

Stager, J. C., A. Ruzmaikin, D. Conway, P. Verburg, and P. J. Mason. 2007. "Sunspots, El Niño, and the Levels of Lake Victoria, East Africa." *Journal of Geophysical Research* 112 (D15): D15106. http://dx.doi.org/10.1029/2006JD008362.

Stager, J. C., D. R. Ryves, B. M. Chase, and F.S.R. Pausata. 2011. "Catastrophic Drought in the Afro-Asian Monsoon Regions During Heinrich Event 1." *Science* 331 (6022): 1299–302. http://dx.doi.org/10.1126/science.1198322.

Stager, J. C., D. Ryves, B. F. Cumming, L. D. Meeker, and J. Beer. 2005. "Solar Variability and the Levels of Lake Victoria, East Africa, During the Last Millennium." *Journal of Paleolimnology* 33 (2): 243–51. http://dx.doi.org/10.1007/s10933-004-4227-2.

Stager, J. C., and M. Thill. 2010. "Climate Change in the Champlain Basin: What Natural Resource Managers Can Expect and Do." Report for The Nature Conservancy.

Stankey, G., B. Bormann, C. Ryan, B. Shindler, V. Sturtevant, R. N. Clark, and C. Philpot. 2003. "Adaptive Management and the Northwest Forest Plan." *Journal of Forestry* 101 (1): 40–6.

Stanturf, J. A., B. J. Palik, and R. K. Dumroese. 2014. "Contemporary Forest Restoration: A Review Emphasizing Function." *Forest Ecology and Management* 331:292–323. http://dx.doi.org/10.1016/j.foreco.2014.07.029.

Stanturf, J. A., B. J. Palik, M. I. Williams, R. K. Dumroese, and P. Madsen. 2014. "Forest Restoration Paradigms." *Journal of Sustainable Forestry* 33 (sup1): S161–94. http://dx.doi.org/10.1080/10549811.2014.884004.

St. Clair, J. B., G. T. Howe, J. Wright, and R. Beloin. 2013. "Center for Forest Provenance Data." Accessed March 19, 2016. http://cenforgen.forestry.oregonstate.edu/.

Ste-Marie, C., E. A. Nelson, A. Dabros, and M. Bonneau. 2011. "Assisted Migration: Introduction to a Multifaceted Concept." *Forestry Chronicle* 87 (6): 724–30. http://dx.doi.org/10.5558/tfc2011-089.

Stein, B. A., L. S. Kutner, and J. S. Adams. 2000. *Precious Heritage: The Status of Biodiversity in the United States*. London: Oxford University Press.

Stein, S. M., J. Menakis, M. A. Carr, S. J. Comas, S. I. Stewart, H. Cleveland, L. Bramwell, and V. C. Radeloff. 2013. "Wildfire, Wildlands, and People: Understanding and Preparing for Wildfire in the Wildland-Urban Interface." Gen. Tech. Rep. RMRS-GTR-299. Fort Collins, CO: U.S. Department of Agriculture,

Forest Service, Rocky Mountain Research Station. http://www.fs.fed.us/rm /pubs/rmrs_gtr299.pdf.

Stephens, S. L., J. K. Agee, P. Z. Fulé, M. P. North, W. H. Romme, T. W. Swetnam, and M. G. Turner. 2013. "Managing Forests and Fire in Changing Climates." *Science* 342 (6154): 41–2. http://dx.doi.org/10.1126/science.1240294.

Stephens, S. L., N. Burrows, A. Buyantuyev, R. W. Gray, R. E. Keane, R. Kubian, S. Liu, F. Seijo, L. Shu, K. G. Tolhurst, et al. 2014. "Temperate and Boreal Forest Mega-fires: Characteristics and Challenges." *Frontiers in Ecology and the Environment* 12 (2): 115–22. http://dx.doi.org/10.1890/120332.

Stephens, S. L., B. M. Collins, and G. Roller. 2012b. "Fuel Treatment Longevity in a Sierra Nevada Mixed Conifer Forest." *Forest Ecology and Management* 285:204–12. http://dx.doi.org/10.1016/j.foreco.2012.08.030.

Stephens, S. L., D. Fry, and E. Franco-Vizcano. 2008. "Wildfire and Forests in Northwestern Mexico: The United States Wishes It Had Similar Fire Problems." *Ecology and Society* 13:10.

Stephens, S. L., R. E. Martin, and N. E. Clinton. 2007. "Prehistoric Fire Area and Emissions from California's Forests, Woodlands, Shrublands, and Grasslands." *Forest Ecology and Management* 251 (3): 205–16. http://dx.doi.org/10.1016/j.foreco .2007.06.005.

Stephens, S. L., J. D. McIver, R.E.J. Boerner, C. J. Fettig, J. B. Fontaine, B. R. Hartsough, P. L. Kennedy, and D. W. Schwilk. 2012a. "Effects of Forest Fuel Reduction Treatments in the United States." *Bioscience* 62 (6): 549–60. http://dx.doi.org /10.1525/bio.2012.62.6.6.

Stephenson, N. L. 1998. "Actual Evapotranspiration and Deficit: Biologically Meaningful Correlates of Vegetation Distribution across Spatial Scales." *Journal of Biogeography* 25 (5): 855–70. http://dx.doi.org/10.1046/j.1365-2699.1998.00233.x.

Sternlieb, F., R. P. Bixler, H. Huber-Stearns, and C. Huayhuaca. 2013. "A Question of Fit: Reflections on Boundaries, Organizations, and Social-Ecological Systems." *Journal of Environmental Management* 130:117–25. http://dx.doi.org/10.1016/j .jenvman.2013.08.053.

Stolzenburg, W. 2009. *Where the Wild Things Were: Life, Death, and Ecological Wreckage in a Land of Vanishing Predators.* New York: Bloomsbury.

Strayer, D. L., V. T. Eviner, J. M. Jeschke, and M. L. Pace. 2006. "Understanding the Long-Term Effects of Species Invasions." *Trends in Ecology & Evolution* 21 (11): 645–51. http://dx.doi.org/10.1016/j.tree.2006.07.007.

Stuart, G. W., and P. J. Edwards. 2006. "Concepts about Forests and Water." *Northern Journal of Applied Forestry* 23 (1): 11–9.

Sturm, M., J. Schimel, G. Michaelson, J. M. Welker, S. F. Oberbauer, G. E. Liston, J. Fahnestock, and V. E. Romanovsky. 2005. "Winter Biological Processes Could Help Convert Arctic Tundra to Shrubland." *Bioscience* 55 (1): 17–26. http://dx.doi .org/10.1641/0006-3568(2005)055[0017:WBPCHC]2.0.CO;2.

Sturrock, R. N., S. J. Frankel, A. V. Brown, P. E. Hennon, J. T. Kliejunas, K. J. Lewis, J. J. Worrall, and A. J. Woods. 2011. "Climate Change and Forest Diseases." *Plant Pathology* 60 (1): 133–49. http://dx.doi.org/10.1111/j.1365-3059.2010.02406.x.

Swanston, C., and M. Janowiak. 2012. "Forest Adaptation Resources: Climate Change Tools and Approaches for Land Managers." Gen. Tech. Rep. NRS-87. Newton Square, PA: USDA, Forest Service, Northern Research Station.

Swanston, C., M. Janowiak, L. Iverson, L. Parker, D. Mladenoff, L. Brandt. P. Butler, M. St. Pierre, A. Prasad, S. Matthews, M. Peters, D. Higgins, and A. Dorland. 2011. "Ecosystem Vulnerability Assessment and Synthesis: A Report from the Climate Change Response Framework Project in Northern Wisconsin." Gen. Tech. Rep. NRS-82. Newtown Square, PA: USDA Forest Service, Northern Research Station.

Swenson, J. J., C. E. Carter, J. C. Domec, and C. I. Delgado. 2011. "Gold Mining in the Peruvian Amazon: Global Prices, Deforestation, and Mercury Imports." *PLoS One* 6 (4): e18875. http://dx.doi.org/10.1371/journal.pone.0018875.

Swetnam, T. W., C. D. Allen, and J. L. Betancourt. 1999. "Applied Historical Ecology: Using the Past to Manage for the Future." *Ecological Applications* 9 (4): 1189–206. http://dx.doi.org/10.1890/1051-0761(1999)009[1189:AHEUTP]2.0.CO;2.

Swetnam, T. W., and C. H. Baisan. 1996. "Historical Fire Regime Patterns in the Southwestern United States Since AD 1700." In *Fire Effects in Southwestern Forests, Proceedings of the Second La Mesa Fire Symposium, Los Alamos, New Mexico, March 29–31, 1994*, ed. C. D. Allen, 11–32. USDA Forest Service General Technical Report RM-GTR-286.

Swetnam, T. W., and C. H. Baisan. 2003. "Tree-Ring Reconstructions of Fire and Climate History in the Sierra Nevada and Southwestern United States." In *Fire and Climatic Change in Temperate Ecosystems of the Western Americas*, ed. T. T. Veblen, W. L. Baker, G. Montenegro, and T. W. Swetnam, 158–195. New York: Springer.

Swetnam, T. W., C. H. Baisan, and J. M. Kaib. 2001. "Forest Fire Histories in the Sky Islands of La Frontera." In *Changing Plant Life of La Frontera: Observations on Vegetation in the United States/Mexico Borderlands*, ed. G. L. Webster and C. J. Bahre, 95–119. Albuquerque: University of New Mexico Press.

Swetnam, T. W., and J. L. Betancourt. 1990. "Fire-southern Oscillation Relations in the Southwestern United States." *Science* 249 (4972): 1017–20. http://dx.doi.org/10.1126/science.249.4972.1017.

Swetnam, T. W., and J. L. Betancourt. 1998. "Mesoscale Disturbance and Ecological Response to Decadal Climatic Variability in the American Southwest." *Journal of Climate* 11 (12): 3128–47. http://dx.doi.org/10.1175/1520-0442(1998)011<3128:MDAERT>2.0.CO;2.

Swetnam, T. W., and P. M. Brown. 2010. "Climatic Inferences from Dendroecological Reconstructions." In *Dendroclimatology: Developments in Paleoenvironmental Research*, ed. M. K. Hughes, T. W. Swetnam, and H. F. Diaz. 11: 263–95.

Swetnam, T. W., D. A. Falk, E. K. Sutherland, P. M. Brown, and T. J. Brown. 2011. "Final Report: Fire and Climate Synthesis (FACS) Project, JFSP 09-2-01-10." Joint Fire Sciences Program. https://www.firescience.gov/projects/09-2-01-10/project/09-2-01-10_final_report.pdf.

Tabor, G. M. 1996. "Yellowstone-to-Yukon: Canadian Conservation Efforts and Continental Landscape/biodiversity Strategy." Kendall Foundation. http://largelandscapes.org/media/publications/Yellowstone-to-Yukon-Canadian-Conservation-Effort.pdf.

Tang, L., G. Shao, Z. Piao, L. Dai, M. A. Jenkins, S. Wang, G. Wu, J. Wu, and J. Zhao. 2010. "Forest Degradation Deepens around and within Protected Areas in East Asia." *Biological Conservation* 143 (5): 1295–8. http://dx.doi.org/10.1016/j.biocon.2010.01.024.

Taylor, P. 2013. "'It's Just Nuts' as Wildfires Drain Budget Yet Again." *Greenwire*, October 30. http://www.eenews.net/greenwire/stories/1059989688.

Taylor, P. D., L. Fahrig, K. Henein, and G. Merriam. 1993. "Connectivity Is a Vital Element of Landscape Structure." *Oikos* 68 (3): 571–3. http://dx.doi.org/10.2307/3544927.

Tele Atlas North America, ESRI. 2012. U.S. and Canada streets Cartographic. Redlands, CA: Tele Atlas.

Tepe, T. L., and V. J. Meretsky. 2011. "Forward-looking Forest Restoration under Climate Change—Are US Nurseries Ready?" *Restoration Ecology* 19 (3): 295–8. http://dx.doi.org/10.1111/j.1526-100X.2010.00748.x.

Terborgh, J., and J. A. Estes. 2010. *Trophic Cascades: Predators, Prey, and the Changing Dynamics of Nature*. Washington, DC: Island Press.

Theobald, D. 2010. "Estimating Natural Landscape Changes from 1992 to 2030 in the Conterminous US." *Landscape Ecology* 25 (7): 999–1011. http://dx.doi.org/10.1007/s10980-010-9484-z.

Thomas, C. D., A. Cameron, R. E. Green, M. Bakkenes, L. J. Beaumont, Y. C. Collingham, B.F.N. Erasmus, M. F. de Siqueira, A. Grainger, L. Hannah, et al. 2004. "Extinction Risk from Climate Change." *Nature* 427 (6970): 145–8. http://dx.doi.org/10.1038/nature02121.

Thomas, C. D., P. K. Gillingham, R. B. Bradbury, D. B. Roy, B. J. Anderson, J. M. Baxter, N.A.D. Bourn, H.Q.P. Crick, R. A. Findon, R. Fox, et al. 2012. "Protected Areas Facilitate Species' Range Expansions." *Proceedings of the National Academy of Sciences of the United States of America* 109 (35): 14063–8. http://dx.doi.org/10.1073/pnas.1210251109.

Thompson, Ian. 2009. "Forest Resilience, Biodiversity, and Climate Change." Convention on Biological Diversity Series No. 43. Montreal: Secretariat of the Convention on Biological Diversity. http://www.cbd.int/doc/publications/cbd-ts-43-en.pdf.

Thompson, I., B. Mackey, S. McNulty, and A. Mosseler. 2009. "Forest Resilience, Biodiversity, and Climate Change: A Synthesis of the Biodiversity/resilience/

stability Relationship in Forest Ecosystems." Tech. Ser. Num. 43. Montreal: Secretariat of the Convention on Biological Diversity.

Thompson, M., A. Ager, M. Finney, D. E. Calkin, and N. M. Vaillant. 2012. "The Science and Opportunity of Wildfire Risk Assessment." In *Novel Approaches and Their Applications in Risk Assessment*, ed. Y. Luo, 99–120. New York: InTech.

Thomson, A. M., K. A. Crowe, and W. H. Parker. 2010. "Optimal White Spruce Breeding Zones for Ontario under Current and Future Climates." *Canadian Journal of Forest Research* 40 (8): 1576–87. http://dx.doi.org/10.1139/X10-112.

Tillery, A. C., M. J. Darr, S. H. Connon, and J. A. Michael. 2011. "Post Wildfire Preliminary Debris Flow Hazard Assessment for the Area Burned by the 2011 Las Conchas Fire in North-Central New Mexico." U.S. Geological Survey Open-file Report 2011-1308.

Tilman, D., C. L. Lehman, and K. T. Thomson. 1997. "Plant Diversity and Ecosystem Productivity: Theoretical Considerations." *Proceedings of the National Academy of Sciences of the United States of America* 94 (5): 1857–61. http://dx.doi.org/10.1073/pnas.94.5.1857.

Tindall, J. R., J. A. Gerrath, M. Melzer, K. McKendry, B. C. Husband, and G. J. Boland. 2004. "Ecological Status of American Chestnut (Castanea dentata) in Its Native Range in Canada." *Canadian Journal of Forest Research* 34 (12): 2554–63. http://dx.doi.org/10.1139/x04-145.

Tingley, M. W., M. S. Koo, C. Moritz, A. C. Rush, and S. R. Beissinger. 2012. "The Push and Pull of Climate Change Causes Heterogeneous Shifts in Avian Elevational Ranges." *Global Change Biology* 18 (11): 3279–90. http://dx.doi.org/10.1111/j.1365-2486.2012.02784.x.

Tingley, M. W., D. A. Orwig, R. Field, and G. Motzkin. 2002. "Avian Response to Removal of a Forest Dominant: Consequences of Hemlock Woolly Adelgid Infestations." *Journal of Biogeography* 29 (10–11): 1505–16. http://dx.doi.org/10.1046/j.1365-2699.2002.00789.x.

TNC (The Nature Conservancy). 2014. *Rio Grande Water Fund: Comprehensive Plan for Wildfire and Water Source Protection*. Santa Fe: The Nature Conservancy.

Torchin, M. E., K. D. Lafferty, A. P. Dobson, V. J. McKenzie, and A. M. Kuris. 2003. "Introduced Species and Their Missing Parasites." *Nature* 421 (6923): 628–30. http://dx.doi.org/10.1038/nature01346.

Torreya Guardians. 2016. "Assisted Migration (Assisted Colonization, Managed Relocation) and Rewilding of Plants and Animals in an Era of Global Warming." Accessed March 19, 2016. www.torreyaguardians.org/assisted-migration.html.

Traveset, A., and D. M. Richardson. 2006. "Biological Invasions as Disruptors of Plant Reproductive Mutualisms." *Trends in Ecology & Evolution* 21 (4): 208–16. http://dx.doi.org/10.1016/j.tree.2006.01.006.

Trenberth, K. 2010. "More Knowledge, Less Certainty." *Nature Reports Climate Change* 4 (1002): 20–1. http://dx.doi.org/10.1038/climate.2010.06.

Turner, M. G., D. C. Donato, and W. H. Romme. 2013. "Consequences of Spatial Heterogeneity for Ecosystem Services in Changing Forest Landscapes: Priorities for Future Research." *Landscape Ecology* 28 (6): 1081–97. http://dx.doi.org/10.1007/s10980-012-9741-4.

Twitchett, R. J. 2006. "The Palaeoclimatology, Palaeoecology, and Palaeoenvironmental Analysis of Mass Extinction Events." *Palaeogeography, Palaeoclimatology, Palaeoecology* 232 (2-4): 190–213. http://dx.doi.org/10.1016/j.palaeo.2005.05.019.

US Department of Agriculture (USDA). 2010. Strategic Plan FY 2010-2015. Washington, DC: Accessed April 4, 2013. www.ocfo.usda.gov/usdasp/sp2010/sp2010.pdf.

US Department of Agriculture (USDA). 2011. "Policy Statement on Climate Change Adaptation." Departmental Regulation 1070–001. Washington, DC: US Department of Agriculture. Accessed July 15, 2013. http://www.ocio.usda.gov/sites/default/files/docs/2012/DR1070-001.pdf.

USDA Forest Service. 2004. "National Strategy and Implementation Plan for Invasive Species Management." http://www.fs.fed.us/invasivespecies/documents/Final_National_Strategy_100804.pdf.

USDA Forest Service. 2008. "Forest Service Strategic Framework for Responding to Climate Change." Version 1. Washington, DC: USDA, Forest Service. Accessed March 25, 2016. http://www.fs.fed.us/climatechange/documents/strategic-framework-climate-change-1-0.pdf.

USDA Forest Service. 2011a. "National Roadmap for Responding to Climate Change." FS-957B. Washington, DC: USDA, Forest Service. Accessed March 25, 2016. http://www.fs.fed.us/climatechange/pdf/Roadmapfinal.pdf.

USDA Forest Service. 2011b. "Navigating the Performance Scorecard: A Guide for National Forests and Grasslands." Version 2, August 2011. Washington, DC: USDA, Forest Service. Accessed March 25, 2016. http://www.fs.fed.us/climatechange/advisor/scorecard/scorecard-guidance-08-2011.pdf.

USDA Forest Service. 2012a. "Future of America's Forest and Rangelands: Forest Service 2010 Resources Planning Act Assessment." Gen. Tech Rep. WO-87. Washington, DC: USDA, Forest Service.

USDA Forest Service. 2012b. "Increasing the Pace of Restoration and Job Creation on Our National Forests." http://www.fs.fed.us/publications/restoration/restoration.pdf.

US Department of Agriculture/US Department of the Interior. 2005. "Wildland Fire Use: Implementation Procedures Reference Guide." In *National Wildfire Coordinating Group*. Boise, ID: National Interagency Fire Center.

US Department of the Interior (USDOI). 2009. "Addressing the Impacts of Climate Change on America's Water, Land, and Other Natural and Cultural Resources." Secretarial Order 3289. Issued September 14, 2009. Accessed March 25, 2016. www.doi.gov/whatwedo/climate/cop15/upload/SecOrder3289.pdf.

US Department of the Interior, Bureau of Reclamation. 2011a. "Colorado River Basin Water Supply and Demand Study." Interim Report No. 1.

US Department of the Interior, Bureau of Reclamation. 2011b. "SECURE Water Act Section 9503(c) – Reclamation Climate Change and Water, Report to Congress."

US Department of the Interior, Bureau of Reclamation. 2013. "San Juan-Chama Project Description." Accessed August 24, 2013. http://www.usbr.gov/projects /Project.jsp?proj_Name=San%20Juan-Chama%20Project.

US Environmental Protection Agency. 2009. "Land Use, Land Use Change and Forestry: Inventory of US Greenhouse Gas Emissions and Sinks, 1990–2007." Washington, DC: US Environmental Protection Agency, Office of Atmospheric Programs.

US Environmental Protection Agency. 2010. "Inventory of US Greenhouse Gas Emissions and Sinks, 1990–2008." EPA 430-R-09-006. Washington, DC: US Environmental Protection Agency, Office of Atmospheric Programs.

US Environmental Protection Agency. 2011. "Climate Change Vulnerability Assessments: Four Case Studies of Water Utility Practices (2011 Final)." Washington, DC: U.S. EPA. EPA/600/R-10/077F, 2011.

US Environmental Protection Agency. 2012. "Climate Change Indicators in the United States." Eastern Research Group. MORE? U.S. Geological Survey. 2013. Rain-on-Snow events. http://wa.water.usgs.gov/projects/rosevents/.

US Geological Survey. 2013. "Rain-on-Snow Events." June. http://wa.water.usgs .gov/projects/rosevents/index.htm.

US Government Accountability Office. 2009. "Wildland Fire Management: Federal Agencies Have Taken Important Steps Forward, but Additional Strategic Action Is Needed." GAO-09-87.

Ukrainetz, N. K., G. A. O'Neill, and B. Jaquish. 2011. "Comparison of Fixed and Focal Point Seed Transfer Systems for Reforestation and Assisted Migration: A Case Study for Interior Spruce in British Columbia." *Canadian Journal of Forest Research* 41 (7): 1452–64. http://dx.doi.org/10.1139/x11-060.

Urban, M. C., J. J. Tewksbury, and K. S. Sheldon. 2012. "On a Collision Course: Competition and Dispersal Differences Create No-analogue Communities and Cause Extinctions During Climate Change." *Proceedings of the Royal Society* 279 (1735): 2072–80. http://dx.doi.org/10.1098/rspb.2011.2367.

Vaks, A., M. Bar-Matthews, A. Ayalon, A. Matthews, L. Halicz, and A. Frumkin. 2007. "Desert Speleothems Reveal Climatic Window for African Exodus of Early Modern Humans." *Geology* 35 (9): 831–4. http://dx.doi.org/10.1130/G23794A.1.

Valachovic, Y. S., C. A. Lee, H. Scanlon, J. M. Varner, R. Glebocki, B. D. Graham, and D. M. Rizzo. 2011. "Sudden Oak Death-Caused Changes to Surface Fuel Loading and Potential Fire Behavior in Douglas-Fir-Tanoak Forests." *Forest Ecology and Management* 261 (11): 1973–86. http://dx.doi.org/10.1016/j.foreco.2011 .02.024.

van de Water, K., and M. North. 2010. "Fire History of Coniferous Riparian Forests in the Sierra Nevada." *Forest Ecology and Management* 260 (3): 384–95. http://dx .doi.org/10.1016/j.foreco.2010.04.032.

van de Water, K., and M. North. 2011. "Stand Structure, Fuel Loads, and Fire Behavior in Riparian and Upland Forests, Sierra Nevada Mountains, USA; A Comparison of Current and Reconstructed Conditions." *Forest Ecology and Management* 262 (2): 215–28. http://dx.doi.org/10.1016/j.foreco.2011.03.026.

van Heeswijk, M., J. S. Kimball, and D. Marks. 1996. "Simulation of Water Available for Runoff in Clearcut Forest Openings During Rain-on-Snow Events in the Western Cascade Range of Oregon and Washington. U.S." Geological Survey Water-Resources Investigations Report: 95–4219.

Vankat, J. L. 2013. *Vegetation Dynamics on the Mountains and Plateaus of the American Southwest.* New York: Springer. http://dx.doi.org/10.1007/978-94-007-6149-0.

van Mantgem, P. J., N. L. Stephenson, J. C. Byrne, L. D. Daniels, J. F. Franklin, P. Z. Fule, M. E. Harmon, A. J. Larson, J. M. Smith, A. H. Taylor, et al. 2009. "Widespread Increase of Tree Mortality Rates in the Western United States." *Science* 323 (5913): 521–4. http://dx.doi.org/10.1126/science.1165000.

van Wagtendonk, J. W., K. A. van Wagtendonk, and A. E. Thode. 2012. "Factors Associated with the Severity of Intersecting Fires in Yosemite National Park, California, USA." *Fire Ecology* 8 (1): 11–31. http://dx.doi.org/10.4996/fireecology.0801011.

Veenhuis, J. E. 2002. "Effects of Wildfire on the Hydrology of Capulin and Rito de los Frijoles Canyons, Bandelier National Monument, New Mexico." US Geological Survey Water-Resources Investigations Report 02–4152.

Venette, R. C., and S. D. Cohen. 2006. "Potential Climatic Suitability for Establishment of Phytophthora Ramorum within the Contiguous United States." *Forest Ecology and Management* 231 (1–3): 18–26. http://dx.doi.org/10.1016/j.foreco.2006.04.036.

Verschuren, D., J. S. Sinninghe Damste, J. Moernaut, I. Kristen, M. Blaauw, M. Fagot, G. H. Haug, B. van Geel, M. De Batist, P. Barker, et al. 2009. "Half-Precessional Dynamics of Monsoon Rainfall near the East African Equator." *Nature* 462 (7273): 637–41. http://dx.doi.org/10.1038/nature08520.

Vitousek, P. M., and W. A. Reiners. 1975. "Ecosystem Succession and Nutrient Retention: A Hypothesis." *Bioscience* 25 (6): 376–81. http://dx.doi.org/10.2307/1297148.

Vitt, P., K. Havens, A. T. Kramer, D. Sollenberger, and E. Yates. 2010. "Assisted Migration of Plants: Changes in Latitudes, Changes in Attitudes." *Biological Conservation* 143 (1): 18–27. http://dx.doi.org/10.1016/j.biocon.2009.08.015.

Vose, J. M., D. L. Peterson, and T. Patel-Weynand, eds. 2012. "Effects of Climatic Variability and Change on Forest Ecosystems: A Comprehensive Science Synthesis for the U.S. Forest Sector." Gen. Tech. Rep. PNW-GTR-870. Portland, OR: USDA, Forest Service, Pacific Northwest Research Station.

Wahungu, G. M., N. W. Gichohi, I. A. Onyango, L. K. Mureu, D. Kamaru, S. Mutisya, M. Mulama, J. K. Makau, and D. M. Kimuyu. 2012. "Encroachment of Open Grasslands and *Acacia drepanolobium* Harms ex B. Y. Sjöstedt Habitats by *Euclea divinorum* Hiern in Ol Pejeta Conservancy, Kenya." *African Journal of Ecology* 51 (1): 130–8. http://dx.doi.org/10.1111/aje.12017.

Wallgren, M., C. Skarpe, R. Bergstrom, K. Danell, A. Bergström, T. Jakobsson, K. Karlsson, and T. Strand. 2009. "Influence of Land Use on the Abundance of Wildlife and Livestock in the Kalahari, Botswana." *Journal of Arid Environments* 73 (3): 314–21. http://dx.doi.org/10.1016/j.jaridenv.2008.09.019.

Wang, T., A. Hamann, A. Yanchuk, G. A. O'Neill, and S. N. Aitken. 2006. "Use of Response Functions in Selecting Lodgepole Pine Populations for Future Climates." *Global Change Biology* 12 (12): 2404–16. http://dx.doi.org/10.1111/j.1365-2486.2006.01271.x.

Wanless, R. M., P. G. Ryan, R. Altwegg, A. Angel, J. Cooper, R. Cuthbert, and G. M. Hilton. 2009. "From Both Sides: Dire Demographic Consequences of Carnivorous Mice and Longlining for the Critically Endangered Tristan Albatrosses on Gough Island." *Biological Conservation* 142 (8): 1710–8. http://dx.doi.org/10.1016/j.biocon.2009.03.008.

Waring, K. M., and K. L. O'Hara. 2005. "Silvicultural Strategies in Forest Ecosystems Affected by Introduced Pests." *Forest Ecology and Management* 209 (1-2): 27–41. http://dx.doi.org/10.1016/j.foreco.2005.01.008.

Warren, R. J., II, and M. A. Bradford. 2014. "Mutualism Fails when Climate Response Differs between Interacting Species." *Global Change Biology* 20 (2): 466–74. http://dx.doi.org/10.1111/gcb.12407.

Warshall, P. 1995. "The Madrean Sky Island Archipelago: A Planetary Overview." In *Biodiversity and Management of the Madrean Archipelago: The Sky Islands of Southwestern United States and Northwestern Mexico*, technical coordinators L. F. DeBano, P. F. Folliott, A. Ortega-Rubio, G. J. Gottfried, H. Robert, and C. B. Edminster, 6–18. USDA Forest Service, Gen. Tech. Rep. RM-GTR-264. Washington, DC: USDA, Forest Service.

Water Research Foundation. 2012. "Report on the Operational and Economic Impacts of Hurricane Irene on Drinking Water Systems." Denver, CO. http://www.waterrf.org/resources/pages/PublicSpecialReports-detail.aspx?ItemID=6

"Water Shortage Blamed for Massive Bird Die-Off At NorCal Wildlife Refuge." 2012. CBS San Francisco, April 21. http://sanfrancisco.cbslocal.com/2012/04/21/water-shortage-blamed-for-massive-bird-die-off-at-norcal-wildlife-refuge/.

Watterson, N. A., and J. A. Jones. 2006. "Flood and Debris Flow Interactions with Roads Promote the Invasion of Exotic Plants along Steep Mountain Streams, Western Oregon." *Geomorphology* 78 (1-2): 107–23. http://dx.doi.org/10.1016/j.geomorph.2006.01.019.

Webb, T., III. 1988. "Eastern North America." In *Vegetation History*, ed. B. Huntley and T. Webb, 385–414. Dordrecht, The Netherlands: Kluwer Academic Publishers. http://dx.doi.org/10.1007/978-94-009-3081-0_11.

Weed, A. S., M. P. Ayres, and J. A. Hicke. 2013. "Consequences of Climate Change for Biotic Disturbances in North American Forests." *Ecological Monographs* 83 (4): 441–70. http://dx.doi.org/10.1890/13-0160.1.

Weidner, E., and A. Todd. 2011. *"From the Forest to the Faucet: Drinking Water and Forests in the US."* Methods Paper. *Ecosystem Services and Markets Program Area. State and Private Forestry.* USDA Forest Service.

Weiss, S. B., D. D. Murphy, and P. H. White. 1988. "Sun, Slope and Butterflies: Topographic Determinants of Habitat Quality for *Euphydryas editha bayensis.*" *Ecology* 69 (5): 1486–96. http://dx.doi.org/10.2307/1941646.

Weng, C., and S. T. Jackson. 1999. "Late Glacial and Holocene Vegetation History and Paleoclimate of the Kaibab Plateau, Arizona." *Palaeogeography, Palaeoclimatology, Palaeoecology* 153 (1–4): 179–201. http://dx.doi.org/10.1016/S0031-0182(99)00070-X.

Wenger, S. J., D. J. Isaak, C. H. Luce, H. M. Neville, K. D. Fausch, J. B. Dunham, D. C. Dauwalter, M. K. Young, M. M. Elsner, B. E. Rieman, et al. 2011. "Flow Regime, Temperature, and Biotic Interactions Drive Differential Declines of Trout Species under Climate Change." *Proceedings of the National Academy of Sciences of the United States of America* 108 (34): 14175–80. http://dx.doi.org/10.1073/pnas.1103097108.

Westerling, A. L. 2009. "Wildfires." In *Climate Change Science and Policy*, ed. S. H. Schneider, A. Rosencranz, M. D. Mastrandrea, and K. Kuntz-Duriseti, 92–103. Washington, DC: Island Press.

Westerling, A. L., T. J. Brown, A. Gershunov, D. R. Cayan, and M. D. Dettinger. 2003. "Climate and Wildfire in the Western United States." *Bulletin of the American Meteorological Society* 84 (5): 595–604. http://dx.doi.org/10.1175/BAMS-84-5-595.

Westerling, A.L., and B. P. Bryant. 2008. "Climate Change and Wildfire in California." *Climatic Change* 87 (1): 231–49.

Westerling, A. L., B. P. Bryant, H. K. Preisler, T. P. Holmes, H. G. Hidalgo, T. Das, and S. R. Shrestha. 2011a. "Climate Change and Growth Scenarios for California Wildfire." *Climatic Change* 109 (S1): 445–63. http://dx.doi.org/10.1007/s10584-011-0329-9.

Westerling, A. L., A. Gershunov, D. R. Cayan, and T. P. Barnett. 2002. "Long Lead Statistical Forecasts of Area Burned in Western U.S. Wildfires by Ecosystem Province." *International Journal of Wildland Fire* 11 (4): 257–66. http://dx.doi.org/10.1071/WF02009.

Westerling, A. L., H. G. Hidalgo, D. R. Cayan, and T. W. Swetnam. 2006. "Warming and Earlier Spring Increase in Western U.S. Forest Wildfire Activity." *Science* 313 (5789): 940–3. http://dx.doi.org/10.1126/science.1128834.

Westerling, A. L., M. G. Turner, E. A. H. Smithwick, W. H. Romme, and M. G. Ryan. 2011b. "Continued Warming Could Transform Greater Yellowstone Fire Regimes by Mid–21st Century." *Proceedings of the National Academy of Science,*

Ecology, Environmental Sciences 108 (32): 13165–70. http://dx.doi.org/10.1073/pnas
.1110199108.

Western Forestry Leadership Coalition. 2010. "The True Cost of Wildfire in the
Western U.S." http://www.wflccenter.org/news_pdf/324_pdf.pdf.

Western States Data. 2007. "Public Land Acreage." Accessed Aug 24, 2013.
http://www.wildlandfire.com/docs/2007/western-states-data-public-land.htm.

Whitlock, C., J. Marlon, C. Briles, A. Brunelle, C. Long, and P. Bartlein. 2008.
"Long-term Relations among Fire, Fuel, and Climate in the Northwestern US
Based on Lake-sediment Studies." *International Journal of Wildland Fire* 17 (1):
72–83. http://dx.doi.org/10.1071/WF07025.

Whitlock, C., S. L. Shafer, and J. Marlon. 2003. "The Role of Climate and Vegetation
Change in Shaping Past and Future Fire Regimes in the Northwestern US and
the Implications for Ecosystem Management." *Forest Ecology and Management* 178
(1-2): 5–21. http://dx.doi.org/10.1016/S0378-1127(03)00051-3.

Wiens, J. A., and D. Bachelet. 2010. "Matching the Multiple Scales of Conservation
with the Multiple Scales of Climate Change." *Conservation Biology* 24 (1): 51–62.
http://dx.doi.org/10.1111/j.1523-1739.2009.01409.x.

Williams, A. P., C. D. Allen, A. K. Macalady, D. Griffin, C. A. Woodhouse, D. M.
Meko, T. W. Swetnam, S. A. Rauscher, R. Seager, H. D. Grissino-Mayer, et al.
2012. "Temperature as a Potent Driver of Regional Forest Drought Stress and
Tree Mortality." *Nature Climate Change* 3 (3): 292–7. http://dx.doi.org/10.1038
/nclimate1693.

Williams, A. P., C. D. Allen, A. K. Macalady, D. Griffin, C. A. Woodhouse, D. M.
Meko, T. W. Swetnam, S. A. Rauscher, R. Seager, H. D. Grissino-Mayer, and J. S.
Dean. 2013. "Temperature as a Potent Driver of Regional Forest Drought Stress
and Tree Mortality." *Nature Climate Change* 3 (3): 292–97.

Williams, A. P., C. D. Allen, C. I. Millar, T. W. Swetnam, J. Michaelsen, C. J. Still,
and S. W. Leavitt. 2010. "Forest Responses to Increasing Aridity and Warmth in
the Southwestern United States." *Proceedings of the National Academy of Sciences
of the United States of America* 107 (50): 21289–94. http://dx.doi.org/10.1073
/pnas.0914211107.

Williams, C. J., E. K. Mendell, J. Murphy, W. M. Court, A. H. Johnson, and S. L.
Richter. 2008. "Paleoenvironmental Reconstruction of a Middle Miocene Forest
from the Western Canadian Arctic." *Palaeogeography, Palaeoclimatology, Palaeoecology* 261 (1-2): 160–76. http://dx.doi.org/10.1016/j.palaeo.2008.01.014.

Williams, J. E., A. L. Haak, H. M. Neville, and W. T. Colyer. 2009. "Potential Consequences of Climate Change to Persistence of Cutthroat Trout Populations."
North American Journal of Fisheries Management 29 (3): 533–48. http://dx.doi.org
/10.1577/M08-072.1.

Williams, J. W., and S. T. Jackson. 2007. "Novel Climates, No-analog Communities,
and Ecological Surprises." *Frontiers in Ecology and the Environment* 5 (9): 475–82.
http://dx.doi.org/10.1890/070037.

Williams, M. I., and R. K. Dumroese. 2013. "Preparing for Climate Change: Forestry and Assisted Migration." *Journal of Forestry* 111 (4): 287–97. http://dx.doi.org/10.5849/jof.13-016.

Williams, S. C., J. S. Ward, and U. Ramakrishnan. 2008. "Endozoochory by White-Tailed Deer (Odocoileus virginianus) across a Suburban/woodland Interface." *Forest Ecology and Management* 255 (3–4): 940–7. http://dx.doi.org/10.1016/j.foreco.2007.10.003.

Williamson, M. A. 2007. "Factors in United States Forest Service District Rangers' Decision to Manage a Fire for Resource Benefit." *International Journal of Wildland Fire* 16 (6): 755–62. http://dx.doi.org/10.1071/WF06019.

Willis, C. G., B. R. Ruhfel, R. B. Primack, A. J. Miller-Rushing, J. B. Losos, and C. C. Davis. 2010. "Favorable Climate Change Response Explains Non-native Species' Success in Thoreau's Woods." *PLoS One* 5 (1): e8878. http://dx.doi.org/10.1371/journal.pone.0008878.

Willis, K. J., and S. A. Bhagwat. 2009. "Biodiversity and Climate Change." *Science* 326 (5954): 806–7. http://dx.doi.org/10.1126/science.1178838.

Willis, K. J., and H. J. B. Birks. 2006. "What Is Natural? The Need for a Long-term Perspective in Biodiversity Conservation." *Science* 314 (5803): 1261–5. http://dx.doi.org/10.1126/science.1122667.

Willis, K. J., L. Gillson, and T. M. Brncic. 2004. "How "Virgin" is Virgin Rainforest?" *Science* 304:402–3. http://dx.doi.org/10.1126/science.1093991.

Wilsey, C. H., J. J. Lawler, E. P. Maurer, D. McKenzie, P. A. Townsend, R. Gwozdz, J. A. Freund, K. Hagmann, and K. M. Hutten. 2013. "Tools for Assessing Climate Impacts on Fish and Wildlife." *Journal of Fish and Wildlife Management* 4 (1): 220–41. http://dx.doi.org/10.3996/062012-JFWM-055.

Winder, R., E. A. Nelson, and T. Beardmore. 2011. "Ecological Implications for Assisted Migration in Canadian Forests." *Forestry Chronicle* 87 (06): 731–44. http://dx.doi.org/10.5558/tfc2011-090.

Wing, S. L., G. J. Harrington, F. A. Smith, J. I. Bloch, D. M. Boyer, and K. H. Freeman. 2005. "Transient Floral Change and Rapid Global Warming at the Paleo-ecene-Eocene Boundary." *Science* 310 (5750): 993–6. http://dx.doi.org/10.1126/science.1116913.

Wippel, G. 2012. "World Heritage Committee Decision on Selous Game Reserve Boundary Changes." Press release, July 30. Uranium-network.org. http://www.uranium-network.org/index.php/africalink/tanzania/253-press-release-re-world-heritage-comittee-decision-on-selous-game-reserve-boundary-changes.

Wittkuhn, R. S., L. McCaw, A. J. Wills, R. Robinson, A. N. Andersen, P. Van Heurck, J. Farr, G. Liddelow, and R. Cranfield. 2011. "Variation in Fire Interval Sequences Has Minimal Effects on Species Richness and Composition in Fire-prone Landscapes of Southwest Western Australia." *Forest Ecology and Management* 261 (6): 965–78. http://dx.doi.org/10.1016/j.foreco.2010.10.037.

Woinarski, J., M. Armstrong, K. Brennan, A. Fisher, A. D. Griffiths, B. Hill, D. J. Milne, C. Palmer, S. Ward, M. Watson, et al. 2010. "Monitoring Indicates Rapid and Severe Decline of Native Small Mammals in Kakadu National Park, Northern Australia." *Wildlife Research* 37 (2): 116–26. http://dx.doi.org/10.1071/WR 09125.

Woodall, C. W., C. M. Oswalt, J. A. Westfall, C. H. Perry, M. D. Nelson, and A. O. Finley. 2009. "An Indicator of Tree Migration in Forests of the Eastern United States." *Forest Ecology and Management* 257 (5): 1434–44. http://dx.doi.org/10 .1016/j.foreco.2008.12.013.

Woodhouse, C. A., D. M. Meko, G. M. MacDonald, D. W. Stahle, and E. R. Cook. 2010. "A 1,200-year Perspective of 21st Century Drought in Southwestern North America." *Proceedings of the National Academy of Sciences of the United States of America* 107 (50): 21283–8. http://dx.doi.org/10.1073/pnas.0911197107.

Woodroffe, R., and J. R. Ginsberg. 1998. "Edge Effects and the Extinction of Populations Inside Protected Areas." *Science* 280 (5372): 2126–8. http://dx.doi.org/10 .1126/science.280.5372.2126.

Work, T. T., D. G. McCullough, J. F. Cavey, and R. Komsa. 2005. "Arrival Rate of Non-Indigenous Insect Species into the United States through Foreign Trade." *Biological Invasions* 7 (2): 323–32. http://dx.doi.org/10.1007/s10530-004-1663-x.

Worrall, J. J., G. C. Adams, and S. C. Tharp. 2010. "Summer Heat and an Epidemic of Cytospora Canker of Alnus." *Canadian Journal of Plant Pathology* 32 (3): 376–86. http://dx.doi.org/10.1080/07060661.2010.499265.

Worrall, J. J., G. E. Rehfeldt, A. Hamann, E. H. Hogg, S. B. Marchetti, M. Michaelian, and L. K. Gray. 2013. "Recent Declines of Populus tremuloides in North America Linked to Climate." *Forest Ecology and Management* 299:35–51. http://dx.doi.org/10.1016/j.foreco.2012.12.033.

Wuebbles, D., G. Meehl, K. Hayhoe, T. R. Karl, K. Kunkel, B. Santer, M. Wehner, B. Colle, E. M. Fischer, R. Fu, et al. 2013. "CMIP5 Climate Model Analyses: Climate Extremes in the United States." *Bulletin of the American Meteorological Society.* http://dx.doi.org/10.1175/BAMS-D-12-00172.1.

Wyborn, C., and R. P. Bixler. 2013. "Collaboration and Nested Environmental Governance: Scale Dependency, Scale Framing, and Cross-scale Interactions in Collaborative Conservation." *Journal of Environmental Management* 123:58–67. http://dx.doi.org/10.1016/j.jenvman.2013.03.014.

Xenopoulos, M., D. Lodge, J. Alcamo, M. Marker, K. Schulze, and D. P. Van Vuuren. 2005. "Scenarios of Freshwater Fish Extinctions from Climate Change and Water Withdrawal." *Global Change Biology* 11 (10): 1557–64. http://dx.doi.org /10.1111/j.1365-2486.2005.001008.x.

Yan, X., L. Zhenyu, W. Gregg, and L. Dianmo. 2001. "Invasive Species in China—An Overview." *Biodiversity and Conservation* 10 (8): 1317–41. http://dx.doi.org/10.1023 /A:1016695609745.

Yates, E. D., D. F. Levia, Jr., and C. L. Williams. 2004. "Recruitment of Three Non-native Invasive Plants into a Fragmented Forest in Southern Illinois." *Forest Ecology and Management* 190 (2–3): 119–30. http://dx.doi.org/10.1016/j.foreco.2003.11.008.

Yohe, G., and R. Leichenko. 2010. "Adopting a Risk-based Approach." *Annals of the New York Academy of Sciences* 1196 (1): 29–40. http://dx.doi.org/10.1111/j.1749-6632.2009.05310.x.

Yuan, D., H. Cheng, R. L. Edwards, C. A. Dyoski, M. J. Kelly, M. Zhang, J. Qing, Y. Lin, Y. Wang, J. Wu, et al. 2004. "Timing, Duration, and Transitions of the Last Interglacial Asian Monsoon." *Science* 304 (5670): 575–8. http://dx.doi.org/10.1126/science.1091220.

Yue, X., L. J. Mickley, and J. A. Logan. 2013a. "Projection of Wildfire Activity in Southern California in the Mid-21st century." *Climate Dynamics*. http://dx.doi.org/10.1007/s00382-013-2022-3.

Yue, X., L. J. Mickley, J. A. Logan, and J. O. Kaplan. 2013b. "Ensemble Projections of Wildfire Activity and Carbonaceous Aerosol Concentrations over the Western United States in the Mid–21st Century." *Atmospheric Environment* 77:767–80. http://dx.doi.org/10.1016/j.atmosenv.2013.06.003.

Yuen, E., S. S. Jovicich, and B. L. Preston. 2013c. "Climate Change Vulnerability Assessments as Catalysts for Social Learning: Four Case Studies in South-Eastern Australia." *Mitigation and Adaptation Strategies for Global Change* 18 (5): 567–90. http://dx.doi.org/10.1007/s11027-012-9376-4.

Zachos, J. C., G. R. Dickens, and R. E. Zeebe. 2008. "An Early Cenozoic Perspective on Greenhouse Warming and Carbon-Cycle Dynamics." *Nature* 451 (7176): 279–83. http://dx.doi.org/10.1038/nature06588.

Zalasiewicz, J., M. Williams, W. Steffen, and P. Crutzen. 2010. "The New World of the Anthropocene." *Environmental Science & Technology* 44 (7): 2228–31. http://dx.doi.org/10.1021/es903118j.

Zerbe, J. 2006. "Thermal Energy, Electricity, and Transportation Fuels from Wood." *Forest Products Journal* 56: 6–14.

Zhang, F., and A. Georgakakos. 2011. "Climate and Hydrologic Change Assessment for Georgia." Proceedings of the 2011 Georgia Water Resources Conference. Athens, GA. April 11–13.

Zhang, J., J. Hudson, R. Neal, J. Sereda, T. Clair, M. Turner, D. Jeffries, P. Dillon, L. Molot, K. Somers, et al. 2010. "Long-term Patterns of Dissolved Organic Carbon in Lakes across Eastern Canada: Evidence of a Pronounced Climate Effect." *Limnology and Oceanography* 55 (1): 30–42. http://dx.doi.org/10.4319/lo.2010.55.1.0030.

Zhang, Y., J. L. Hanula, and S. Horn. 2012. "The Biology and Preliminary Host Range of Megacopta cribraria (Heteroptera: Plataspidae) and Its Impact on Kudzu Growth." *Environmental Entomology* 41 (1): 40–50. http://dx.doi.org/10.1603/EN11231.

Zhu, K., C. W. Woodall, and J. S. Clark. 2012. "Failure to Migrate: Lack of Tree Range Expansion in Response to Climate Change." *Global Change Biology* 18 (3): 1042–52. http://dx.doi.org/10.1111/j.1365-2486.2011.02571.x.

Ziska, L. H., J. B. Reeves, and B. Blank. 2005. "The Impact of Recent Increases in Atmospheric CO_2 on Biomass Production and Vegetative Retention of Cheatgrass (Bromus tectorum): Implications for Fire Disturbance." *Global Change Biology* 11 (8): 1325–32. http://dx.doi.org/10.1111/j.1365-2486.2005.00992.x.

Zommers, Z., and D. W. MacDonald. 2012. "Protected Areas as Frontiers for Human Migration." *Conservation Biology* 26 (3): 547–56. http://dx.doi.org/10.1111/j.1523-1739.2012.01846.x.

Contributors

EDITORS

R. Patrick Bixler is a research fellow with the RGK Center for Philanthropy and Community Service in the Lyndon B. Johnson School of Public Affairs at the University of Texas, Austin, and a senior fellow at the Pinchot Institute for Conservation.

Char Miller is a senior fellow at the Pinchot Institute for Conservation and the W. M. Keck Professor of Environmental Analysis at Pomona College, Claremont, CA.

V. Alaric Sample is the president of the Pinchot Institute for Conservation in Washington, DC.

CONTRIBUTING AUTHORS

Craig D. Allen is a research ecologist at the US Geological Survey, Jemez Mountains Field Station, Los Alamos, NM.

Mark Anderson is the director of conservation science at The Nature Conservancy, Arlington, VA.

Susan Beecher is a research fellow at the Pinchot Institute for Conservation, Washington, DC.

R. Travis Belote is a forest ecologist at The Wilderness Society, Bozeman, MT.

Timothy J. Brown is a research professor at the Desert Research Institute, Reno, NV.

Anne A. Carlson is a climate associate at The Wilderness Society, Bozeman, MT.

Tim Caro is a professor of wildlife biology in the Department of Wildlife, Fish and Conservation Biology, University of California, Davis.

Grace K. Charles is a PhD scholar in the Department of Plant Sciences, University of California, Davis.

David Cleaves is the climate change advisor at the USDA Forest Service, Office of the Chief, Washington, DC.

Dena J. Clink is a PhD scholar in the Department of Anthropology, University of California, Davis.

Ayesha Dinshaw is a research analyst for World Resources Institute's Vulnerability and Adaptation Initiative, Washington, DC.

R. Kasten Dumroese is a research plant physiologist with the Grassland, Shrubland, and Desert Ecosystems Science Program for the US Forest Service Rocky Mountain Research Station, Moscow, ID.

Jonas Epstein is an ORISE economic research fellow, US Forest Service, Washington, DC.

Alexander M. Evans is the research director at the Forest Guild, Santa Fe, NM.

Todd Gartner is a senior associate for the World Resources Institute's Food, Forests and Water Program and manager of WRI's Natural Infrastructure for Water Program, Washington, DC.

Jessica E. Halofsky is a research ecologist, University of Washington, College of the Environment, School of Environmental and Forest Sciences, Seattle, WA.

Nels Johnson is the deputy state director for the Pennsylvania Chapter of The Nature Conservancy.

Linda A. Joyce is a quantitative ecologist with the Human Dimensions Research Program, US Department of Agriculture Forest Service, Rocky Mountain Research Station, Fort Collins, CO.

Paige Lewis is the deputy state director and director of Conservation Programs for The Nature Conservancy in Colorado.

Laura Falk McCarthy is the director of conservation programs for The Nature Conservancy, Santa Fe, NM.

Heather McGray is the director for World Resources Institute's Vulnerability and Adaptation Initiative, Washington, DC.

Constance I. Millar is a research paleoecologist with the US Department of Agriculture, Pacific Southwest Research Station, Albany, CA.

James Mulligan is the executive director at Green Community Ventures, a 501c3 nonprofit, Washington, DC.

Chadwick Dearing Oliver is Pinchot Professor of Forestry and Environmental Studies and director of the Global Institute of Sustainable Forestry, School of Forestry and Environmental Studies, Yale University, New Haven, CT.

David L. Peterson is a research biological scientist with the US Department of Agriculture, Forest Service, Pacific Northwest Research Station, Seattle, WA.

Will Price is the director of conservation programs at the Pinchot Institute for Conservation in Washington, DC.

Janine M. Rice is a research associate with the University of Colorado, Western Water Assessment, and the US Department of Agriculture Forest Service, Rocky Mountain Research Station, Fort Collins, CO.

Jason Riggio is a PhD scholar, Department of Wildlife, Fish and Conservation Biology, University of California, Davis.

Tania Schoennagel is a research scientist in the Geography Department and the Institute of Arctic and Alpine Research (INSTAAR), University of Colorado, Boulder.

Mark L. Shaffer is a biodiversity conservationist with the US Fish and Wildlife Service, Arlington, VA.

Curt Stager is a professor of natural sciences, Paul Smith's College, and an adjunct research associate at the Climate Change Institute, University of Maine, Orono.

Scott L. Stephens is a professor of fire science in the Department of Environmental Policy, Science, and Management, University of California, Berkeley.

Thomas W. Swetnam is a regents' professor of dendrochronology and director of the Laboratory of Tree-Ring Research at the University of Arizona, Tucson.

Gary M. Tabor is the executive director of the Center for Large Landscape Conservation, Bozeman, MT.

Christopher Topik is the director of the Restoring America's Forests Initiative at The Nature Conservancy, Arlington, VA.

Monica G. Turner is Eugene P. Odum Professor of Ecology, Department of Zoology, University of Wisconsin, Madison.

Thomas T. Veblen is a professor of geography at University of Colorado, Boulder.

Alexandra M. Weill is a PhD scholar, Department of Plant Sciences, University of California, Davis.

Anthony L. Westerling is an associate professor of environmental engineering and geography, Sierra Nevada Research Institute, University of California, Merced.

Carolyn Whitesell is a PhD scholar, Graduate Group in Ecology, University of California, Davis.

Mary I. Williams is a postdoctoral research assistant with Michigan Technological University, living in Salt Lake City, UT.

Index

Page numbers in italic indicate illustrations.